Springer Series in Advanced Manufacturing

Series Editor

Professor D.T. Pham
Manufacturing Engineering Centre
Cardiff University
Queen's Building
Newport Road
Cardiff CF24 3AA
UK

Sergej Fatikow
Editor

Automated Nanohandling
by Microrobots

 Springer

Sergej Fatikow, Professor, Dr.-Ing. habil.
Department of Computing Science
University of Oldenburg
26111 Oldenburg
Germany

ISBN 978-1-84628-977-4   e-ISBN 978-1-84628-978-1

Springer Series in Advanced Manufacturing ISSN 1860-5168

British Library Cataloguing in Publication Data
Fatikow, S. (Sergej), 1960-
  Automated nanohandling by microrobots. - (Springer series
  in advanced manufacturing)
  1. Microfabrication 2. Microelectromechanical systems
  3. Robotics 4. Nanostructured materials 5. Robots, Industrial
  I. Title
  620.5
ISBN-13: 9781846289774

Library of Congress Control Number: 2007933584

Printed on acid-free paper

9 8 7 6 5 4 3 2 1

springer.com

Preface

"What I want to talk about is the problem of manipulating and controlling things on a small scale" stated Richard P. Feynman at the beginning of his visionary talk "There's Plenty of Room at the Bottom", given on December 29th 1959 at the annual meeting of the American Physical Society at the California Institute of Technology. Today, almost half a century after this first insight into unlimited opportunities on the nanoscale level, we still want – and have to – talk about the same issue. The problem identified by Feynmann turned out to be a very difficult one due to a lack of understanding of the underlying phenomena in the nanoworld and a lack of suitable nanohandling methods. This book addresses the second issue and tries to contribute to the tremendous effort of the research community in seeking proper solutions in this field.

Automated robot-based nanomanipulation is one of the key challenges of microsystem technology and nanotechnolgy, which has recently been addressed by a rising number of R&D groups and companies all over the world. Controlled, reproducible assembly processes on the nanoscale will enable high-throughput manufacturing of revolutionary products and open up new application fields. The ultimate goal of these research activities is the development of automated nanomanipulation processes to build a bridge between existing precise handling strategies for micro- and nanoscale objects and aspired high-throughput fabrication of micro- and nanosystems. These activities include, amongst others, the development of new nanohandling robots; the investigation of application-specific nanohandling strategies; the construction of new application-specific tools; the development of advanced control approaches; as well as the investigation of suitable sensing technologies. Real-time sensory feedback and fast and precise control systems are of particular importance for automated nanohandling, so the book will take a thorough look at these issues.

Despite the growing interest in automated nanomanipulation, there is hardly any publication that treats this research in a coherent and comprehensive way. This book is an attempt to provide the researcher with an overview of some important aspects of this rapidly expanding technology. The other main purpose of this book is to inform the practicing engineer and the engineering student about automation on the nanoscale as well as the promising fields of application. The latter can be of a very different nature as nanohandling is strongly interdisciplinary in character so that the borders between established scientific and technical disciplines fade.

The idea of the book originates from the lecture courses on microrobotics and nanohandling which have been given to students of computer sciences and physics at the University of Oldenburg since 2001. At the same time, the book is a comprehensive summary of research work that has been performed by my teams at the Division of Microrobotics and Control Engineering at the same university as well as at the Division of Microsystems Technology and Nanohandling at the Oldenburg Institute for Information Technology (OFFIS) for the last six years. All the contributors are – or were for a long time – members of the Divisions' research staff.

It is obviously impossible to pick up every idea and every piece of research work on nanohandling and automation on the nanoscale that has been discussed in the literature. A representative selection of them was made in the overview section of each chapter, and the authors believe that most relevant results have been covered. Many of the nanohandling approaches and devices presented in the book are at the forefront of technology. Eventually, they will reach maturity and open up a mega-market for nanotechnology products. The market penetration and success will be caused to a great extent by the innovators who are currently experimenting with automated handling on the nanoscale. It is the strong wish of the authors' team that this work will help to generate an awareness of this new, diversified technology and to guide the interested reader.

This work was done by the team of researchers involved in quite a few international and German joint research projects. Any active researcher would understand how difficult it is to spare the time for serving the research community by writing a book. For this reason, my strongest vote of thanks goes to all the authors who have contributed to this book. I especially want to thank Professor Duc Truong Pham, the Director of the Manufacturing Engineering Centre at Cardiff University and the scientific editor of the Springer book series on Advanced Manufacturing, for triggering the idea of writing a book about my field of research. The linguistic proofreading was done by Nicholas Pinfield and Christian Fatikow. We are indebted to them for many suggestions that have improved the book a great deal. We appreciate the support by Professor Sylvain Martel, the Director of the NanoRobotics Lab at Montreal University, who read the manuscript and made a lot of valuable comments. We are grateful to the colleagues who provided us with graphs and pictures which make it much easier to understand the text. The book team had much help with the time-consuming drawing of the artwork: we are indebted to Sascha Fatikow for his excellent work. Dr. Markus Kemper deserves our sincere thanks for his time and effort with the meticulous preparation of the final manuscript for printing. Our thanks also go to Daniel Jasper and Dr. Kwangsoo Kim, who helped us with error checking and correction in the final manuscript.

Oldenburg, March 2007

Sergej Fatikow

Contents

List of Contributors

Volkmar Eichhorn
University of Oldenburg,
Department of Computing Science,
Division of Microrobotics and
Control Engineering,
26111 Oldenburg, Germany.

Stephan Fahlbusch
EMPA,
Laboratory for Mechanics of
Materials and Nanostructures,
Feuerwerkerstr. 39,
3602 Thun, Switzerland.

Sergej Fatikow
University of Oldenburg,
Department of Computing Science,
Division of Microrobotics and
Control Engineering,
26111 Oldenburg, Germany.

Saskia Hagemann
OFFIS e.V.
Division of Microsystems
Technology and Nanohandling,
Escherweg 2,
26121 Oldenburg, Germany.

Helge Hülsen
University of Oldenburg,
Department of Computing Science,
Division of Microrobotics and
Control Engineering,
26111 Oldenburg, Germany.

Marco Jähnisch
OFFIS e.V.
Division of Microsystems
Technology and Nanohandling,
Escherweg 2,
26121 Oldenburg, Germany.

Iulian Mircea
OFFIS e.V.
Division of Microsystems
Technology and Nanohandling,
Escherweg 2,
26121 Oldenburg, Germany.

Torsten Sievers
University of Oldenburg,
Division of Microrobotics and
Control Engineering,
Department of Computing Science,
26111 Oldenburg, Germany.

Albert Sill
University of Oldenburg,
Department of Computing Science,
Division of Microrobotics and
Control Engineering,
26111 Oldenburg, Germany.

Christian Stolle
OFFIS e.V.
Division of Microsystems
Technology and Nanohandling,
Escherweg 2,
26121 Oldenburg, Germany.

Thomas Wich
University of Oldenburg,
Department of Computing Science,
Division of Microrobotics and
Control Engineering,
26111 Oldenburg, Germany.

1

Trends in Nanohandling

Sergej Fatikow

Division of Microrobotics and Control Engineering,
Department of Computing Science,
University of Oldenburg, Germany

1.1 Introduction

The handling of micro- and nanoscale objects is an important current trend in robotics. It is often referred to as **nanohandling**, having in mind the range of aspired positioning accuracy. The Greek word "nanos" (dwarf) refers to the physical unit of a nanometer = 1 nm = 10^{-9}m. In this book, we understand nanohandling as the manipulation of microscale and nanoscale objects of different nature with an accuracy in the (sub-) nanometer range, which may include their finding, grasping, moving, tracking, releasing, positioning, pushing, pulling, cutting, bending, twisting, *etc*. Additionally, different characterization methods, *e.g.*, indenting or scratching on the nanoscale, measurement of different features of the object, requiring probe positioning with nanometer accuracy, structuring or shaping of nanostructures, and generally all kinds of changes to matter at nanolevel could also be defined as nanohandling in the broadest sense.

Obviously, not all conceivable nanohandling operations are based on robotics, *e.g.*, the so-called self-assembly, which will be introduced later. This book does not attempt to cover the whole palette of nanohandling options and will confine itself to the approaches that can be implemented and eventually automated with the help of microrobots with nanohandling capabilities. As in the field of industrial robotics, where humans leave hard, unacceptable work to robots, microrobots can help humans to handle extremely small objects with very high accuracy. Drastically miniaturized robots, or **microrobots**, are able to operate in extremely constricted work spaces, *e.g.*, under a light microscope or in the vacuum chamber of a scanning electron microscope (SEM). In particular, microsystem technology (MST) and nanotechnology require this kind of robot, since humans lack capabilities in manipulation at those scales. Automated nanohandling by microrobots will have a great impact in both these technologies in the near future and will contribute to the development of high-throughput nanomanipulation processes.

Microsystem technology aims at producing miniature systems involving micro-machined components or structures. Such systems enable new functions and new applications as well as having cost benefits. [1] provides a good overview of the present and future applications of MST. The emphasis of microsystem technology is clearly on the system aspect. However, the vast majority of today's MST products are components, which have to be further integrated into complete micro-systems. This integration often requires the robot-based handling of microscale objects with an accuracy in the nanometer range.

Nanotechnology is a new approach that refers to understanding and mastering the properties of matter at the nanoscale as well as to the miniaturization of devices down to the nanometer level. At this level, matter exhibits different and often amazing properties, which leads to a revolutionary potential in terms of possible impact on industrial production routes. Nanotechnology offers possible solutions to many current problems by means of smaller, lighter, faster, and better performing materials, components, and systems. A good overview of the most promising appli-cations of nanotechnology is given in [2, 3]. Nanotechnology is now just at the beginning of its commercial development. There are still many challenges to be solved, in order to be able to control single atoms, molecules, or nanoscale par-ticles in the best possible way. The ultimate goal of nanohandling here is nano-assembly – the organization of nanoscale objects into ordered structures – with precise control over the individual objects' relative positions.

The transfer of classical robotic "know-how" from our macroscopic world to the world of microscale and nanoscale objects being handled is, however, a huge technological challenge for the robotics research community. Some critical issues, *e.g.*, parasitic surface forces derogating the positioning accuracy, are regarded in Chapter 2. New, advanced actuator and sensor technologies which are suitable for nanohandling have to be investigated. Another crucial issue is the development of control architectures and methods tailored to the demands of automated nano-manipulation. The state of the art for nanohandling control approaches includes teleoperated and semi-automated control strategies. The reader will find a good review of current work on these approaches in [4]. Here, the operator controls the nanohandling robot directly or sends task commands to the robot controller, using vision, force or tactile feedback to control the nanohandling process. However, it is usually rather slow and not repeatable. In the automatic control approach, the robot has closed-loop control using sensory information without any user intervention. The latter approach is very challenging, especially due to the difficulty in getting available real-time nanoscale visual feedback and the lack of advanced control strategies able to deal with changing and uncertain physical parameters and disturbances [5, 6]. This book tries to show some solutions to these problems and to present promising applications, where tremendous benefit can be gained from the controlled handling of matter on the nanoscale and where smart microrobots may play an important role both as a high-throughput automated nanohandling technology as well as a complementary process to other techniques.

1.2 Trends in Nanohandling

There are several ways to classify nanohandling approaches. The following three approaches are being pursued by the majority of the nanohandling research groups, and they seem to be most promising and versatile for future developments in this field:

- **top-down approach** utilizing serial nanohandling by microrobot systems. The main goal is the miniaturization of robots, manipulators, and their tools as well as the adaptation of the robotic technology (sensing, handling, control, automation, *etc.*) to the demands of MST and nanotechnology. This approach is the topic of this book and will be shortly introduced in Section 1.3.
- **bottom-up approach** or **self-assembly** utilizing parallel nanohandling by autonomous organization of micro- and nanoobjects into patterns or structures without human intervention.
- the use of a **scanning probe microscope (SPM)** as a nanohandling robot. In this approach, the (functionalized) tip of an atomic force microscope (AFM) probe or of a scanning tunneling microscope (STM) probe acts as a nanohandling tool affecting the position or the shape of a nanoscale part. Actually, SPM-based nanohandling has to be pigeonholed into the **top-down** drawer, regarding a microscope scanner as a nanomanipulator and microscope probe as a nanohandling tool. However, it is worth taking a more precise look at this fascinating technology.

Several other approaches, *e.g.*, the use of optical tweezers [7–10] or electrophoresis [11, 12] might also be adapted for automated nanohandling. They are primarily used for the manipulation of fragile biological samples because of the low grasping forces lying in the pN range. The latter is clearly one of the limitations of these non-contact manipulators.

1.2.1 Self-assembly

Self-assembly can be considered as a new strategy for **parallel** nano- and microfabrication, which draws its inspiration from numerous examples in nature: self-assembly is one of the most important strategies used by nature for the development of complex, functional structures. Recent technological advances have led to the development of many novel "bottom-up" self-assembly strategies capable of creating ordered structures with a wide variety of tunable properties. In this context, self-assembly can be defined as the spontaneous formation of higher ordered structures from basic units. This approach is increasingly being exploited to assemble systems at the micro- and nanoscale. Especially at the nanoscale and when the assembly process deals with a large number of parts, the ability to efficiently manipulate single parts gradually diminishes, as the size of objects decreases, and the need for a parallel manipulation method arises.

Recent research activities on self-assembly in the microscale and nanoscale have been reviewed in [13, 14]. Generally, the self-assembly process involves recognition and making connections to the other parts of the system. For this reason, each part has to be equipped with a mechanism supporting its process of self-assembly, *i.e.*, the ability to recognize (self-assembly programming mechanism) and connect (self-assembly binding/driving force) to the proper adjacent part or template. Additionally, an external agitation mechanism is often needed to drive the system to the global energy minimum that corresponds to the correct self-assembly.

The goal of self-assembly in the **microscale** is usually the exact planar positioning of parts onto a substrate (2D self-assembly) or the creation of 3D microstructures which cannot be fabricated by existing micromachining methods. Self-assembly is enabled by pre-programming the structure and the functions of parts during their synthesis so that parts self-assemble into ordered 2D and 3D architectures under appropriate conditions [15, 16]. Typically, a large number of parts to be self-assembled are put into a fluid on the substrate surface. The parts in the fluid flow are "looking" for the suitable binding sites and spontaneously build an ordered structure on the substrate. To guide the self-assembly, *e.g.*, gravitational, magnetic, or capillary forces can be utilized. A typical application of gravity includes the agitation of parts to make them move on the substrate surface until they "find" suitable binding sites – particularly shaped recesses in the substrate – and get stuck in them [17, 18]. Capillary forces are increasingly used to guide 2D self-assembly [19-21]. Usually, the self-assembly is performed by exploitation of the hydrophobic and/or hydrophilic features of substrate and microparts, which can be modulated in different ways to improve controllability and selectivity. Electrostatic forces can also be utilized to build ordered planar microstructures [22–24]. The main advantage of this approach is the ability to dynamically control the self-assembly process by modulating the electric force. In comparison to 2D self-assembly, the 3D approaches are just at the very beginning of their active investigation. Again, a liquid is usually used as the medium for guided self-assembly and several promising approaches have been demonstrated [25–27].

In the **nanoscale**, self-assembly may enable many of the most difficult steps in nanofabrication, including those involving atomic-level modifications of structure. As a result, ordered structures with sub-nanometer precision can be expected, both in 2D and 3D architectures. Typically, the self-assembly of nanoobjects, *e.g.*, nanocrystals, nanowires, or carbon nanotubes (CNT), is triggered using chemistry and exploits biologically inspired interaction paradigms such as shape complementarity, van der Waals forces, hydrogen bonding, hydrophobic interactions, or electrostatic forces. Maybe the best-known example of self-assembled nanostructures is the so-called self-assembled monolayers (SAM) which are built from organic molecules that chemically bind to a substrate and form an ordered lattice [28-30]. SAM can be used for the modulation of surface-dependent phenomena [30, 31], which is of interest for different applications of nanotechnology, especially for nanoelectronics and nanooptics. Also, 3D self-assembled nanostructures are possible [32, 33], *e.g.*, utilizing a molecular recognition process for binding complementary DNA strands. This approach also enables part-to-part and part-to-substrate self-assembly by using DNA hybridization [34–35], which is a highly selective programmable process for generating 3D structures with nanoscale precision. Self-assembly of nanowires and CNT has recently attracted significant attention. The reason is to pursue many promising applications both in nanoelectronics and nanooptics as well as in nano–micro interface technologies. The assembly of nanowires and CNT is a challenging task due to their shape anisotropy, which makes their proper integration into a device difficult [13, 14]. Electric fields between the electrodes on a substrate are widely used in dealing with this task and to trigger the self-alignment of rod-shaped nanoobjects [11, 36–38]. The above-mentioned SAM approach is another option for self-assembling CNT, which is based on the fabrication of binding sites through SAM patterning [39].

It is obvious that self-assembly has the potential to radically change the automated fabrication of microscale and nanoscale devices, as it enables the parallel handling of very different objects in a very selective and efficient way. However, despite promising results achieved up to now, this technology still remains on the level of basic research. Critical **challenges** in the development of future devices through self-assembly are the limited availability of suitable integration tools that enable automatic, site-specific, localization and integration of parts into the system, especially when the number of sites is very large, as well as the increasing complexity of parts due to the necessary fabrication steps for the implementation of binding features. The study of defects in self-assembled systems and the introduction of fault-tolerant architectures like in biological systems will also play a prominent role in transferring self-assembly from research laboratories to device manufacturing [13]. These challenges are currently being addressed by the self-assembly research community.

Usually, robot-based assembly (top-down) and self-assembly (bottom-up) are investigated separately. However, **hybrid approaches,** using the advantages of both serial and parallel technologies, seem to be a promising solution that is worth pursuing for different applications in order to achieve higher complexity or productivity [40–43]. A major European research project that started in 2006 aims at combining ultra-precision robots with innovative self-assembly technologies, with the goal of developing a new versatile 3D automated production system with a positioning accuracy of at least 100 nm for complex microscale products [43]. The combination of serial robot-based assembly and parallel self-assembly has never yet been achieved at the industrial scale, and the project team is anxious to prove the viability of this new production concept.

To sum up, self-assembly is a fascinating research field attracting a rapidly increasing number of research groups from multiple disciplines. There is a clear indication that self-assembly can be exploited as a supporting technology, and it will be able to contribute to automated robot-based assembly approaches. The ability to make a complete device by only using self-assembly steps and to become one of the key assembly approaches for the products of MST and nanotechnology remains to be seen.

1.2.2 SPM as a Nanohandling Robot

SPMs can deliver high-resolution images of a wide class of hard and soft samples, which are used, *e.g.*, for materials and surface sciences, bioscience research, or nanotechnology. Additionally, these devices can be used to interact with nanoscale parts, which results in a change of their position or their shape. The first nanomanipulation was reported in 1990 by IBM researchers, who "wrote" the IBM logo with xenon atoms by nanomanipulation with an STM [44]. It was the beginning of the active investigation of this novel nanohandling approach, especially by using AFMs, which offer the widest range of applications in the SPM field.

The nanohandling capabilities of the AFM were discovered rather by accident during AFM imaging scans; some particles just could not be found on the substrate, because they were moved apart by the AFM tip in the previous scan. Controllable positioning of nanoscale parts by an AFM acting as nanohandling robot has been actively investigated in the last decade [45–50]. The ultimate goal is to **automatically assemble nanoscale objects** in nanosystems in ambient conditions, aiming at the rapid prototyping for nanodevices.

To control the movement of a part, it has first to be localized on the substrate by an imaging scan performed in dynamic mode. In the second step, the AFM tip is brought by the AFM control system to the immediate vicinity of the part and is moved afterwards – staying in dynamic mode without AFM feedback in the z-direction – to the centres of the part towards a predetermined location. As a result, the part is pushed in a "blind" feed-forward way by repulsive forces [45, 48]. The precondition is the AFM's ability to perform one-line scans in any direction on the substrate surface. The re-imaging of the area of interest afterwards reveals the results of the manipulation, which are often not satisfying and require frequent trial-and-error experiments. Current research work aims at developing a **high-level AFM control system** to perform predictable nanohandling operations, which might open the door to high-throughput automated nanomanipulation processes [47–52]. Also, several other SPM modes have been used for pushing nanoparticles or molecules [46, 53, 54]. The whole variety of operational modes of SPM [55] has not been fully investigated in regard to nanomanipulation, which often makes trial-and-error experiments necessary for a given task.

Besides the positioning of nanoscale parts, the SPM tip can also be used to **modify surfaces** with nanometer resolution or to change the object shape, *e.g.*, by scratching, indenting, cutting, dissecting, *etc.* [56–62]. A destructive interaction between tip and sample is usually an unwanted effect while imaging. However, for nanomachining purposes, the SPM tip can be exploited as a nanohandling tool, *e.g.*, milling cutter, nanoscalpel, or nanoindenter. Nanoscratching can be implemented by moving an AFM tip on a surface and applying a high load force on the tip. The technique can be used amongst others for mask-free lithography on the nanoscale level. Biological specimens can also be handled in this way. The chromosomal microdissection by AFM can *e.g.* be used for isolating DNA [63, 64]. The AFM is applied first in non-contact mode or in tapping mode for the localization of the cut site in the genetic material. After that, a DNA chromosome is extracted by one AFM linescan and picked up by the AFM tip through hydrophilic attraction.

"Writing" on a substrate surface by the AFM tip is another interesting option for shaping on the nanoscale. The mask-free nanolithography mentioned above can be implemented not only by nanoscratching but also by anodic oxidation [65, 66] or by the so-called dip-pen lithography (DPN) [67, 68]. A line width of a few tens of nanometers can be achieved by both approaches. To perform nanostructuring by **anodic oxidation**, a nanometer-thin metal layer is deposited on the substrate surface, and a voltage is applied between the metal and the conductive AFM tip. Since the metal surface is moistened in ambient atmosphere, an electrolytic process is triggered by the voltage, resulting in a tiny metal oxide dot on the surface. By proper process control, these dots can form a sophisticated nanoscale pattern.

As the name of the approach implies, **dip-pen lithography** works in a manner analogous to that of a dipped pen. Chemical reagents ("ink") are transported from the AFM tip to nanoscopic regions of a target substrate by using capillary forces. This direct-writing process enables the building of nanoscale structures and patterns on different surfaces by literally drawing molecules onto a substrate. The AFM tip is coated with the ink that is to be deposited, and the molecules of the ink are delivered to the surface through a solvent meniscus forming between the tip and the substrate under ambient conditions. This simple method of directly depositing molecules onto a substrate has recently become an attractive tool for nanoscientists, especially because of its versatility. The approach enables molecular deposition of virtually any material (hard and soft) on any substrate. However, ink/substrate combinations must be chosen carefully, so that the ink does not agglomerate or diffuse. Additionally, the ink molecules have to be able to anchor themselves to their deposition location (molecular "glue"). These challenges are the subject of the current basic research activities in this field.

Going back to nanomanipulation and taking into account the primary concern of this book, **automated nanohandling**, the main drawback of AFM-based nanohandling is the lack of real-time visual feedback. The same AFM tip cannot be simultaneously used for both imaging and handling, so that the results of nanohandling have to be frequently visualized by an AFM scan to verify the performance. This procedure makes the nanohandling process inefficient, rather unsuitable for high throughput, and includes uncertainties due to being "blind" during the actual manipulation. Several research groups are trying to overcome this problem by modeling the nanohandling task, which might enable nanomanipulation in open-loop mode, without visual feedback. Having a valid model of the nanopart behavior, including all relevant interactions between tip, part and substrate, it might be possible to mathematically simulate the behavior of the nanoobjects during manipulation and to calculate the expected position of the part in real-time. Such a model is the basis of the so-called **augmented reality** systems "translating" the nanoworld into virtual reality and delivering calculated "visual feedback" during manipulation [69, 70]. This approach, however, requires exact knowledge of nanomanipulation phenomena, which is not available in the current state of the nanosciences. The usability, therefore, of augmented reality systems for automated nanomanipulation, especially in regard to reliability and reproducibility, is currently limited due to a lack of understanding of what exactly is going on during nanomanipulation.

Another problem in regard to nanohandling automation arises when accuracies in the sub-nanometer range are required [71]. Most commercial devices cannot offer a reliable position feedback at this level, and spatial uncertainty in AFM – because of the thermal drift of AFM components, creep and hysteresis of piezo-actuators, and other **variant effects and nonlinearities** – cannot be taken care of in a direct way. Some solutions to this problem have been addressed in [50, 72, 73].

From the automation point of view, a combination of the AFM as a nano-handling robot with other imaging techniques that supply independent visual feedback from the work scene during nanomanipulation by the AFM seems to be most promising. For positioning accuracies of about 0.5 μm and worse, the manipulation can be monitored by a **light microscope**. This approach is frequently used in bioscience research to provide multimodal imaging capabilities for yielding extensive information on biomolecules and biological processes [74].

SPM–SEM hybrid systems are currently attracting rising interest. Here, an SPM head is integrated into the vacuum chamber of an SEM, so that both microscopy methods are used in a complementary fashion to analyze the sample properties, building a sophisticated nanocharacterization device [75–77]. To exploit such a system for nanohandling, SEM can be used as a sensor for visual feedback during nanomanipulation or modification of sample surface by the AFM. Important issues to be addressed are the synchronizing of both microscopes and proper system engineering enabling the AFM tip to act in the SEM's field of view. The latter usually requires a large tilt of the SPM against the electron beam [76]. The use of an environmental SEM (ESEM) may open up new applications as vacuum-incompatible samples such as biological cells can also be analyzed or handled this way.

The latter concept already builds a bridge to the topic of this book, which is introduced in the next section. Indeed, if we exclude the visualization feature of AFM and just think of the nanopositioning capability of the AFM scanner, then we are left with a nanopositioning module carrying a tiny cantilever with a (functionalized) nanoscale tip – a three degrees-of-freedom (DoF) nanohandling robot that can be used for diverse applications in the vacuum chamber of the SEM.

1.3 Automated Microrobot-based Nanohandling

Different concepts are being followed to carry out micro- and nanomanipulation for specific classes of application:

- Purely **manual manipulation** is the most often used method today. It is a common practice, *e.g.*, in medicine and biological research. Even in industry, such tasks are often carried out by specially trained technicians, who position the parts with tiny hammers and tweezers under a microscope and finally fasten them in the desired position. However, with progressive part miniaturization, the tolerances have become smaller and smaller, and the capabilities of the human hand are no longer adequate.
- The application of **teleoperated manipulation** systems, which transform the user's hand motions into the finer 3D motions of the system manipulators by a sophisticated man–machine user interface. Here, special effort is devoted to the development of methods which allow the transmission of feedback information from the work scene (images, forces, noises) in a user-friendly form. The user interface can include a haptic device, providing tactile information that helps the user to operate in a more intuitive way. The haptic device might also be integrated into a virtual reality environment

that is based on the mathematical modeling of the application-relevant phenomena in the nanoworld. However, the fundamental problems of resolution of the fine motion and of speed as well as of repeatability remain, since the motion of the tool is a direct imitation of that of the user's hand.

- The use of **automated nanohandling desktop stations** supported by miniaturized nanohandling robots, which exploit direct drives typically implemented by using piezoelectric, electrostatic, or thermal microactuators. The flexibility of such a microrobot can be enhanced by dividing the actuator system into a **coarse positioning** module (*e.g.*, a mobile platform) and a **fine positioning** module or nanomanipulator carrying an application-specific tool. There is no direct connection between the user's hands and the robot. The user commands are given through a graphical user interface to the station's control system, which generates corresponding commands for the robot actuators. The degree of abstraction of the user instructions is determined by the capabilities of the control system. Several microrobots can be active at the same time to deal with a handling task.

Different aspects of the latter approach are discussed in this book, along with promising applications in MST, nanotechnology, biotechnology and material science.

Microrobotics for handling microscale and nanoscale parts has been established as a self-contained research field for nearly 15 years [78–107]. In recent years, a trend towards the microrobot-based automation of nanohandling processes emerged, and different concepts are currently being investigated [81, 104, 108–127]. **Process feedback**, *i.e.*, the transmission of information from the nanoworld to the macroworld to facilitate the control of the handling process, has emerged as the most crucial aspect of nanohandling automation. Vision feedback and force feedback are the two information channels to be used for automation purposes.

With present technology, it is rather difficult to obtain reliable **force information**, while handling microscale and especially nanoscale parts. Real-time force feedback is, on the other hand, the inherent feature of the AFM, so the use of an AFM probe for nanohandling may provide the necessary force feedback [80]. As optical detection based on a reflected laser beam and utilized in most AFMs imposes serious limitations on the robot's mobility, piezoresistive AFM probes seem to be a more practical solution, even though this offers worse resolution compared to the laser beam [80]. A few promising approaches are introduced in Chapter 6.

Nevertheless, **vision feedback** is often the only way to control a nanohandling process. The capability of a light microscope rapidly decreases with the parts being scaled down to the nanoscale level. Scanning near-field optical microscopy (SNOM) may, however, be exploited for nanoscale manipulations in an ambient environment [86]. The vacuum chamber of an SEM is for many applications the best place for a nanohandling robot. It provides an ample work space, very high resolution up to 1 nm, and a large depth of field (see Chapter 2 for more information). Quite a few research groups have recently been investigating different aspects of nanohandling in SEM, *e.g.*, [78, 88, 89, 93, 94, 102, 108, 116, 121, 128, 129]. However, real-time visual feedback from changing work scenes in the SEM containing moving microrobots is a challenging issue, which is thoroughly analyzed in Chapter 4.

Figure 1.1 presents a generic concept of the **automated microrobot-based nano-handling station (AMNS)**, first introduced in [130] and further gradually developed at the University of Karlsruhe and the University of Oldenburg [120–127].

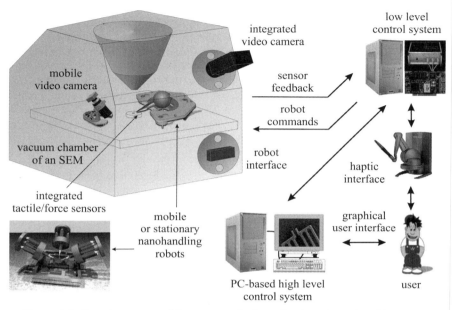

Figure 1.1. Generic concept of the automated microrobot-based nanohandling station

Positioning with nanometer precision is the first precondition for the development of an AMNS. Typically, the microrobots are driven by piezoactuators with resolutions down to sub-nm ranges. The travel range is comparatively large with several tens of millimeters for **stationary** microrobots and with almost no limitation for **mobile** microrobot platforms [128, 131, 132]. The mobile robots of the station have a micromanipulator integrated in their platform, which makes them capable both of moving over longer distances and of manipulating with nanometer accuracy. The latter leads to more flexibility, as the robots can be deployed everywhere inside the SEM vacuum chamber. Stationary robots are, on the other hand, easier to control, which makes them more suitable for high-throughput automation. The robots usually have to be tethered and get power, *e.g.*, driving voltages for the robots' piezoactuators, over the vacuum-sealed mechanical **robot interface** integrated into the SEM wall.

The flexibility of the system can additionally be enhanced by accommodating **several robots** in the station, which can cooperate and carry out handling tasks as a team. Moreover, one of the robots can act as a "cameraman", carrying a miniature video microscope. Such a mobile microscope can deliver images from virtually any point of view in the work space of the nanohandling robots. The combination of a mobile video microscope and SEM with an integrated video camera can contribute to the station's flexibility and versatility, providing a smooth transition between the different magnifications during a nanohandling process.

Various **tools** can be attached to the manipulator and exchanged according to the task to be accomplished. The already-mentioned use of a (functionalized) **AFM probe** with its extremely sharp tip (10–20nm) as a robot tool is currently one of the most actively pursued approaches [4, 80, 86, 133].

SEM, video cameras, force sensors, as well as – if available – position sensors integrated into the robot axes, build the **sensor system of the AMNS**. SEM delivers near-field sensor information for fine positioning of the robot tool, and video cameras provide necessary far-field information for the coarse positioning of the robot. The sensor data are sent to the station's control system for **real-time signal processing**. Its task is to calculate the positions of the robots and their tools as well as the positions of the parts to be handled or other objects of interest. The calculated positions serve as input data for the closed-loop robot control.

Even though the AMNS is designed for automated nanohandling, teleoperated work is also possible. The latter is performed by using a **haptic interface** and/or a **graphical user interface (GUI)**. Teleoperation is often the first step on the way to automation, as it helps the user to learn more about the nanohandling task to be implemented. A good overview of teleoperation techniques and applications is given in [134].

The positioning accuracy of the microrobots during automated nanohandling is affected by several factors, so that a powerful position control approach is required. The latter is run on the low level of the control system. The demands on the **low-level control system** and the implementation results are discussed in detail in Chapter 3. The **high-level control system** is responsible amongst others for path planning, error handling, and the parallel execution of tasks. Both user interfaces, GUI and haptic interface, are supported by the high-level control system as well. An advanced control system architecture that is currently being implemented [135] and tailored for nanohandling automation in the SEM is introduced in Section 7.6.

The AMNS concept has been implemented [136–140] or is currently being implemented [141–145] in different application fields in which (semi-) automated nanomanipulation is required. The implementation results are presented and ongoing work is discussed in Chapter 2 and Chapters 6–10.

1.4 Structure of the Book

The following is a brief summary of the topics covered in the following chapters.

Chapter 2 introduces fundamental considerations for the application of a scanning electron microscope as position sensor of an automated microrobot-based nanohandling station. Due to its resolution, its image acquisition times and scalability, the SEM is the preferred sensor for nanohandling and nanoassembly processes. In order to successfully automate such processes, parasitic forces have to be taken care of. Furthermore, the automatic detection of contact between objects or their height difference is a major problem. Possible solutions to these problems are discussed. Nanohandling and nanoassembly processes are described as a combination of simple tasks and subtasks (primitives). Based on this representation, approaches for the optimization of reliability and throughput of the process can be defined. One of the first implementations of the AMNS concept, a two-robot station for the (semi-) automated handling of silicon lamellae is presented.

Chapter 3 deals with the closed-loop pose controller for the mobile microrobots of the AMNS. It contains a trajectory controller, a motion controller, and an actuator controller. The actuator controller is implemented as a Self-Organizing Locally Interpolating Map (SOLIM), resulting in a learning direct inverse model controller based on Self-Organizing Maps (SOMs). The performance of the SOLIM approach is validated by the learning of an inverse model of a virtual two-joint serial kinematic link. Finally, both an automatically and a manually trained SOLIM actuator controller are applied to a mobile microrobot platform of the AMNS presented in Chapter 2.

Chapter 4 covers the SEM in combination with real-time image processing algorithms, which is proposed as the near-field sensor for the automation of nanohandling tasks. The specific properties of an SEM as image sensor are explained, and the requirements for real-time image processing methods are outlined. Then, the integration of an SEM into an image processing system is demonstrated. The latter enables real-time image access and electron beam control. The main part of the chapter is devoted to the implementation and validation of real-time tracking algorithms in the vacuum chamber of an SEM, using cross-correlation and active contours with region-based minimization.

Chapter 5 deals with a crucial issue of 3D vision feedback for nanohandling in an SEM. For precise handling, it is often necessary to know the position of objects and tools in all three dimensions of space, *i.e.*, 3D visual information is to be acquired. Basic concepts such as stereoscopic imaging in nature are analyzed. In addition, approaches for depth detection in the SEM are illustrated, and a 3D imaging system, tailored for nanohandling tasks, is presented. The application of this system is shown for the handling of two different kinds of nanoobjects.

Chapter 6 gives an overview of the fundamentals and principles of micro/nano force measurement. The main emphasis is placed on the special requirements in force feedback for nanohandling by microrobots. Near visual process monitoring using force feedback is the most important source of sensor data in nanohandling. The main challenge is the measurement of forces in the range of micro- and nanonewtons, to control the interaction between parts and tools as well as for nanomechanical characterization of, *e.g.*, nanowires, carbon nanotubes, or biological cells. The integration of force feedback provides essential process information for both the operator working with telemanipulation devices and the control system operating in automatic mode. The state of the art in force microsensors and force sensing for robot- and AFM-based nanohandling systems is described.

Chapter 7 outlines different issues around the characterization and handling of carbon nanotubes. The basics of CNTs are mentioned, followed by an analysis of structural, electronical, and mechanical properties of CNTs. Fabrication techniques and possible applications for CNTs are explained. Characterization techniques are outlined, and the advantages of CNT characterization in an SEM are demonstrated. The latter requires SEM-tailored nanohandling robot systems; their state of the art is discussed. The developed AMNS for the handling and characterization of CNTs is introduced, and preliminary implementation results are shown. Finally, the novel control system architecture for automated CNT nanohandling is introduced.

Chapter 8 deals with the characterization and manipulation of biological objects by an atomic force microscope. An overview of relevant parameters for AFM measurements in liquids and of soft samples is given. After a short introduction of the biological background, the current state of the art in AFM-based characterization is analyzed, ranging from imaging tasks in biological processes, through conductance measurements on DNA, to activation and measurements of stress-activated ion channels. The first implementation steps of the AMNS for cell characterization are presented, and preliminary results of the experiments with measuring cell elasticity are introduced.

Chapter 9 presents the application of nanohandling microrobotics for nano-mechanical characterization. An emphasis is placed on the characterization of very thin coatings by using nanoindentation. The first section outlines the theoretical background of the well-established method of instrumented indentation for the determination of material hardness and Young's modulus. The second section demonstrates the use of an AMNS for the nanoindentation of epoxy-based electrically conductive adhesives (ECA). The current setup of the station is introduced, and the necessary calibration steps for the apparatus are analyzed. Finally, the first experimental results of hardness measurements on ECA samples are discussed.

Chapter 10 covers current research work on Electron Beam-induced Deposition (EBiD) inside an SEM. EBiD is relevant for the fabrication of nano-mechanical elements, *e.g.*, pins, flexible hinges, or more complicated structures, as well as for nanoassembly. The latter can be accomplished by EBiD of a suitable precursor material between two parts (nanosoldering). Based on the interactions between electron beam and substrate, the rate equation model for EBiD is investigated, and the relevant parameters are analyzed. The molecular flux density has been identified as a crucial parameter for the optimization of the growth rate of the resulting nanostructures. This parameter can be modulated by using an advanced gas injection system (GIS), so design considerations and control methods for microrobot-based GIS are discussed. Basic principles of EBiD process control are illustrated, and promising methods are introduced. Necessary mechanical data is offered from the appropriate literature, and own experimental results are presented.

1.5 References

[1]   Maluf, N. 2000, *An Introduction to Microelectromechanical Systems Engineering*, Artech House., Boston.
[2]   *Nanotechnology: Innovation for Tomorrow's World*, 2004. European Commission, Research DG.
[3]   Holister, P. 2001, *Nanotech: the Tiny Revolution,* CMP Cientifica.
[4]   Sitti, M. 2003, 'Teleoperated and automatic nanomanipulation systems using atomic force microscope probes', *IEEE Conference on Decision and Control*, Maui, Hawaii.
[5]   Sitti, M. & Hashimoto, H. 1999, 'Teleoperated nanoscale object manipulation', in *Recent Advances on Mechatronics*, Springer-Verlag, p. 322.

[6]   Hülsen, H., Fatikow, S., Pham, D. T. & Wang, Z. 2006, 'Self-organising locally interpolating map for the control of mobile microrobots', *Intelligent Production Machines and Systems*, Elsevier, pp. 595–600.

[7]   Holmlin, R. E., Schiavoni, M., Chen, C. Y., Smith, S. P., Prentiss, M. G. & Whitesides, G. M. 2000, 'Light-driven microfabrication: Assembly of multicomponent, three-dimensional structures by using optical tweezers', *Angewandte Chemie-International Edition*, vol. 39, pp. 3503–3506.

[8]   Sinclair, G. , Jordan, P., Laczik, J., Courtial, J. & Padgett, M. 2004, 'Semi-automated 3-dimensional assembly of multiple objects using holographic optical tweezers', *SPIE's Optical Trapping and Optical Micromanipulation*, vol. 5514, pp. 137–142.

[9]   Yu, T., Cheong, F.-C. & Sow, C.-H. 2004, 'The manipulation and assembly of CuO nanorods with line optical tweezers', *Nanotechnology*, vol. 15, pp. 1732–1736.

[10]   Ashkin, A. 1997, 'Optical trapping and manipulation of neutral particles using lasers', *National Academy of Scienc., USA*, vol. 94, pp. 4853–4860.

[11]   Yamamoto, K., Akita, S. & Nakayama, Y. 1996, 'Orientation of carbon nanotubes using electrophoresis', *Japanese Journal of Applied Physics*, vol. 35, pp. L917–L918.

[12]   Cristofanelli, M., De Gasperis, G., Zhang, L., Hung, M.-C., Gascoyne, P. R. C. & Hortobagyi, G. N., 2002, 'Automated electrorotation to reveal dielectric variations related to HER2/neu overexpression in MCF-7 sublines', *Clinical Cancer Research*, vol. 8, pp. 615–619.

[13]   Parviz, B. A., Ryan, D. & Whitesides, G. M. 2003, 'Using self-assembly for the fabrication of nano-scale electronic and photonic devices', *IEEE Transactions on Advanced Packaging*, vol. 26, no. 3, pp. 233–241.

[14]   Morris, C. J., Stauth, S. A. & Parviz, B. A. 2005, 'Self-assembly for microscale and nanoscale packaging: steps toward self-packaging', *IEEE Transactions on Advanced Packaging*, vol. 28, no. 4, pp. 600–611.

[15]   Srinivasan, U., Liepmann, D. & Howe, R. T. 2001, 'Microstructure to substrate self-assembly using capillary forces', *J. Microelectromech. Sys.*, vol. 10, pp. 17–24.

[16]   Palacin, S., Hidber, P. C., Bourgoin, J. P., Miramond, C., Fermon, C. & Whitesides, G. M. 1996, 'Organization of nanoscale objects via polymer demixing', *Chemical Materials*, vol. 8, p. 1316.

[17]   Verma, A. K., Hadley, M. A., Yeh, H.-J. & Smith, J. S. 1995, 'Fluidic self-assembly of silicon microstructures', *45th Electronic Components and Technology Conf.*, Las Vegas, NV, pp. 1263–1268.

[18]   Soga, I., Ohno, Y., Kishimoto, S., Maezawa, K. & Mizutani, T. 2003, 'Fluidic assembly of thin GaAs blocks on Si substrates', *Japanese Journal of Applied Physics*, vol. 42, pp. 2226–2229.

[19]   Böhringer, K. F., Srinivasan, U. & Howe, R. T. 2001, 'Modeling of capillary forces and binding sites for fluidic self-assembly', *IEEE Conf. Micro Electro Mechanical Systems*, Interlaken, Switzerland, pp. 369–374.

[20]   Morris, C. J., Stauth, S. & Parviz, B. A. 2005, 'Using capillary forces for self-assembly of functional microstructures', *2nd Annual Conference Foundations of Nanoscience*, pp. 51–56.

[21]   Gyorvary, E., O'Riordan, A., Quinn, A. J., Redmond, G., Pum, D. &, Sleytr, U. 2003, 'Biomimetic nanostructure fabrication: nonlithographic lateral patterning and self-assembly of functional bacterial S-layers at silicon supports', *Nano Letters*, vol. 3, no. 3, pp. 315–319.

[22]   Böhringer, K. F., Goldberg, K., Cohn, M., Howe, R. T. & Pisano, A. 1998, 'Parallel microassembly using electrostatic force fields', *IEEE International Conference Robotics Automation*, Leuven, Belgium, pp. 1204–1211.

[23] Yang, M., Ozkan, C. S. & Gao, H. 2003, 'Self assembly of polymer structures induced by electric field', *J. Assoc. Lab. Automation*, vol. 8, pp. 86–89.

[24] Tien, J., Terfort, A. & Whitesides, G.M. 1997, 'Microfabrication through electrostatic self-assembly', *Langmuir*, vol. 13, p. 5349.

[25] Clark, T. D., Tien, J., Duffy, D. C., Paul, K. E. & Whitesides, G. M. 2001, 'Self-assembly of 10-microm-sized objects into ordered three-dimensional arrays', *Journal of the American Chemical Society*, vol. 123, pp. 7677–7682.

[26] Onoe, H., Matsumoto, K. & Shimoyama, I. 2004, 'Three-dimensional microself-assembly using hydrophobic interaction controlled by self-assembled monolayers', *Journal of Microelectromechanical Systems*, vol. 13, pp. 603–611.

[27] Breen, T. L., Tien, J., Oliver, S. R. J., Hadzic, T. & Whitesides, G. M. 1999, 'Design and self-assembly of open, regular, 3-D mesostructures', *Science*, vol. 284, pp. 948–952.

[28] Ulman, A. 1996, 'Formation and structure of self-assembled monolayers', *Chemical Reviews*, vol. 96, pp. 1533–1554.

[29] Love, J. C., Wolfe, D. B., Haasch, R., Chabinyc, M. L., Paul, K. E., Whitesides, G. M. & Nuzzo, R. G. 2003, 'Formation and structure of self-assembled monolayers of alka-nethiolates on palladium', *Journal of the American Chemical Society*, vol. 125, pp. 2597–2609.

[30] Maboudian, R., Ashurst, W. R. & Carraro, C. 2000, 'Self-assembled monolayers as anti-stiction coatings for MEMS: Characteristics and recent developments', *Sensors & Actuators*, vol. 82, pp. 219–223.

[31] Li, Y., Moon, K.-S. & Wong, C. P. 2004, 'Electrical property of anisotropically conductive adhesive joints modified by self-assembled monolayer (SAM)', *Electronic Components and Technology Conference*, pp. 1968–1974.

[32] Niemeyer, C. M., Adler, M., Lenhert, S., Gao, S., Fuchs, H. & Chi, L. F. 2001, 'Nucleic acid supercoiling as a means for ionic switching of DNA nanoparticle networks', *ChemBioChem*, vol. 2, pp. 260–264.

[33] Fritz, J., Baller, M. K., Lang, H. P., Rothuizen, H., Vettiger, P., Meyer, E., Guntherodt, H. J., Gerber, C. & Gimzewski, J. K. 2000, 'Translating biomolecular recognition into nanomechanics', *Science*, vol. 288, pp. 316–318.

[34] McNally, H., Pingle, M., Lee, S. W., Guo, D., Bergstrom, D. E. & Bashir, R. 2003, 'Self-assembly of micro- and nano-scale particles using bio-inspired events', *Applied Surface Science*, vol. 214, pp. 109–119.

[35] Wang, C.-J., Lin, L. Y., Dong, J.-C. & Parviz, B. A. 2005, Self-assembled waveguides by DNA hybridization, *Nanotech Insight Conference*, Luxor, Egypt, pp. 75–76.

[36] Smith, P. A., Nordquist, C. D., Jackson, T. N., Mayer, T. S., Martin, B. R., Mbindyo, J. & Mallouk, T. E. 2000, 'Electric-field assisted assembly and alignment of metallic nanowires', *Applied Physics Letters*, vol. 77, pp. 1399–1401.

[37] Nagahara, L. A., Amlani, I., Lewenstein, J. & Tsui, R. K. 2002, 'Directed placement of suspended carbon nanotubes for nanometer-scale assembly', *Applied Physics Letters*, vol. 80, pp. 3826–3829.

[38] Kamat, P. V., Thomas, K. G., Barazzouk, S., Girishkumar, G., Vinodgopal, K. & Meisel, D. 2004, 'Self-assembled linear bundles of single wall carbon nanotubes and their alignment and deposition as a film in a dc field', *Journal of the American Chemical Society*, vol. 126, pp. 10757–10762.

[39] Rao, S. G., Huang, L., Setyawan, W. & Hong, S. 2003, 'Nanotube electronics: Large-scale assembly of carbon nanotubes', *Nature*, vol. 425, pp. 36–37.

[40] Dual, J. 1996, 'Nanorobotics in an optical microscope', *Microsystem Technologies - 5th International Conference on Micro Electro, Opto, Mechanical Systems and Components*, Potsdam.

[41] Codourey, A., Rodriguez, M. & Pappas, I. 1997, 'A Task-oriented teleoperation system for assembly in the microworld', *International Conference on Advanced Robotics*, Monterey, USA.

[42] Shen, W.-M., Will, P. & Khoshnevis, B. 2003, 'Self-assembly in space via self-reconfigurable robots', *Multi-Robot Systems: the Second NATO Workshop*, Kluwer Academic.

[43] *Hydromel 2006*, Hydromel-Integrated Project under the Sixth Framework Programme of the European Community (2002–2006), <NMP-2-CT-2006-026622>.

[44] Eigler, D. M. & Schweizer, E. K. 1990, 'Positioning single atoms with a scanning tunnelling microscope', *Nature*, vol. 344, pp. 524–526.

[45] Junno, T., Deppert, K., Montelius, L. & Samuelson, L. 1995, 'Controlled manipulation of nanoparticles with an atomic force microscope', *Applied Physics Letters*, vol. 66, pp. 3627–3629.

[46] Ramachandran, T. R., Baur, C., Bugacov, A., Madhukar, A., Koel, B. E., Requicha, A. & Gazen, C. 1998, 'Direct and controlled manipulation of nanometer-sized particles using the non-contact atomic force microscope', *Nanotechnology*, vol. 9, pp. 237–245.

[47] Baur, C., Bugacov, A., Koel, B.E, Madhukar, A., Montoya, N., Ramachandran, T. R., Requicha, A., Resch, R. & Will, P. 1998, 'Nanoparticle manipulation by mechanical pushing: underlying phenomena and real-time monitoring, *Nanotechnology*, vol. 9, pp. 360–364.

[48] Requicha, A. A. G., Meltzer, S., Teran Arce, F. P., Makaliwe, J. H., Siken, H., Hsieh, S., Lewis, D., Koel B. E. & Thompson, M. 2001, 'Manipulation of nanoscale components with the AFM: principles and applications', *IEEE International Conference on Nanotechnology*, Maui.

[49] Arbuckle, D. J., Kelly J. & Requicha, A. A. G. 2006, 'A high-level nanomanipulation control framework', *International Advanced Robotics Programme Workshop on Micro and Nano Robotics*, Paris, France.

[50] Mokaberi, B., Yun, J., Wang, M. & Requicha, A. A. G. 2007, 'Automated nano-manipulation with atomic force microscopes', *Proc. IEEE International Conference on Robotics & Automation*, Italy.

[51] Chen, H., Xi, N. & Li, G. 2006, 'CAD-guided automated nanoassembly using atomic force microscopy-based nanorobotics', *IEEE Trans. on Automation Science & Engineering*, vol. 3, no. 3, pp. 208–217.

[52] Makaliwe, J. H. & Requicha, A. A. G. 2001, 'Automatic planning of nanoparticle assembly tasks', *IEEE International Symp. on Assembly and Task Planning*, Fukuoka, Japan, pp. 288–293.

[53] Sheehan, P. E. & Lieber, C. M. 1996, 'Nanotribology and nanofabrication of MoO$_3$ structures by atomic force microscopy', *Science*, vol. 272, p. 1158.

[54] Schaefer, D. M., Reifenberger, R., Patil, A. & Andres, R. P. 1995, 'Fabrication of two-dimensional arrays of nanometer-size clusters with the atomic force microscope', *Applied Physics Letters*, vol. 66, p. 1012.

[55] Meyer, E., Jarvis, S. P. & Spencer, N. D. 2004, 'Scanning probe microscopy in materials science', *MRS Bulletin*, pp. 443–445.

[56] Quate, C. F. 1992, *Manipulation and Modification of Nanometer-scale Objects with the STM*, Highlights in condened matter physics and future prospects, NATO series, Plenum Press, New York, pp. 573–630.

[57] Schimmel, Th., von Blanckenhagen, P. & Schommers, W. 1999, 'Nanometer-scale structuring by application of scanning probe microscopes and self-organisation Processes', *Applied Physics*, vol. A68, p. 263.

[58] von Blanckenhagen P. 1999, *Applications of Scanning Probe Microscopy in Materials Science: Examples of Surface Modifications and Quantitative Analysis*, Atomic Force Microscopy/Scanning Tunneling Microscopy, vol. 3, Springer, New York.

[59] Heckl, W. M. 1997, *Visualization and Nanomanipulation of Molecules in the Scanning Tunnelling Microscope*, Pioneering Ideas for the Physical and Chemical Sciences, Plenum Press, New York.

[60] Klehn, B. & Kunze, U. 1999, 'Nanolithography with an atomic force microscope by means of vector-scan controlled dynamic plowing', *Journal of Applied Physics,* vol. 85, no. 7, pp. 3897–3903.

[61] Fang, T.-H., Weng C.-I. & Chang, J.-G. 2000, 'Machining characterization of the nano-lithography process using atomic force microscopy', *Nanotechnology*, vol. 11, pp. 181–187.

[62] M. Villarroya, *et al.* 2004, 'AFM lithography for the definition of nanometre scale gaps: application to the fabrication of a cantilever-based sensor with electrochemical current detection', *Nanotechnology*, vol. 15, pp. 771–776.

[63] Lü, J., Li, H., An, H., Wang, G., Wang, Y., Li, M., Zhang, Y., & Hu, J., 2004, 'Positioning isolation and biochemical analysis of single DNA molecules based on nanomanipulation and single-molecule PCR', *Journal of Am. Chem. Soc.*, vol. 126, no. 36, pp. 11136–11137.

[64] Heckl, W. M. 1998, 'The combination of AFM nanodissection with PCR', *BIOforum International*, vol. 2, p. 133.

[65] Held, R., Vancura, T., Heinzel, T., Ensslin, K., Holland, M., & Wegscheider, W. 1998, 'In-plane gates and nanostructures fabricated by direct oxidation of semiconductor heterostructures with an atomic force microscope', *Applied Physics Letters*, vol. 73, p. 262.

[66] Wouters, D. & Schubert, U. S. 2004, 'Nanolithography and nanochemistry: probe related patterning techniques and chemical modification for nanometer-sized devices', *Angewandte Chemie, International Edition*, vol. 43, pp. 2480–2495.

[67] Piner, R. D., Zhu, J., Xu, F., Hong, S. & Mirkin, C. A. 1999, 'Dip-pen nanolithography', *Science*, vol. 283, pp. 661–663.

[68] Ginger, D. S., Zhang, H. & Mirkin, C. A. 2004, 'The evolution of dip-pen nanolithography', *Angewandte Chemie, International Edition,* vol. 43, pp. 30–45.

[69] Li, G. Y., Xi, N., Chen, H. P., Pomeroy, C. & Prokos, M. 2005, 'Videolized AFM for interactive nanomanipulation and nanoassembly', *IEEE Transactions on Nanotechnology*, vol. 4, pp. 605–615.

[70] Li, G. Y., Xi, N. & Yu, M. 2004, 'Development of augmented reality system for AFM based nanomanipulation', *IEEE/ASME Transactions on Mechatronics*, vol. 9, no. 2, pp. 199–211.

[71] Requicha, A. A. G. 1993, 'Nanorobots, NEMS and nanoassembly', *Proceedings of the IEEE*, vol. 91, no. 11, pp. 1922–1933.

[72] Mokaberi, B. & Requicha, A. A. G. 2006, 'Drift compensation for automatic nanomanipulation with scanning probe microscopes', *IEEE Transactions on Automation Science and Engineering*, vol. 3, no. 3, pp. 199–207.

[73] Schitter, G., Stark, R. W. & Stemmer A. 2004, 'Fast contact-mode atomic force microscopy on biological specimen by model-based control', *Ultramicroscopy*, vol. 100, pp. 253–257.

[74] Rubio-Sierra, F. J., Heckl, W. M. & Stark, R. W. 2005, 'Nanomanipulation by atomic force micrscopy', *Advanced Engineering Materials*, vol. 7, no. 4, pp. 193–196.

[75] Troyton, M., Lei, H. N., Wang, Z. & Shang, G. 1998, 'A scanning force microscope combined with a scanning electrone microscope for multidimensional data analysis', *Scanning Microscopy*, vol. 12, no. 1, pp. 139–148.

[76] Joachimsthaler, J., Heiderhoff, R. & Balk, L. J. 2003, 'A universal scanning-probe-microscope-based hybrid system', *Measurement Science and Technology*, vol. 14, pp. 87–96.

[77] Kikukawa, A., Hosaka, S., Honda, Y. & Koyanagi, H. 1993, 'Magnetic force microscope combined with a scanning electrone microscope', *Journal of Vacuum Science Technology*, vol. A11, pp. 3092–3098.

[78] Aoyama, H., Iwata, F. and Sasaki, A. 1995, 'Desktop flexible manufacturing system by movable miniature robots', *International Conference on Robotics and Automation*, pp. 660–665.

[79] Brufau-Penella, J., Puig-Vidal, M., López-Sánchez, J., Samitier, J., Driesen, W., Breguet, J.-M., Gao, J., Velten, T., Seyfried, J., Estaña, R. & Wörn, H. 2005, 'MICRoN: small autonomous robot for cell manipulation applications', *IEEE International Conference on Robotics and Automation*.

[80] Sitti, M. & Hashimoto, H. 2000, 'Two-dimensional fine particle positioning under optical microscope using a piezoresistive cantilever as a manipulator', *Journal of Micromechatronics*, vol. 1, no. 1, pp. 25–48.

[81] Codourey, A., Zesch, W., Büchi, R. 1995, 'A robot system for automated handling in mirco-world', *IEEE/RSJ International Conference on Intelligent Robots and Systems*, Pittsburgh, Pennsylvania, pp. 185–190.

[82] Dario, P., Valleggi, R., Carrozza, M. C., Montesi, M. C. & Cocco, M. 1992, 'Microactuators for microrobots: a critical survey', *Journal of Micromechanics and Microengineering*, vol. 2, pp. 141–157.

[83] Driesen, W., Varidel, T, Regnier, S. & Breguet, J.-M. 2005, 'Micro manipulation by adhesion with two collaborating mobile micro robots', *Journal of Micromechanics and Microengineering*, vol. 15, pp. 259–267.

[84] Fatikow, S., Magnussen, B. & Rembold, U. 1995, 'A piezoelectric mobile robot for handling of microobjects', *International Symposium on Microsystems, Intelligent Materials and Robots*, Sendai, pp. 189–192.

[85] Fujita, H. 1993, 'Group work of microactuators', *International IARP-Workshop on Micromachine Technologies and Systems*, Tokyo, pp. 24–31.

[86] Fukuda, T., Arai, F. & Dong, L. 2003, 'Assembly of nanodevices with carbon nanotubes through nanorobotic manipulations', *Proceedings of the IEEE*, vol. 91 no. 11, pp. 1803–1818.

[87] Gengenbach, U. 1996, 'Automatic assembly of microoptical components', *International Symposium on Intelligent Systems & Advanced Manufacturing*, vol. 2906, pp. 141–150.

[88] Hatamura, Y., Nakao, M. and Sato, T. 1995, 'Construction of nano manufacturing world', *Microsystem Technologies*, vol. 1, pp.155–162.

[89] Johansson, St. 1995, 'Micromanipulation for micro- and nano-manufacturing', *INRIA/IEEE Conference on Emerging Technologies and Factory Automation*, Paris, vol. 3, pp. 3–8.

[90] Magnussen, B., Fatikow, S. & Rembold, U. 1995, 'Actuation in microsystems: problem field overview and practical example of the piezoelectric robot for handling of microobjects', *INRIA/IEEE Conference on Emerging Technologies and Factory Automation*, vol. 3, pp. 21–27.

[91] Mitsuishi, M., Kobayashi, K., Nagao, T., Hatamura, Y., Sato, T. & Kramer, B. 1993, 'Development of tele-operated micro-handling/machining system based on information transformation', *IEEE/RSJ International Conference on Intelligent Robots and Systems*, Yokohama, pp. 1473–1478.

[92] Morishita, H. & Hatamura, Y. 1993, 'Development of ultra precise manipulator system for future nanotechnology', *International IARP Workshop on Micro Robotics and Systems*, Karlsruhe, pp. 34–42.

[93] Mazerolle, S., Rabe, R., Fahlbusch, S., Michler, J. & Breguet, J.-M. 2004, 'High precision robotics system for scanning electron microscopes', *Proceedings of the IWMF*, vol. 1, pp. 17–22.

[94] Sato, T., Kameya, T., Miyazaki, H. & Hatamura, Y. 1995, 'Hand-eye System in nano manipulation world', *International Conference on Robotics and Automation*, Nagoya, pp. 59–66.

[95] Arai, T. & Tanikawa, T. 1997, 'Micro manipulation using two-finger hand', *Proc. of the Int. Workshop on Working in the Micro- and Nano-Worlds: Systems to Enable the Manipulation and Machining of Micro-Objects, IEEE/RSJ International Conference on Intelligent Robots and Systems*, France, pp. 12–19.

[96] Breguet, J.-M. & Renaud, Ph. 1996, 'A 4-degrees-of-freedom microrobot with nanometer resolution', *Robotica*, vol. 14, pp. 199–203.

[97] Fearing, R. S. 1992, 'A miniature mobile platform on an air bearing', *3rd International Symposium on Micro Machine and Human Science*, Nagoya, pp. 111–127.

[98] Fukuda, T. & Ueyama, T. 1994, 'Cellular robotics and micro robotic systems', *World Scientific*, Singapore.

[99] Hesselbach, J., Pittschellis, R. & Thoben, R. 1997, 'Robots and grippers for micro assembly', *9th International Precision Engineering Seminar*, Braunschweig, pp. 375–378.

[100] Menciassi, A., Carozza, M. C., Ristori, C., Tiezzi, G. & Dario, P. 1997, 'A workstation for manipulation of micro objects', *IEEE International Conference on Advanced Robotics*, Monterey, California, pp .253–258.

[101] Rembold, U. & Fatikow, S. 1997, 'Autonomous microrobots', *Journal of Intelligent and Robotic Systems*, vol. 19, pp. 375–391.

[102] Weck, M., Hümmler, J. & Petersen, B. 1997, 'Assembly of hybrid micro systems in a large-chamber scanning electron microscope by use of mechanical grippers', *International Conference on Micromachining and Microfabrication*, Austin, Texas, pp. 223–229.

[103] Sitti, M. 2001, 'Survey of nanomanipulation systems', *1st IEEE Conference on Nanotechnology*, Maui, HI, USA, pp. 75–80.

[104] Martel S., Madden P., Sosnowski L., Hunter I. & Lafontaine S. 1999, 'NanoWalker: a fully autonomous highly integrated miniature robot for nano-scale measurements', *European Optical Society and SPIE International Symposium on Environsense, Microsystems Metrology and Inspection*, vol. 3825, Munich, Germany.

[105] Bourjault, A. & Chaillet, N. 2002, *La microrobotique*, Hermes.

[106] Ferreira, A., Fontaine, J-G. & Hirai, S. 2001, 'Virtual reality-guided microassembly desktop workstation', *5th Japan-France Congress on Mecatronics*, Besançon, France, pp. 454–460.

[107] Montane, E., Miribel, P., Puig-Vidal, M., Bota, S. A. & Samitier, J. 2001, 'High voltage smart power circuits to drive piezoceramic actuators for microrobotic applications', *IEE Circuits Devices and Systems*, vol. 148, pp. 343–347.

[108] Schmoeckel, F., Fahlbusch, St., Seyfried, J., Buerkle, A. & Fatikow, S. 2000, 'Development of a microrobot-based micromanipulation cell in an SEM', *SPIE's International Symposium on Intelligent Systems & Advanced Manufacturing: Conference on Microrobotics and Microassembly*, Boston, MA, USA, pp. 129–140.

[109] Fatikow, S. 1996, 'An automated micromanipulation desktop-station based on mobile piezoelectric microrobots', *SPIE's International Symp. on Intelligent Systems &*

Advanced Manufacturing, Boston, MA, vol. 2906: Microrobotics: Components and Applications, pp. 66–77.

[110] Fatikow, S. & Rembold, U. 1997, *Microsystem Technology and Microrobotics*, Springer-Verlag, Berlin.

[111] Fatikow, S., Rembold, U. & Wörn, H. 1997, 'Design and control of flexible microrobots for an automated microassembly desktop-station', *SPIE's International Symposium on Intelligent Systems & Advanced Manufacturing*, vol. IS02: Microrobotics and Microsystem Fabrication, Pittsburgh, PA, pp. 66–77.

[112] Fatikow, S., Munassypov, R. & Rembold, U. 1998, 'Assembly planning and plan decomposition in an automated microrobot-based microassembly desktop station', *Journal of Intelligent Manufacturing*, vol. 9, pp. 73–92.

[113] Fahlbusch, St., Buerkle, A. & Fatikow, S. 1999, 'Sensor system of a microrobot-based micromanipulation desktop-station', *International Conference on CAD/CAM, Robotics and Factories of the Future*, Campinas, Brazil, vol. 2, no. RW4, pp. 1–6.

[114] Yang, G., Gaines, J. A. & Nelson, B. J. 2003, 'A supervisory wafer-level 3D microassembly system for hybrid MEMS fabrication', *Journal of Intelligent and Robotic Systems*, vol. 37, pp. 43–68.

[115] Bleuler, H., Clavel, R., Breguet, J.-M., Langen, H. & Pernette, E. 2000, 'Issues in precision motion control and microhandling', *International Conference on Robotics & Automation*, San Francisco, USA.

[116] Kasaya, T., Miyazaki, H., Saito, S. & Sato, T. 1999, 'Micro object handling under SEM by vision-based automatic control', *International Conference on Robotics and Automation*, Detroit, USA, pp. 2736–2743.

[117] Clevy, C., Hubert, A., Agnus, J. & Chaillet, N. 2005, 'A micromanipulation cell including a toll changer', *Journal of Micromechanics and Microengineering*, vol. 15, pp. 292–301.

[118] Tanikawa, T., Kawai, M., Koyachi, N., Arai, T., Ide, T., Kaneko, S., Ohta, R. & Hirose, T. 2001, 'Force control system for autonomous micromanipulation', *International Conference on Robotics and Automation*, Seoul, pp. 610–615.

[119] Fatikow, S. 2000, *Microrobotics and Microassembly* (in German), Teubner Verlag, Stuttgart.

[120] Fatikow, S., Seyfried, J., Fahlbusch, St., Buerkle, A. & Schmoeckel, F. 2000, 'A flexible microrobot-based microassembly station', *Journal of Intelligent and Robotic Systems*, Kluwer, Dordrecht, vol. 27, pp. 135–169.

[121] Fatikow, S., Fahlbusch, St., Garnica, St., Hülsen, H., Kortschack, A., Shirinov, A. & Sill, A. 2002, 'Development of a versatile nanohandling station in a scanning electron microscope', *3rd International Workshop on Microfactories*, Minneapolis, Minnesota, USA, September 16-18, pp. 93–96.

[122] Hülsen, H., Trüper, T., Kortschack, A., Jähnisch, M. & Fatikow, S. 2004, 'Control system for the automatic handling of biological cells with mobile microrobots', *American Control Conference*, Boston, MA, USA, pp. 3986–3991.

[123] Fatikow, S., Kortschack, A., Hülsen, H., Sievers, T. & Wich, Th. 2004, 'Towards fully automated microhandling', *4th International Workshop on Microfactories*, Shanghai, China, vol. 1, pp. 34–39.

[124] Wich, Th., Sievers, T., Jähnisch, M., Hülsen, H. & Fatikow, S. 2005, 'Nanohandling automation within a scanning electron microscope', *IEEE International Symposium on Industrial Electronics*, Dubrovnik, Croatia, pp. 1073–1078.

[125] Sievers, T. & Fatikow, S. 2005, 'Visual servoing of a mobile microrobot inside a scanning electron microscope', *IEEE International Conference on Intelligent Robots and Systems*, Edmonton Canada, 2-6, pp. 1682–1686.

[126] Fatikow, S., Eichhorn, V., Wich, Th., Hülsen, H., Hänßler, O. & Sievers, T. 2006 'Development of an automatic nanorobot cell for handling of carbon nanotubes', *IARP/IEEE-RAS/EURON International Workshop on Micro/Nano Robotics*.

[127] Fatikow, S., Wich, Th., Hülsen, H., Sievers, T. & Jähnisch, M. 2007, 'Microrobot system for automatic nanohandling inside a scanning electron microscope', *IEEE-ASME Transactions on Mechatronics*.

[128] Nakajima, M., Arai, F., Dong, L., Nagai, M. & Fukuda, T. 2004, 'Hybrid nanorobotic manipulation system inside scanning electron microscope and transmission electron microscope', *IEEE/RSJ International Conference on Intelligent Robots and Systems*, Sendai, Japan, pp. 589–594.

[129] Misaki, D., Kayano, S., Wakikaido, Y., Fuchiwaki, O. & Aoyama, H. 2004, 'Precise automatic guiding and positioning of microrobots with a fine tool for microscopic operations', *IEEE/RSJ International Conference on Intelligent Robots and Systems*, Sendai, Japan, pp. 218–223.

[130] Fatikow, S., Buerkle A. & Seyfried J. 1999, 'Automatic control system of a micro-robot-based microassembly station using computer vision', *SPIE's International Symposium on Intelligent Systems & Advanced Manufacturing*, Conference on Microrobotics and Microassembly, Boston, MA, USA, pp. 11–22.

[131] *Klocke Nanotechnik* 2005, http://www.nanomotor.de/.

[132] Kortschack, A. & Fatikow, S. 2004, 'Development of a mobile nanohandling robot', *Journal of Micromechatronics*, vol. 2, no. 3, pp. 249–269.

[133] Mircea, J. & Fatikow, S. 2007, 'Microrobot-based nanoindentation of an epoxy-based electrically conductive adhesive', *IEEE NANO-Conference*, Hong Kong.

[134] Ferreira, A. & Mavroidis, C. 2006, 'Virtual reality and haptics for nanorobotics', *IEEE Robotics & Automation Magazine*, pp. 78–92.

[135] Stolle, Ch. & Fatikow, S. 2007, 'Control system of an automated nanohandling robot cell', *22nd IEEE International Symposium on Intelligent Control*, Singapore.

[136] Trüper, T., Kortschack, A., Jähnisch, M., Hülsen, H. & Fatikow, S. 2004, 'Transporting cells with mobile microrobots', *IEE Proceedings - Nanobiotechnology*, vol. 151, no. 4, pp. 145–150.

[137] Hülsen, H., Trüper, T., Kortschack, A., Jähnisch, M. & Fatikow, S. 2004, 'Control system for the automatic handling of biological cells with mobile microrobots', *American Control Conference*, Boston, MA, USA, pp. 3986–3991.

[138] Wich, Th., Sievers, T. & Fatikow, S. 2006, 'Assembly inside a scanning electron microscope using electron beam induced deposition', *IEEE International Conference on Intelligent Robots and Systems*, Beijing, China, pp. 294–299.

[139] Sievers, T., Garnica, S., Tautz, S., Trüper, T. & Fatikow, S. 2007, 'Microrobot station for automatic cell handling', *Journal on Robotics & Autonomous Systems*.

[140] Fatikow, S., Eichhorn, V., Tautz, S. & Hülsen, H. 2006, 'AFM probe-based nanohandling robot station for the characterization of CNTs and biological cells', 5th *International Workshop on Microfactories*, Besancon, Frankreich.

[141] Hagemann, S., Krohs, F. & Fatikow, S. 2007, 'Automated characterization and manipulation of biological cells by a nanohandling robot station', *International Conference Nanotech Northern Europe*, Helsinki, Finland.

[142] Fatikow, S., Eichhorn, V., Krohs, F., Mircea, J., Stolle, Ch. & Hagemann, S. 2007, 'Development of an automated microrobot station for nanocharacterization', *SPIE's International. Conference Microtechnologies for the New Millennium*, Maspalomas, Gran Canaria.

[143] Luttermann, T., Wich, Th., Stolle, Ch. & Fatikow S. 2007, 'Development of an automated desktop station for EBiD-based nano-assembly', *2nd International Conference on Micro-Manufacturing*, Greenville, South Carolina, USA.

[144] Jasper, D. & Fatikow, S. 2007, 'CameraMan – nanohandling robot cell inside a scanning electron microscope with flexible vision feedback', *SPIE's International Symposium on Optomechatronic Technologies*, Lausanne, Switzerland.
[145] Eichhorn, V., Carlson, K., Andersen, K.N., Fatikow, S. & Boggild, P. 2007, 'Nanorobotic manipulation setup for pick-and-place handling and non-destructive characterization of carbon nanotubes', *20th International Conference on Intelligent Robots and Systems*, San Diego, California.

Robot-based Automated Nanohandling

Thomas Wich and Helge Hülsen

Division of Microrobotics and Control Engineering,
Department of Computing Science,
University of Oldenburg, Germany

2.1 Introduction

Within the last ten years, the interest of industry and research and development institutes in the handling of micro- and nanometer-sized parts has grown rapidly [1]. Micro- and nanohandling has become a very common task in the industrial field and in research in the course of ongoing miniaturization. Typical applications include the manipulation of biological cells under an optical light microscope, the assembly of small gears for miniaturized gearboxes, the handling of lamellae cut out of a silicon wafer in the semiconductor industry, and the chemical and physical characterization of nanoscale objects. The number of applications for nanohandling and nanoassembly is expected to grow rapidly with the development of nanotechnology. The handling process is the precursor of the assembly process, hence, in this chapter, these expressions are used equally where not explicitly stated.

Often, a distinction is made between macro-, micro-, and nanoscale assembly with respect to the part size, where the part dimensions are larger than 1 mm for macroscale, smaller than 1 mm for microscale, and smaller than 1 μm for nanoscale handling [2]. This distinction should even be tightened, because the interaction between the handling system and the handled parts is mostly determined by its smallest dimension, which determines the necessary **positioning accuracy** (Chapter 1). A typical example for parts with very exotic aspect ratios, but which are still considered as nanometer-sized parts, are nanofibers or nanotubes. They can, *e.g.*, be produced by electro-spinning [3], which results in lengths in the cm-range. However, most of these handling processes are still accomplished by means of manual operation [4-6]. Very often, this leads to either very long process durations with high reliability or to shorter durations with low reliability.

The handling process itself can be distinguished by the number of parts handled at a time, *i.e.,* when only one part is handled at a time the expression **serial**

approach is used, in contrast to **parallel approach** for simultaneous handling of multiple parts [2]. These two approaches are based on very different considerations: the serial approach is the more conservative one, where the principles of handling known from the macroscale are adapted to the micro- and nanoscale. Naturally, special considerations have to be taken into account when downscaling, which is one of the main issues discussed in this chapter.

The other approach is the parallel handling of micro- and nanometer-sized parts, where force fields are used to position and orientate objects. The aim here is to maintain the advantages of batch processes, as applied in the MEMS (micro-electro-mechanical systems) and semiconductor industries.

Handling processes can be evaluated in respect of two parameters: **throughput** and **reliability**. (Massively) parallel handling or manufacturing aims at very high throughputs, *e.g.,* assembly of dies in the semiconductor industry. In contrast, the serial approach handles only one part at a time, where high reliability is the main requirement because of the special value of the handled parts. A typical example is the handling of TEM (transmission electron microscope) lamellae that are small slices (approx. 20 μm × 10 μm × 100 nm) cut out of a processed silicon wafer by a focused ion beam (FIB). These lamellae are then transferred to a TEM for inspection, *i.e,.* the TEM lamellae are the micrographs of the semiconductor industry. This approach is very important for discovering failures in semiconductor processes, and high reliability of handling is required. In general, the criteria to be considered when distinguishing between a serial and a parallel approach are the number of parts to be handled or assembled, the complexity of the process, and the individuality of the single parts.

The given examples for the serial and parallel approach represent two applications with very different demands. Still, the goal is always to maximize reliability and throughput for every handling system, independent of the chosen approach, but sometimes reliability is more important than throughput, and vice versa. This chapter focuses on automation issues in the field of **nanohandling for the serial approach**.

The handling of nanoscale objects usually takes place in a special environment necessary for observation, *e.g,.* under optical microscopes or scanning electron microscopes (SEM). The advantages and disadvantages of the single **vision sensors** and resulting consequences for the handling of objects with respect to automation will be discussed in Section 2.2. When the size of the handled objects is reduced, the relationship between surface and volume changes dramatically, resulting in a stronger influence of **parasitic forces** on the objects compared to the macroworld. These forces have to be overcome in order to successfully automate handling and assembly processes (more information in Section 2.3). Another major issue for the process automation discussed in Section 2.3 is the **contact detection**, which is the detection of height distances between objects. Critical issues regarding handling processes and the **planning** of these processes by a combination of simple tasks and subtasks will be discussed in Section 2.4. Based on these, measures and approaches for **optimizing reliability and throughput** of handling and assembly processes will be described and discussed in Section 2.5. The setup and results achieved with an **automated microrobot-based nanohandling station (AMNS)**, implemented for the handling of TEM lamellae, will be described in Section 2.6.

2.2 Vision Sensors for Nanohandling Automation

Nanohandling can be seen as the continuation of macrorobotics to the nanometer scale, taking several new issues into account. One of these issues is the need for a near-field vision sensor (Chapter 1), providing visual information about the handling process, *i.e.*, making the nanoworld accessible to the human eye.

Although nanohandling tasks could be performed without any visualization by relying on measurement data from the handling tools like forces, velocity, and time, it is much more precise to measure geometric values (position, length, distance, *etc.*) directly using vision-based sensors. As will be shown in Section 2.4, especially for nanohandling tasks the continuous gathering of geometric information is very important to achieve high reliability. There are three main reasons for this:

1. The **parasitic forces** result in an apparently unpredictable behavior of objects, as many parameters for these parasitic forces are either unknown or change continuously. A typical example is the release of a small micro- or nanoscale object from a gripper. Opening the gripper jaws does not necessarily lead to dropping the object due to gravitation; instead the object often sticks to one of the jaws due to parasitic forces.
2. The sensors that are used in nanohandling and manufacturing to determine an object's state (*e.g.,*. "gripped"), position (*e.g.,* "distance to the gripper") or orientation ("upright") are usually bigger in size than the objects them- selves. By contrast, when handling parts in the macroscale, the sensors are smaller or in the same range as the tools. For example, modern grippers can easily handle an egg without damaging it, due to integrated force sensors, whereas the integration of force sensors into a gripper with a jaw cross-section of a few micrometers is extremely challenging (Chapter 6). Thus, on the nanoscale the **sensor density**, *i.e.*, the amount of sensors per handling tool, is much smaller than on the macroscale.
3. The near-field vision sensors (Chapter 1) considered are **global vision sensors**, *i.e.*, they measure a scene based on a global coordinate system.

The combination of the first two circumstances – apparently undetermined beha- vior of the objects due to hardly determinable parasitic forces and a significantly lower sensor density – are the major challenges for any nanohandling process.

Vision sensors, however, provide a tremendous amount of information, because objects can be recognized and their relationship to each other can thus be qualified and quantified. A typical example is the gripping of a small glass sphere, where it is necessary to know if the sphere is between the gripper jaws. By evaluating this information using object recognition, it is possible, *e.g.*, to operate without a force sensor on the gripper jaws.

2.2.1 Comparison of Vision Sensors for Nanohandling Automation

Of further interest are the **geometric scales**, which have to be bridged during nanohandling tasks. Consider a typical robot used for assembly tasks in the automotive industry. The range of the robot is several meters ($\sim 10^0$ m), and typical position accuracies are in the range of a millimeter ($\sim 10^{-3}$ m); the dimension of the geometric scale is four orders of magnitude. As a comparison, the range of the AMNS described in Section 2.6 is from one decimeter ($\sim 10^{-1}$ m) to a position accuracy of 100 nm ($\sim 10^{-7}$ m), thus seven orders of magnitude are passed through. For most applications, it is preferable to have a vision sensor that can be zoomed seamlessly, in order to cover the full range of geometric scale.

The resolution of the imaging sensor is defined by the distance between two objects needed to recognize them as separated. Therefore, the resolution indicates the size of the object, which the imaging sensor can track in a handling process. However, it must be kept in mind that among other factors, the resolution in the scanning microscopy is strongly dependent on the **image acquisition time**, *e.g.*, the slower an image is scanned, the better the resolution becomes.

The image acquisition time has a major influence on the automation process, not only with respect to the image quality. It also determines the maximum velocity with which objects or tools can be moved under observation. This topic will be discussed in Section 2.2.3.

Of considerable interest for object recognition purposes and for the user is the **information** contained in the image acquired from the sensor. Images based on light from the visible spectrum can be colored (light optical microscope) with the colors giving information about the geometric surface of the object. By contrast, images gathered by an SEM using an energy dispersive X-ray detector (EDX-detector) contain information about the material from a depth of up to 3 µm from the object's surface. Images gathered by a scanning probe microscope (SPM), *e.g.*, an atomic force microscope (AFM), contain information about the tip-sample interaction, *i.e.*, the distinction between two objects lying on each other is hardly possible without previous knowledge. With regard to automation of handling tasks, the necessary information is based on geometric conditions (*e.g.*, "position", "orientation" and "distance"). However, the information contained in an image can be of different quality, *e.g.*, material, material contrast, conductivity, or atomic forces. Thus the mapping from the image information to the geometric information about an object condition can in many cases only be fulfilled with previous knowledge.

The **interactions** between sensor medium and object have also to be considered as the medium influences the object. For example, in the SEM, the electron beam used for scanning the object can lead to electrical charging of the object or even to damage. The tip of an AFM used for scanning an object can move the object due to parasitic forces and thus accidentally interfere with a handling process.

Very important for handling tasks is the **dimensionality** of the gathered image. Two-dimensional images are common, *e.g.*, in light microscopy or SEM. Necessary for handling tasks is often the determination of the geometric condition of the object in three dimensions, which has to be done when only 2D images are available. Specialized methods are discussed in Section 2.3.2 and in Chapter 5.

Other issues are the environmental requirements and the constraints imposed by the vision sensor. Typical restrictions are given in installation space, vacuum compatibility, and electromagnetic shielding.

Three typical vision sensors used for nanohandling are given in Table 2.1. The **light microscope** has its major domain as a vision sensor for the handling of micro-sized parts, due to its comparatively low resolution. Still, the frame rate is only determined by the quality of the camera grabbing the images and not by the medium itself, as it is not a scanning vision sensor. Furthermore, the geometric information can be directly evaluated. The 2D images with a low depth of focus are well suited for automation purposes, as the height difference between two objects can easily be quantified (Section 2.3.2).

Within this book the focus regarding vision sensors is on **scanning electron microscopes**, whose major advantages are their high resolution combined with a high range of magnification. The information contained in the images about the objects is mapped to geometric conditions, even with the commonly used Everhart-Thornley SE detector. Further advantages can be drawn from imaging with specialized detectors (*e.g.,* object recognition through material identification). A drawback for automation is certainly the comparatively low frame rate. The consequences of this issue are considered in Section 2.3, and possible solutions are presented in Chapter 4. Additionally, the high depth of focus of the 2D images complicates the determination of height distances between objects. Solutions for this problem will be discussed in Section 2.3.3. The electron beam scanning the objects and tools can also lead to electric charging and thus undetermined parasitic forces.

Atomic force microscopes become more and more interesting for the automation of nanohandling tasks where the object's size is only a few nm. Due to its very high resolution, the AFM – or more generally the SPM – is the only option. However, its very low frame rate prevents the automation of processes at reasonable speeds; this might change when the first high-speed AFMs become commercially available [7]. From an automation aspect, the generation of image information that can easily be transferred into three-dimensional views of the handling process is very advantageous, although a reference level (usually the substrate, on which the objects are placed) has to be present.

Recapitulating the issues discussed above, the conclusion can be drawn that the light microscope has the most advantages regarding microhandling. The low requirements with regard to the environment and the sensor medium make it a comparatively cheap and flexible sensor. Although light microscopes can open the door to the nanometer range, automation of nanohandling is hardly possible due to a lack of geometric information. For example, silicon nanowires with a diameter of a couple of hundred nanometers and a length of several micrometers are visible under the light microscope as interference.

The application of AFMs for nanohandling tasks is reasonable for extremely small objects. The gap between light microscopy and AFM is best bridged using SEMs.

Table 2.1. Comparison of light microscope, SEM, and AFM as vision sensors for nanoscale automation

	Light microscope	**Scanning electron microscope (SEM)**	**Atomic force microscope (AFM)**
Resolution	Several hundred nm	1-3 nm for thermionic electron guns, approx. 0.1 nm for field emission guns	$\ll 1$ nm
Orders of magnitude in scale (oom) = difference between field length at least magnification and resolution	$10^{-2}\,\text{m} - 10^{-7}\,\text{m}$ 6 oom	$10^{-3}\,\text{m} - 10^{-10}\,\text{m}$ 8 oom	$10^{-4}\,\text{m} - 10^{-10}\,\text{m}$ 7 oom
Frame rate or image acquisition time	Approx. 25 frames per second, $T = 40$ ms	$T = 0.1 - 100$ s, depending on image quality and size	$T = 10 - 100$ s, 15 ms have been reported for high speed AFMs [7]
Image information	• Surface condition • Color • Geometric shape	• Topography (SE detector) • Material contrast (SE/BSE detector) • Material (EDX detector)	• Tip-sample interaction, force, height
Interaction between vision medium and object	Visible light, interactions used for optical trapping	Electron beam, can lead to charging, heating, and damage	Force between AFM tip and object is measured, surface can be damaged, object can interfere with the tip
Image dimension	2D low depth of focus	2D high depth of focus	3D
Environment	Ambient atmosphere	High vacuum or reduced pressure	Ambient atmosphere

2.2.2 Zoom Steps and Finding of Objects

Although the handled objects are in the nanometer to micrometer range, they must often be transported over several centimeters (*e.g.,* handling of TEM lamellae). For assembly applications, micro- to nanometer-sized parts must mostly be joined onto parts with a size of several millimeters (*e.g.,* bonding of carbon nanotubes on AFM tips).

Therefore, several orders of magnitude on the geometric scale have to be gone through, mostly starting on the centimeter scale and zooming in multiple steps. Typical examples for the different zoom steps are given in Figures 2.14 and 2.15. Usually, after every zoom step a positioning step follows, in order to center the object for the next zoom step.

Quite often, different actuators and/or sensors are used for different magnifications, *e.g.,* a DC motor for coarse positioning at low magnifications and a piezo drive for fine positioning at high magnifications. Consequently, every combination of a zoom step and its subsequent positioning step has to be seen as a task of its own in the process chain with its individual reliability (Section 2.4). Hence, the number of **zoom-and-center (ZAC) steps** should be minimized. It is therefore of considerable interest to calculate how many ZAC steps are at least necessary, until an object is sufficiently magnified, starting from a defined imaged area.

For determining the number of steps, the following approach is sensible: the imaged area should be square-shaped with an edge length k_n; the minimum edge length of the next (k_{n+1}) magnification step can then be determined by the following equation:

$$k_{n+1} = H \cdot s + u_{Pos} + \frac{u_{Pixel}}{A_{Pixel}} \cdot k_n .$$ (2.1)

In Figure 2.1, the single terms are illustrated. The term $H \cdot s$ represents the minimum desired image size, consisting of the structural size s of the object and the hull factor H. The structural size s is the size of an object (*e.g.,* diameter of a carbon nanotube, CNT) or the distance between object and tool in a handling process. The hull factor H determines the image size compared to the object size needed for the handling process.

After every zoom step, the object is centered again. The actuator performing this positioning step has an accuracy of u_{Pos}, which has to be added to the hull. Furthermore, the edge length k_{n+1} of the $(n+1)$th zoom step depends on the accuracy of the object recognition of the n-th step, *i.e.,* defining the maximum difference between measured and real position. The term $u_{Pixel} \cdot k_n / A_{Pixel}$ reflects the object recognition accuracy, consisting of the number of pixels A_{Pixel} per edge length k_n and the accuracy of the object recognition u_{Pixel} given in pixels.

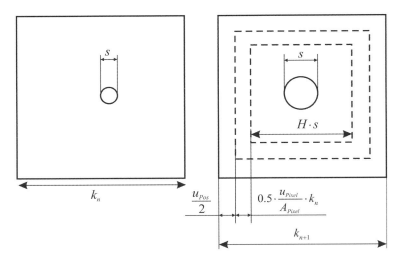

Figure 2.1. Two subsequent zoom-and-center steps during magnification of the object with diameter s. The left image has a lower magnification compared to the right one, *i.e.,* $k_n > k_{n+1}$. In the right image, the parameters used in Equation 2.2 are shown. The inner square is used as the area of movement for the object with a diameter s.

Based upon the above equation, the edge length of the n-th zoom step can be calculated by transferring the recurrence relation in Equation 2.1 into an explicit expression for k_n:

$$k_n = \left(k_0 - \frac{H \cdot s + u_{Pos}}{1 - u_{Pixel} / A_{Pixel}} \right) \cdot \left(\frac{u_{Pixel}}{A_{Pixel}} \right)^n + \frac{H \cdot s + u_{Pos}}{1 - u_{Pixel} / A_{Pixel}}. \tag{2.2}$$

Assuming that the edge size of the n-th frame should be equal to the ε-tolerated sum of the hull and inaccuracies through positioning and object recognition, Equation 2.2 can be written as

$$(1 + \varepsilon) \cdot (H \cdot s + u_{Pos}) \cdot \left(1 + \frac{u_{Pixel}}{A_{Pixel}} \right) =$$
$$\left(k_0 - \frac{H \cdot s + u_{Pos}}{1 - u_{Pixel} / A_{Pixel}} \right) \cdot \left(\frac{u_{Pixel}}{A_{Pixel}} \right)^n + \frac{H \cdot s + u_{Pos}}{1 - u_{Pixel} / A_{Pixel}}. \tag{2.3}$$

Equation 2.3 can then be solved for n, thus returning the minimum number of ZAC steps needed for automated switching from low magnification to high magnification.

As a further restriction, the length reflected by one pixel in the lowest magnification (k_0 / A_{Pixel}), has to be at least the same as the structural size s. Assuming that the ε-factor is 10%, the positioning accuracy should be half the size s.

Furthermore, if the object recognition accuracy is about 1% of the number of pixels ($A_{Pixel} = 500$), then the number of ZAC steps is the ceiling of the value calculated with Equation 2.3, thus $n = 2$. The first magnification step ($n = 1$) is then used for magnifying approx. 50 times, the second step ($n = 2$) for magnifying again approx. two times. From this typical example, it can be concluded that for applications where in the first magnification the object is just recognized, two ZAC steps are generally necessary to reach the desired magnification. With very good parameter sets, zooming and centering can be achieved in one step. For poor object recognition accuracies and number of pixels, the number of zoom steps increases to approximately three. Hence, in typical nanohandling applications, the **number of ZAC steps** is smaller than or equal to three.

Consequently, it should be considered that nanoscale handling processes contain **more tasks** compared to macroscale processes, as zoom-and-center steps occur more frequently.

2.2.3 SEM-related Issues

In the last section, the necessary number of ZAC steps was discussed. In the work space of SEMs, an increase in magnification implies a higher resolution, thus opening the possibility of substituting on-board position sensors by higher resolution SEM image acquisition and object recognition. This method has been used widely for the automation of nanohandling tasks [8, 9]. However, image acquisition takes longer than reading a sensor value; shorter image acquisition and processing times result in noisier images. Thus, the magnification where the on-board position sensor is replaced by object recognition has to be chosen taking the resolution enhancement as well as the delays into account.

2.2.3.1 Sensor Resolution and Object Recognition
The SEM is a high-resolution image acquisition unit and thus, as mentioned above, can be used as a sensor for closed-loop position control, substituting or supplementing an on-board position sensor (Figure 2.2). SEM object recognition challenges are discussed in Chapter 4. Issues regarding the position controller are considered in Chapter 3.

In Figure 2.3 a comparison between the resolution of a common on-board sensor for a linear axis and the achievable resolution using image processing and object recognition are plotted against the magnification of the SEM. It is clearly visible that already at magnification higher than 300 times in this case, the resolution achieved through object recognition is better than the on-board sensor resolution. Hence, with respect to image resolution, switching between the on-board position sensor and object recognition can occur when

$$\frac{k}{A_{Pixel}} \cdot u_{Pixel} < u_{Sensor}, \qquad (2.4)$$

where A_{Pixel} is the number of pixels for the imaged square with edge length k, u_{Pixel} is the accuracy of object recognition in pixels, and u_{Sensor} is the sensor accuracy.

Thus object recognition is preferably used compared to the on-board sensors, if the resolution of the object recognition system is significantly better. Modern piezoactuators, stick-slip, or continuous actuators accomplish step widths of 10 to 20 nm. In most cases, therefore, the **sensor resolution is the bottleneck** rather than the actuator resolution.

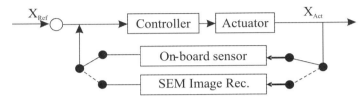

Figure 2.2. Typical control schematic for an actuator used for nanohandling automation. The on-board position sensor is substituted by SEM object recognition where the resolution is significantly better.

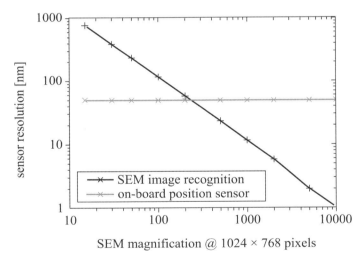

SEM magnification @ 1024 × 768 pixels

Figure 2.3. Comparison between the achievable sensor resolution using the on-board sensor of linear stick-slip axes and SEM object recognition. The on-board sensor is of type Numerik Jena L4 [10] and has an interpolated resolution of 50 nm, independent of the magnification. The resolution of SEM object recognition was measured according to the left side of Equation 2.4, using a recognition accuracy of 1 pixel.

2.2.3.2 Noise

The SEM as a scanning image sensor has a resolution of approx. 0.1 nm for field emission guns. These high resolutions are only achievable with comparatively long image acquisition times, when the signal-to-noise ratio is maximized. In general, the image quality is improved when scanning speed is reduced. However, for most automation processes, **high scanning speeds** combined with **fast object recognition** are desirable. Therefore, a compromise between these two aspects has to be chosen. The typical noise of an SEM image plotted against the scanning time is shown in Figure 2.4.

Obviously, it is necessary to increase the scan time if object recognition fails due to noisy images. For automation purposes, an update rate for sensor poses of 2 per second is tolerable, but a lower rate significantly slows down the process. A reduction of noise by the factor 0.5 leads to approx. 10 times longer scan times, whereas a reduction of the scan time to one third leads to three times more noise in the image. Thus, specialized methods for high update rates at high recognition reliability are necessary, which will be described in Chapter 4.

2.2.3.3 Velocity and Image Acquisition Time

The scanning speed also limits the maximum travel speed of an actuator, when object recognition is used as sensor feedback. Consider a point-like object under observation of a vision sensor with a frame refresh rate f_S, respectively a frame acquisition time T_S and a resolution A_{Pixel}, given in pixels. The area under observation is assumed to have an edge length of k. Then the object recognition accuracy $u_{recognition}$ is determined by the following equation:

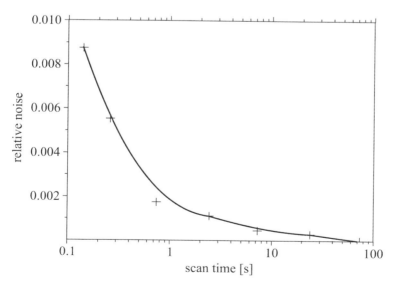

Figure 2.4. Relative noise of an SEM image for different scanning times at an image size of 512 × 442 pixels. The relative noise was calculated as mean-square error relative to the error at a maximum scan time of 73 s.

$$u_{recognition} = \frac{k}{A_{Pixel}} \cdot u_{Pixel} , \qquad (2.5)$$

where u_{Pixel} is a constant depending on the system and taking into account how precisely the position should be recognized. Typical values range from 1 to 10, *e.g.,* a value of 2, using the sampling theorem. The scan process is simplified here to be a process where the image is scanned line by line and the column-wise scanning within a line is neglected. Then, two cases have to be considered for estimating the maximum allowed velocity v_{max} of an object, if it has to be recognized at least in two subsequent images until the scanned area is left:

- The object is **moving orthogonal** to the scanning direction, *i.e.,* the time Δt between two occurrences of the object can be considered approx. T_S.
- The object is moving in the same direction as the scan is running. Then again two cases have to be considered: In the first, object movement and scan direction are **anti-parallel**. Then the time Δt until an object occurs in the following frame is shorter than the frame time T_S. For the second, the object's movement is **parallel** to the scan direction, resulting in longer time Δt between the occurrences in two successive frames. Then the maximum allowed velocity v_{max} can be calculated from the intersection of two lines, resulting in

$$v_{max} = \frac{1 - \dfrac{u_{Pixel}}{A_{Pixel}}}{2 \cdot T_S} \cdot k , \qquad (2.6)$$

where T_S is the image acquisition and object recognition time. Taking for example a scan time of 0.5 s (*i.e.,* a sensor refresh rate of 2 Hz) at an edge length of the scan field of 50 μm and a recognition accuracy of 10%, then the speed limit according to the above equation would be $v_{max} = 45$ μm/s.

2.3 Automated Nanohandling: Problems and Challenges

2.3.1 Parasitic Forces

The expression "parasitic forces" is commonly used as the collective term for surface forces that have major relevance in the micro- and nanoscale, *i.e.,* van der Waals, electrostatic, and capillary forces.

Electrostatic forces refer to forces due to electric charging of objects. Two objects that are charged with the same polarity repel each other, whereas oppositional polarity leads to attraction. Typical causes for electrostatic forces in handling processes are contact electrification, triboelectrification, and direct charging through the electron beam in the SEM. Charging through the electron beam of an

SEM can lead to repelling forces, resulting in objects floating around on the substrate surface. Other observed effects are electrostatic actuators in the SEM driven by the electron beam or image artifacts occurring due to the electrostatic deflection of the primary beam. In [11], the equation for estimating the electrostatic force between a sphere of radius r with a charge q and a conductive plane in distance is given by

$$F_{el} = \frac{q^2}{4\pi\varepsilon(2r)^2} \, ,$$

$$(2.7)$$

where ε is the dielectric permittivity. In practice, the force is hard to estimate because in general neither the charge nor the dielectric permittivity are known. In the SEM, special measures preventing or minimizing charging through the electron beam can be taken, *e.g.*, observation in low-vacuum modus or optimization of the beam parameters.

Van der Waals forces denote the (attractive) forces between atoms and molecules due to interatomic forces and can be calculated for a sphere on a plane by the following equation [11, 12]:

$$F_{vdW} = \frac{h \cdot r}{8\pi z^2} \, ,$$

$$(2.8)$$

where h is the Lifshitz-van der Waals constant and z is the atomic distance between the sphere and the plane. In [12], values for the Lifshitz-van der Waals constant are given for several material combinations, although this term is in general hard to estimate for handling processes.

Capillary forces are due to liquid films between two objects. Even in high-vacuum chambers, *e.g.*, when an SEM is used as vision sensor, liquid films on the surface of objects cannot be avoided. Due to water films (condensation), oil films (pump oil), *etc.*, the surfaces of objects, tools, and substrate can never be considered dry. Estimations for the capillary force between a sphere and a plane are given in [11, 13], resulting in the following equation:

$$F_{cap} = 4\pi r\gamma \, ,$$

$$(2.9)$$

where γ is the surface tension. However, the capillary force between object and gripper has also been used successfully for gripping, thus using the parasitic effect for handling objectives [14-16].

Fearing surveyed the parasitic forces and their influence on parts below 1 mm size. In [11], he suggested the following actions, among others, for reducing the influence of adhesion forces:

1. Usage of conductive materials for reducing charging effects. In micro- and nanotechnology, silicon is, however, a very common material for handling

tools, which forms an insulating oxide. A work-around for this problem is covering the tools with a **conductive layer**, *e.g.,* gold, where possible.

2. **Rough gripper jaw surfaces** in order to minimize the contact area. This measure should even be adopted for the design of gripper jaws, which allow a minimum of point-point contacts between object and gripper. Furthermore, the gripper geometry should always be adapted to the object to be handled.

It can be concluded from the survey of parasitic forces that most of them are very hard to estimate. Many factors in a handling system are either unknown or hardly measurable, *e.g.,* the capillary forces due to condensed water in an SEM's vacuum chamber. The complexity regarding geometry and interactions between multiple objects in a handling system additionally complicate the calculation of parasitic forces. Furthermore, these forces are time-variant, *i.e.,* they can change dramatically during a handling process. For example, a silicon gripper charged through the SEM's electron beam can be discharged through contact with the substrate surface.

Thus the parasitic forces are the major problem for the automation of handling tasks on the micro- and nanoscale due to their uncertainty and time-variance [16]. Experimental observations [17, 18] proved this conclusion.

2.3.2 Contact Detection

One of the major issues in the handling and assembly of nanoscale parts is the detection of contact between two objects. The issue arises when 2D images, *e.g.,* from a light microscope or SEM, are used as global vision sensors determining the out-of-plane positions of objects relative to each other. Object recognition can be used for contact detection within the observed plane, but out-of-plane contact detection is not possible. A typical example is the detection of whether a gripper touches a probe surface. If this scene is observed from above, *e.g.,* with an SEM providing high depth of focus, it is hardly possible to distinguish with common image recognition tools if the gripper touches the surface or not. Possible approaches for solving this problem are presented briefly in the following paragraphs.

Depth from focus: The depth-from-focus method is described in [19, 20] for measuring the height difference between two objects by means of a focus sweep. For the series of images, two regions of interest containing both objects are defined. For every region and image, the variance is determined. The variance shows a local minimum in the variance function over the changing focus, where the object's sharpness is at its maximum. Based on this method, the difference in height between two (object) surfaces can be determined.

Touchdown sensor: This sensor [21-23] provides a method for measuring contact between two objects by means of a resonance method. The resonator consists of a piezoelectric actuator and a piezoelectric sensor, on which the tool, *e.g.,* a gripper, is mounted. The piezoelectric actuator oscillates very close to the resonance frequency of the system. The system's mechanical oscillation induces an electrical current in the piezosensor. This current oscillates at the same frequency and with a certain amplitude. When the tool touches another object or a surface, the resonance frequency of the system changes, leading to a drop in the measured amplitude.

3D vision: The idea of creating a 3D image out of two SEM images recorded from different angles has been the subject of research for several years. Especially with regard to automation processes, it is obligatory to deflect the electron beam instead of the probe. A promising approach will be presented in Chapter 5.

Vision-based force measurement: Vision-based force measurement quantifies the deformation of a stressed object by means of object recognition [24, 25]. The algorithm can be applied for measuring forces – and thus contact – between two objects. For calculating the forces applied on an object, *a priori* knowledge is necessary, whereas for simple contact detection the deformation is evidence enough. However, this method is best applied where the object's stiffness is low, *e.g.,* measuring of contact or force between a gripper and a nanotube [26].

2.4 General Description of Assembly Processes

In this section, the process design and considerations with regard to serial assembly tasks on the nanometer scale will be given. Figure 2.5 provides an overview of the typical tasks that are necessary to accomplish an assembly process.

A description of the single tasks will be given in the next section. Based on this, further consideration regarding the reliability of assembly processes is given in the section after. Basically, simple handling tasks are the separation of an object, its transportation to another position and its release. Assembly processes comprehend (multiple) handling tasks, but are extended by joining processes and eventual inspection processes for quality assurance.

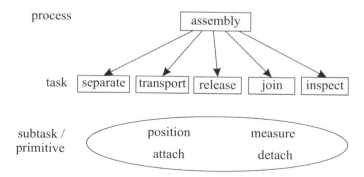

Figure 2.5. Overview of the tasks required in an assembly process. A typical process consists of the tasks "Separate", "Transport", and "Release". Every task can again be separated into subtasks, which can be seen as primitives.

2.4.1 Description of the Single Tasks

The tasks and subtasks forming a process can be best described as a change between two quantifiable states A and B. Especially for nanohandling tasks, it is very important that these states are quantifiable by means of measurement values, *e.g.,* giving a position in coordinates or describing the contact between two objects. Hence, ideally, the process is always in a definite state. In the following sub-sections, elementary tasks will be described with respect to the special requirements of the nano- and the macroworld.

Separation: The separation task can be considered as the occasion when the object's attachment status is changed, *i.e.,* in the beginning its condition is "connected to substrate" and at the end of the process it is connected to a tool or an intermediate product. The task itself can be achieved by several methods, *e.g.,* gripping and lifting, gluing and etching. Already from these examples it is obvious that the separation task includes at least two subtasks, *i.e.,* connecting the object to the tool and releasing the object from the substrate.

The separation of objects is one of the most difficult tasks in the field of nanohandling (Figure 2.6). The main reasons are the influences of parasitic forces and the comparatively weak forces which can be exerted by grippers or similar tools. An overview of strategies for lifting off small objects on the nanoscale, *e.g.,* nanowires, is given in [6, 17].

Transport: The transport task differs from macroscale transport tasks simply by the possible number of orders of magnitude in geometric scale, but is conceptually the same: the gripped or previously fixed object is transported from position A to a position B. Care has to be taken that the object is not released by accident during transport, *e.g.,* caused through vibrations of the actuators. A further description of positioning issues and position control will be given in Chapter 3 and Section 2.6.

Release: The "release" task is the reverse of "separation", *i.e.,* at first the object is attached to the substrate and then the object is detached from the tool. The parasitic forces cause an object on the nanoscale to stick to handling tools until the forces between the surface, where the object should be placed, are higher than the forces between the tool and the object. A reduction in sticking forces can be achieved using two principal ways:

1. The sticking force is reduced by **reducing the contact size**, *e.g.,* through specially formed gripper jaws or by reducing the influence of the parasitic forces, *e.g.,* through special coatings on the jaws. These measures usually require a considerable technical effort.
2. Instead of adapting the gripper to the gripping task, a **special technique** for releasing the object can be applied, *e.g.,* wiping off [17] or shaking off. Both techniques aim at shifting the balance between gripper-object forces and object-substrate forces to the substrate side.

Joining: The joining of objects is the central task for assembly processes. In principle, three different methods can be used for joining objects:

1. **Material closure:** two objects are connected using a material connection between both, *e.g.,* welding, gluing, or soldering. In Chapter 10, electron beam induced deposition (EBiD) will be explained in more detail, as it is a very promising method for joining parts through material closure inside the SEM.
2. **Force closure:** two objects are connected by a force, which can also be a parasitic force [27].
3. **Form closure:** two objects are joined by their geometry. This approach is not very common in nanoassembly tasks.

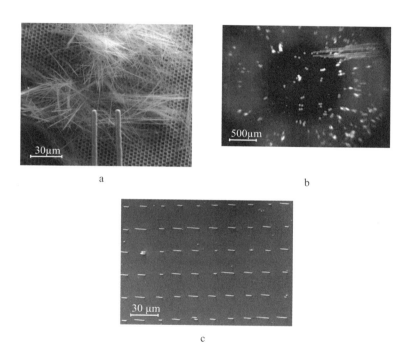

a

b

c

Figure 2.6. Typical objects to be manipulated in micro- and nanohandling tasks. a. A gripper trying to grab one out of a bunch of silicon nanowires in the SEM. The nanowires with a diameter between 200 and 500 nm and a length of several micrometers have been put on the substrate simply by peeling off. This separation task is very hard to automate, due to the parasitic forces holding the nanowires together. b. A glass ball (diameter approx. 30 μm) has to be gripped by a silicon gripper in the light microscope. This task can be automated, because the balls are split up and do not adhere to the surface due to the reduced contact area. c. CNTs grown in a matrix on a silicon wafer. This is a good starting point for automation processes, as the nanowires are separated and orientated on the wafer. However, they have to be detached from the substrate by breaking or etching.

Inspection: The inspection task serves for quality assurance. Many parameters can be tested to prove that the assembly process has been successful: stressing the connection up to a threshold force, chemical analysis of deposited material, electrical characterization of the bonding through resistance measurements.

2.4.2 General Flowchart of Handling Processes

Based on the simple tasks described above, it is possible to set up more complex processes, *e.g.,* handling and assembly by combining these to a **linked task chain**. Generally, every process, task, and subtask can be described in the process flow chart as a change from condition A to condition B. The conditions can be described as a vector containing the position data of single components, *i.e.,* objects and tools, and their relation to each other, *e.g.,* "part 1 connected to part 2". Based on these measurable values, it is possible to **trace failures** that would result in a process failure. Figure 2.7 shows the main tasks needed for bonding a CNT to an AFM tip. Additionally, the number of subtasks needed for successfully fulfilling a task is given.

2.5 Approaches for Improving Reliability and Throughput

2.5.1 Improving Reliability

The reliability of the overall process $R_{process}$ for a series of subtasks can be calculated by multiplying the single subtask reliabilities, *i.e.,*

$$R_{process} = \prod_{i=1}^{n} r_{subtask\_i} = \overline{r_{subtask}}^{n} ,$$

(2.10)

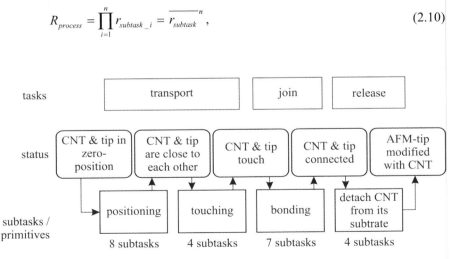

Figure 2.7. Typical process layout for bonding a single CNT to an AFM tip, showing the conditions between the tasks, the tasks and the number of subtasks for every task

where $r_{subtask\_i}$ is the reliability of the i-th subtask and n is the amount of subtasks. For example, a mean reliability $r_{subtask}$ of 98% for the 23 subtasks shown in Figure 2.7, leads to an over-all reliability for the process of 63%. A further decrease of the mean reliability for the subtasks down to 95% reduces the over-all reliability to 31%. This example shows the importance of maximizing the subtask reliability on the one hand and minimizing the number of subtasks on the other.

Another means, which is of special interest for serial assembly processes, is the definition of fallback markers in the process chain. If a task or a subtask fails, the system should be brought into a defined state, from where the chain can be continued. This method is referred to as **failure analysis with non-ambiguous retrace**.

Minimizing the number of subtasks: For minimizing the number of subtasks, several measures can be adopted:

1. Skillfully planning the handling tasks. This can be achieved through reducing the number of tools needed. One very important consideration, which should be taken into account for assembly tasks, is, *e.g.,* the omission of a gripper. This reduces the number of tasks substantially, because intermediate tasks, which are not directly concerned with connecting two objects to each other, can be left out. For example, connecting CNTs to AFM tips is a process where the gripper can be left out, if the CNTs come in a suitable pre-packaged orientation (Figure 2.6c).
2. Optimizing the number of subtasks needed for fulfilling a task. A typical example is minimizing the number of necessary ZAC steps (Section 2.2.2). Further improvements can be achieved if, *e.g.,* position sensors providing very high resolution over a wide scale are used. This prevents the switching of sensors and thus leads to a reduction of subtasks for the same positioning task.

Maximizing subtask reliability: subtask reliability can be increased using the following measures:

1. Continuous application of sensors, ideally setting up of closed-loop control systems, in order to trap exceptions. *In situ* measurement methods are of special interest for controlling subtasks.
2. Attaching and detaching subtasks are of special interest in terms of reliability. Based on the indetermination of parasitic forces, form and force closure should be substituted through material closure where possible, *e.g.,* bonding TEM lamellae to tips instead of gripping them with mechanical grippers.

2.5.2 Improving Throughput

The throughput D of the process can be defined as the inverse of the mean time needed for one process $T_{process}$, *i.e.,*

$$D = \frac{1}{\overline{T_{process}}} \, . \tag{2.11}$$

For estimating the influence of single parameters, the following assumptions are made: a process consists of n subtasks, with a mean duration $\overline{T_{subtask}}$ and a mean reliability $\overline{r_{subtask}}$ each. Every subtask is repeated until it has succeeded and thus the whole process is successful. For a large number of equal processes, the mean process duration can then be calculated from

$$\overline{T_{process}} = \frac{n \cdot \overline{T_{subtask}}}{\overline{r_{subtask}}} \, . \tag{2.12}$$

From this equation, the influence of the single parameters on the throughput can be qualified. For maximizing the throughput of a process, it is thus necessary to minimize the number of subtasks and to maximize the subtask reliability. Both measures have already been discussed in the section above. However, for minimizing the mean duration $\overline{T_{subtask}}$ of a subtask, the following measures can be taken:

1. **Optimizing the travel speed** of actuated parts through fast sensors. Additionally, as scanning vision sensors are widely used for nanohandling, the optimization of image acquisition and recognition has to be taken into account.
2. Especially for handling tasks, where the separation task is often very critical, the duration can be minimized by **optimizing the layout of the stored objects**.
3. Contact detection can be one of the very time-consuming tasks, because the travel speed of the actuators has to be reduced, in order to prevent hard crashes. Therefore, **vision-based methods** for the determination of the distance between object and tool, respectively substrate, are preferred. One approach, 3D vision sensors, is discussed in Chapter 5.
4. **Optimizing controllers for speed**, as discussed in Chapter 3.

2.6 Automated Microrobot-based Nanohandling Station

The **automated microrobot-based nanohandling station** for TEM lamellae handling was one of the first implementations of the generic AMNS concept presented in Chapter 1. The station was developed by the Division Microrobotics and Control Engineering (AMiR), University of Oldenburg, in the framework of the EU project ROBOSEM (grant number GRD1-2001-41861). The main purpose of the project was the integration of microrobots as well as position and force sensors into the vacuum chamber of an SEM. The client-server-based control system supports the user during nanohandling processes, where the objects' sizes range from some hundred μm to some hundred nm. A good example that has evoked interest from industrial partners is the handling of silicon lamellae that are to be evaluated in a TEM.

2.6.1 AMNS Components

2.6.1.1 Setup

The setup of the nanohandling station (schematic in Figure 2.8 and picture in Figure 2.9) consists of one microrobot that positions the sample to be handled ("sample robot"), and one microrobot that positions the end-effector, performing the actual handling task ("handling robot") [9, 28, 29]. Besides the main sensor, the SEM with image processing, CCD cameras with image processing, a position sensor, and a sensor for contact detection ("touchDown sensor") are employed to support the user and to allow for automatic positioning. The setup is attached to an exchangeable door of the SEM, so that it can be assembled, maintained and tested outside the vacuum chamber with a light microscope as SEM replacement. This reduces valuable SEM time, avoids pump time for the vacuum and generally allows for easy access to all components.

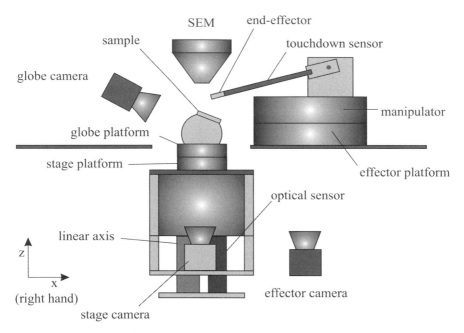

Figure 2.8. Schematic of the nanohandling station. The sample robot consists of the stage platform with a sphere carrying the specimen. It can position the sample in all six DoF. The handling robot consists of the effector platform with the manipulator carrying the touchdown sensor and a gripper. It can position the end-effector in the three DoF of a horizontal plane. The two translational DoF have a high actuation resolution.

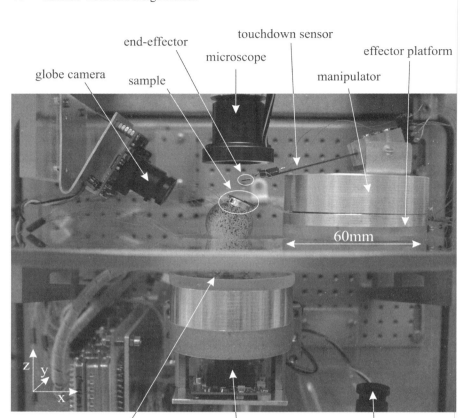

Figure 2.9. Setup of the nanohandling station. For development purposes, a light microscope is used and removed when the station is used in the SEM.

2.6.1.2 Actuators

The **sample robot** consists of two single-disk mobile platforms (diameter: 30 mm) and a linear axis. One mobile platform (**stage platform** in Figure 2.8) moves on a horizontal glass plate and carries another platform (**globe platform**), which is mounted upside down to rotate the sphere-shaped sample holder in all three rotational degrees of freedom (DoF). The working principle of the mobile platforms is explained below. The mobile platforms holding the sample can be moved vertically with a piezo-based **linear axis** from [30], which is fixed to the SEM door. Using all components of the sample robot, the sample can be positioned in all six DoF.

The **handling robot** consists of a triple-disk mobile platform (diameter: 60 mm), which carries a manipulator with an end-effector, *e.g.,* a gripper. The mobile platform (**effector platform** in Figure 2.8) moves on a separate horizontal glass plate around the sample robot to coarse-position the end-effector, and the **manipulator** then positions the end-effector with higher resolution to its desired (x, y) -position. The manipulator consists of two piezo stack actuators, which drive leverage arms fixed by flexible hinges. The maximum stroke of the table is about 40 μm. The **end-effector** can thus be positioned in x and y with high resolution, and rotated around the z-axis. The end-effector itself can be passive, like an STM (scanning tunneling microscope) tip, or an AFM cantilever, or active like a microgripper from [31], or from the Technical University of Denmark, Lyngby, Denmark (Chapter 7).

2.6.1.3 Mobile Microrobots

Two different implementations of a mobile microrobot platform are integrated into the AMNS [32-34]. The **triple-disk platform** is actuated by three piezodisks, which are each segmented into three parts (Figure 2.10a and c). A small ruby bead is glued to each segment, and each three-tuple of ruby beads drives one of three metal or sapphire spheres, which support the mobile platform. Instead of three piezodisks with three segments each, the **single-disk platform** consists of one piezodisk with nine segments (Figure 2.10b and d). In the setup, the triple-disk platform implements the effector platform while the single-disk platform implements the stage platform and the globe platform.

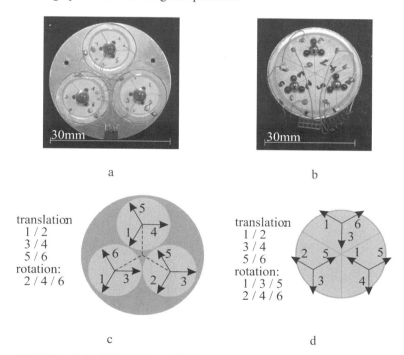

a   b

c   d

Figure 2.10. The triple-disk platform with three piezodisks: a. bottom view and c. channel configuration. The single-disk platform with one piezodisk: b. bottom view and d. channel configuration.

The developed platforms make use of the **stick-slip principle** by applying a voltage signal, which consists of a part with a gentle slope and a part with a steep slope, such that the segments are bent correspondingly slowly and fast (Figure 2.11). During the slow bending, the small ruby bead moves the large sapphire sphere, leading to a small rotation. This is called the stick phase. During the fast bending, the small ruby bead slides over the sapphire sphere, which therefore keeps its orientation. This is called the slip phase.

The number of control channels is reduced from nine to six by electrically connecting each of three piezo segment pairs. The configurations given in Figure 2.10c and Figure 2.10d yield three principal translation directions and one principle rotation direction, such that the microrobot platforms can move in all three degrees of freedom.

2.6.1.4 Sensors

The main high-resolution sensor of the nanohandling station is a LEO 1450 **SEM** [35], in combination with image processing, which provides "poses" of micro- and nanoobjects and end-effectors with resolutions down to 2 nm (magnification

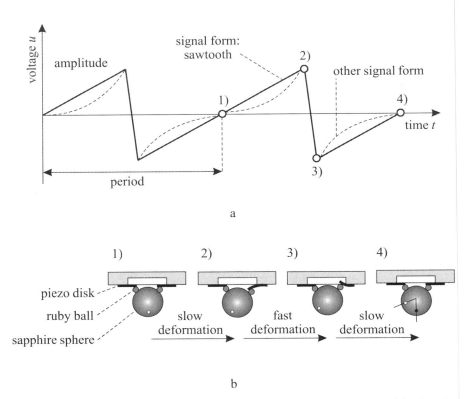

Figure 2.11. a. Typical voltage signal that is applied to a piezo segment. b. Stick phase by slow deformation and slip phase by fast deformation of the piezo segment.

50,000×, fast scanning). The generated pictures are acquired using the digital image acquisition unit offered by [36] and are processed using algorithms that have been developed at the AMiR [37-39]. For the coarse positioning of the mobile platforms, three corresponding **CCD cameras** are mounted on the SEM door. Together with the image processing software developed at the AMiR, they measure the mobile platform's poses with resolutions of about 60 µm (stage camera) and 170 µm (effector camera).

Measurements in the vertical direction are performed with two sensors. An optical **position sensor** from [40] measures the z-position of the linear axis with respect to a fixed reference (resolution below 1 µm). In addition, a **touchdown sensor** detects when the end-effector touches another object, *e.g.,* a microobject to be handled [9]. The touchdown sensor is a bimorph piezo-bending actuator, which is attached to the manipulator, and acts as a cantilever holding the end-effector. One ceramic layer is driven by an AC voltage with small amplitude (5 mV), and the other layer measures the amplitude of the resulting mechanical oscillation (approx. 50 nm). A contact between end-effector and microobject then results in a considerable and distinct drop in the measured amplitude.

2.6.1.5 Control Architecture

The control system is set up as client-server architecture with communication over TCP/IP to allow for flexible use of the control and sensor modules in different applications (Figure 2.12).

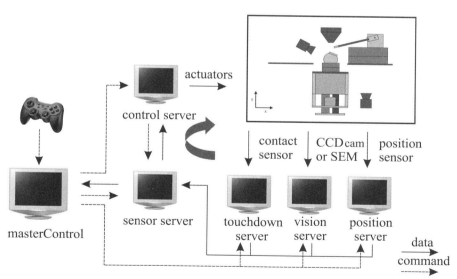

Figure 2.12. Control system architecture of the nanohandling station (a detailed view of the inlay figure in given in Figure 2.8). The touchdown server, the vision server, and the position server send their measurement data to the sensor server. The control server requests the measurement data when needed to control the actuators. The super-client master control controls all servers remotely and provides an interface to the user.

On the sensor side, there is a **vision server** that is responsible for the acquisition and processing of images from the SEM and from the CCD cameras. Its task is to detect features of microobjects, end-effectors, or microrobots and to determine their pose in a global frame (Chapter 4). Also, the optical position sensor and the touchdown sensor have their own server applications (**touchdown server** and **position server**). All servers continuously send their data to a **sensor server**, which stores the most recent data of each sensor and which provides that data to any client requesting it. On the actuation side, a low-level **control server** is responsible for the access of the actuation hardware and for the execution of control primitives with sensor feedback via the sensor server. The most common process primitive is positioning an object like an end-effector with feedback from a microscope. The **master control** module serves as a super-client and controls all servers remotely. For the vision servers, it remotely chooses between different tracking models and input sources, and starts and stops the tracking and defines regions of interest. The touchdown server and the position server can be switched on and off remotely. For monitoring, the master control module continuously requests data from the sensor server about the microrobots' poses. On the control server, it remotely starts and stops the execution of low-level control primitives and forwards data from a teleoperation device. The user can control the microrobots via teleoperation or in a semi-automated way by triggering process primitives. The user receives status and position feedback from the graphical user interface of the high-level control module or visual feedback from the vision servers.

2.6.1.6 User Interface
The user can interact with the AMNS via a **graphical user interface** (GUI) and via a **teleoperation device**. In addition, an **emergency stop button** stops all actuators by disconnecting them from the power supply.

The **GUI** is the main part of the master control module and allows control of the connection to the servers and automation, supports teleoperation and gives information for monitoring (Figure 2.13). The **connection** part effects connection to, and disconnection from, the servers. The servers are identified by their IP addresses and their constant port numbers. The **automation** part effects the triggering of a predefined sequence of process primitives. Depending on the current automation state, it is possible to start this sequence at different intermediate states. An additional **calibration** provides a comfortable read-out and saving of different desired poses. For example, the pose of the effector platform, at which the connected gripper is in the SEM's field of view, can be reached by teleoperating the platform accordingly. The current pose can then be used by the automation module as the desired pose when coarse-positioning the gripper. The **pose** part can show the pose values of all sensors, which are connected to the sensor server. The **teleoperation and sensor control** part shows the current sensor and the current actuator and allows changing them. Controlled by a **cordless gamepad**, the current actuator can be moved in any desired direction and with any desired velocity. The maximum velocity can be adjusted in a wide range from constant movement to a single step. Furthermore, the current sensor and actuator can be changed by the gamepad. Finally, in the **system messages** part, there is a textual output about the current and the last tasks.

2.6.2 Experimental Setup: Handling of TEM Lamellae

The control system can be seen as a tool to support user-performed nanohandling tasks in a teleoperated or in an automatic mode. Within the framework of the ROBOSEM project, the nanohandling station has been used to demonstrate the **semi-automated handling of lamellae**.

The handling sequence comprises the automated and teleoperated positioning of mobile microrobots. The names of actuators and sensors of the following sequence steps refer to the description above:

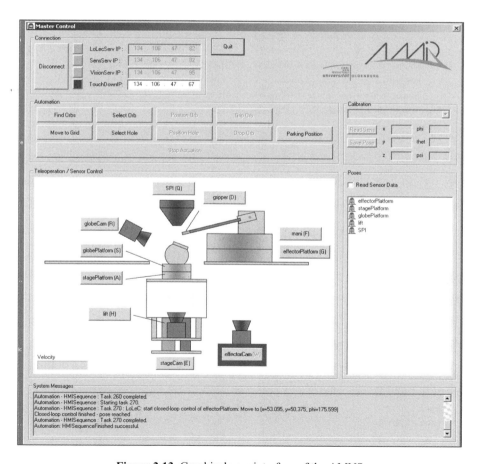

Figure 2.13. Graphical user interface of the AMNS

- **Lamella selection.** The sample, a silicon chip with four lamellae, is placed on the sphere, which can be rotated by the globe platform. First, the sample is coarse-positioned into the SEM's field of vision by moving the stage platform to a predefined pose with respect to the stage camera. The user then selects one of the lamellae, which is detected in the SEM image (Figure 2.14a).
- **Automatic lamella positioning.** The selected lamella is successively fine-positioned at a pose that allows good gripping with SEM feedback at magnifications of 50×, 300×, and 1000× (Figure 2.14b, c, and d).
- **Automatic gripper positioning.** To avoid a collision during the gripper positioning, the lamella is lowered by some 100 μm using the linear axis with feedback from the optical sensor. The gripper is then coarse-positioned into the SEM's field of vision by moving the effector platform to a predefined pose with respect to the effector camera, followed by a successive fine-positioning with SEM feedback at magnifications of 50× and 300× (Figure 2.15a and b). The most accurate positioning of the gripper is then achieved by using the manipulator with feedback from the SEM. For gripping, the lamella is lifted up again, until the touchdown sensor detects a contact between gripper and specimen (Figure 2.15c).
- **Lamella gripping.** The lamella is gripped in teleoperation mode since the mismatch of the gripper's stiffness and the lamella's pull-off force leads to low gripping reliability (Figure 2.15d).

The lamella can then be transported to a TEM grid, which is placed on the same sample holder. The sequence is similar to the gripping sequence, *i.e.,* the positioning of TEM grid and gripper can be done automatically with feedback from

Table 2.2. Positioning accuracies during different tasks when handling lamellae

Task	Actuator	Meas. frame	Positioning accuracy			
			x [μm]	y [μm]	z [μm]	φ [°]
lamella coarse positioning	stage platf.	Stage camera	20	20		0.5
lamella fine positioning	stage platf.	SEM (world)	2	2		
gripper coarse positioning	effector platf.	Effector camera	20	20		0.5
gripper fine positioning	effector platf.	SEM (world)	1	1		
gripper fine positioning	manipulator	SEM (world)	0.5	0.5		
lamella coarse positioning	linear axis	optical sensor			1	

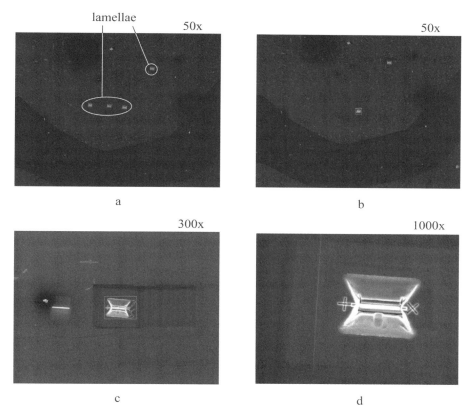

Figure 2.14. a. Selection of one of four lamellae. Successive positioning of the selected lamella with SEM feedback with magnifications of b. 50×, c. 300× and d. 1000×. Size of the lamella: 20 μm × 10 μm × 100 nm.

cameras, SEM, and touchdown sensor, while the actual release process must be carried out by teleoperation.

Table 2.2 lists the achieved positioning accuracies during different tasks when handling TEM lamellae. The accuracies are given with respect to the sensors which provide the feedback. For example, the gripper that is attached to the effector platform is positioned with an accuracy of 1 μm with respect to the SEM (gripper fine positioning in Table 2.2). The world frame is set to be the SEM image frame such that poses for the gripper and for the lamella can be used without any trans-formations. Moreover, the positioning accuracies that are given in Table 2.2 are the thresholds for the controller to stop the corresponding positioning process. They are chosen rather conservatively to increase the overall robustness and speed of the handling process.

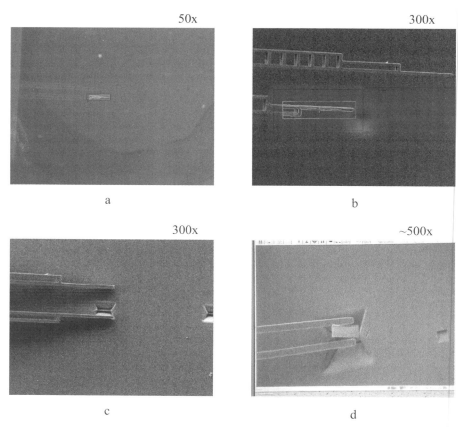

Figure 2.15. Positioning of the gripper with SEM feedback with magnifications of a. 50× and b. 300×. c. lift-up of the lamella with feedback from touchdown sensor. d. teleoperated gripping of the lamella. Gripper opening: 50 μm.

2.7 Conclusions

Vision sensors are essential for nanohandling tasks. A reasonable choice due to its good resolution, reasonable image acquisition times, and scalability is the **scanning electron microscope**. The SEM provides the possibility of substituting on-board position sensors and thus extending the closed-loop positioning resolution of a sensor-actuator system. The seamless zooming over several orders of magnitude is especially advantageous for automation, because typically more than one zoom-and-center step is necessary for accomplishing a handling or assembly task on the nanoscale. The number of necessary ZAC steps can be quantified to determine and optimize the overall number of process tasks. When applying SEM-based pose measurement in a closed-loop control structure, however, there is a trade-off between long image acquisition times and noisy images. The image acquisition time determines the maximum allowable velocity of the actuators and the noise determines the reliability of the object tracking.

There are two major challenges regarding automated nanohandling. Firstly, inevitable and time-variant **parasitic forces** result in ambiguous behavior of objects. Possible approaches are the avoidance of grippers and the performance of handling tasks with material closure between object, tool, and substrate. Secondly, a reliable method for the **automatic detection of contact** between tools, objects, and substrate must be found. Possible approaches are depth-from-focus methods, the touchdown sensor concept, 3D SEM vision, and vision-based force measurement.

For the **optimization** of automated nanohandling and assembly **processes**, the **reliability** of each task can be **improved**. This can be accomplished by reducing the number of tasks and subtasks, by maximizing each subtask's reliability (closed-loop control, constant monitoring, *in situ* measurement), and by applying material closure for attaching and detaching subtasks. Furthermore, the throughput can be improved by prearranging parts for an optimized separation, by using a fast method for contact detection, and by maximizing the velocity of actuated parts.

Finally, the concept of an **automated microrobot-based nanohandling station** has been demonstrated for the handling of TEM lamellae, which integrates all essential components named above. An SEM and CCD cameras, together with dedicated image processing, are used as position sensors, which provide the feedback for the closed-loop control of different microrobots. Mobile platforms are applied for coarse positioning and piezostack actuators for fine positioning. A touchdown sensor detects contact between gripper and substrate. The communication framework for the sensors and actuators is a designed TCP/IP-based distributed control system. The separation, transportation, and release tasks are implemented, although separation proved to be difficult due to the parasitic forces, to the low gripping force, and to the low stiffness of the gripper.

The future activities aim at **improving reliability and throughput** for automated handling on the nanoscale. EBiD will be used as a reliable joining technique, avoiding grippers where possible. A very important step towards reliable processes on the nanoscale is the application of subtask **failure analysis and non-ambiguous retrace**, as well as a final inspection task for the quality assurance. Different contact detection methods will be evaluated with respect to reliability and speed. The communication framework is currently being redesigned and will use the common object request broker architecture (CORBA), which allows the integration of modules on different platforms and written in different programming languages. Finally, a script-language-based high-level controller allows flexible execution of different nanohandling processes with the same system (Chapter 7).

2.8 References

[1]   Clevy, C., Hubert, A. & Chaillet, N. 2006, 'Micromanipulation and micro-assembly systems', *International Advanced Robotics Programme (IARP) 2006*.

[2]   Böhringer, K. F., Fearing, R. S. & Goldberg, K. Y. 1999, 'Microassembly', *Handbook of Industrial Robotics*, 2nd edn, Shimon Y. Nof (ed.), pp. 1045–1066.

[3]   Dzenis, Y. 2004, 'Spinning continuous fibers for nanotechnology', *Science*, vol. 304, no. 5679, pp. 1917–1919.

[4]   Nelson, B. & Yu Zhou Vikramaditya, B. 1998, 'Sensor-based microassembly of hybrid MEMS devices', *Control Systems Magazine, IEEE*, vol. 18, no. 6, pp. 35–45.

[5]   Brufau-Penella, J., Puig-Vidal, M., López-Sánchez, J., Samitier, J., Driesen, W., Breguet, J.-M., Gao, J., Velten, T., Seyfried, J., Estaña, R. & Woern, H. 2005, 'MICRoN: small autonomous robot for cell manipulation applications', *Proceedings of IEEE International Conference on Robotics and Automation (ICRA)*.

[6]   Fukuda, T. & Arai, F. D. 2003, 'Assembly of nanodevices with carbon nanotubes through nanorobotic manipulations', *Proceedings of the IEEE*, vol. 91, no 11, pp. 1803-1818

[7]   Humphris, A. D. L., Miles, M. J. & Hobbs, J. K. 2005, 'A mechanical microscope: High-speed atomic force microscopy', *Applied Physics Letters*, vol. 86, no. 3, p. 034106.

[8]   Sievers, T. 2006, 'Global sensor feedback for automatic nanohandling inside a scanning electron microscope', *Proceedings of IPROMS NoE Virtual International Conference on Intelligent Production Machines and Systems*, pp. 289–294. Received the Best Presentation Award.

[9]   Fatikow, S., Wich, T., Hülsen, H., Sievers, T. & Jähnisch, M. 2007, 'Microrobot system for automatic nanohandling inside a scanning electron microscope', *IEEE-ASME Transactions on Mechatronics*, accepted.

[10] Numerik Jena GmbH Germany, 2007, 'Datasheet for Encoder Kit L4', Online: http://numerik.itool4.net/frontend/files.php4?dl_mg_id=221&file=dl_mg_114422%6420.pdf.

[11] Fearing, R. 1995, 'Survey of sticking effects for micro parts handling', *iros*, vol. 2, p. 2212.

[12] Krupp, H. & Sperling, G. 1966, 'Theory of adhesion of small particles', *Journal of Applied Physics*, vol. 37, no. 11, pp. 4176–4180.

[13] Hecht, L. 1990, 'An introductory review of particle adhesion to solid surfaces', *Journal of the IES*.

[14] Lambert, P. D. 2005, 'A study of capillary forces as a gripping principle', *Assembly Automation*, vol. 25, no. 4, pp. 275–283.

[15] Driesen, W., Varidel, T., Régnier, S. & Breguet, J.-M. 2005, 'Micro manipulation by adhesion with two collaborating mobile microrobots', *Journal of Micromechanics and Microengineering*, vol. 15, pp. 259–267.

[16] Zhou, Q. 2006, 'More confident microhandling', *Proc. Int. Workshop on Microfactories (IWMF'06)*, Besancon, France.

[17] Mølhave, K., Wich, T., Kortschack, A. & Boggild, P. 2006, 'Pick-and-place nanomanipulation using microfabricated grippers', *Nanotechnology*, vol. 17, no. 10, pp. 2434–2441.

[18] Fatikow, S., Wich, T., Hülsen, H., Sievers, S. & Jähnisch, M. 2006, 'Microrobot system for automatic nanohandling inside a scanning electron microscope', *Proceedings of IEEE International Conference on Robotics and Automation (ICRA)*.

[19]   Estana, R., Seyfried, J., Schmoeckel, F., Thiel, M., Buerkle, A. & Woern, H. 2004, 'Exploring the micro- and nanoworld with cubic centimetre-sized autonomous microrobots', *Industrial Robot*, vol. 31, no. 2, pp. 159–178.

[20]   Watanabe, M., Nayar, S. K. & Noguchi, M. N. 1996, 'Real-time computation of depth from defocus', *Proc. of SPIE: Three-Dimensional and Unconventional Imaging for Industrial Inspection and Metrology*, vol. 2599, pp. 14–25.

[21]   Arai, F., Motoo, K., Kwon, P., Fukuda, T., Ichikawa, A. & Katsuragi, T. 2003, 'Novel touch sensor with piezoelectric thin film for microbial separation', *IEEE International Conference on Robotics and Automation, 2003. Proceedings. ICRA '03.*, Vol. 1, pp. 306–311.

[22]   Motoo, K., Arai, F., Fukuda, T., Matsubara, M., Kikuta, K., Yamaguchi, T. & Hirano, S. 2005, 'Touch sensor for micromanipulation with pipette using lead-free $(K,Na)(Nb,Ta)O_3$ piezoelectric ceramics', *Journal of Applied Physics*, vol. 98, no. 9, p. 094505.

[23]   Wich, T. & Fatikow, S. 2007, 'Assembly in the SEM', *Robotics Science and Systems Conference* - http://www.me.cmu.edu/faculty1/sitti/RSS06/RSSWorkshop.htm.

[24]   Wang, X., Ananthasuresh, G.K. & Ostrowski, J.P, 20 November 2001, 'Vision-based sensing of forces in elastic objects', *Sensors and Actuators A: Physical*, vol. 94, pp. 142–156(15).

[25]   Greminger, M. A. & Nelson, B. J. 2004, 'Vision-based force measurement', *IEEE Transactions on Pattern Analysis and Machine Intelligence*, vol. 26, no. 3, pp. 290–298.

[26]   Wich, T., Sievers, T. & Fatikow, S. 2006, 'Assembly inside a scanning electron microscope using electron beam induced deposition', *Proc. Int. Conf. on Intelligent Robots and Systems (IROS'06)*, Beijing, China, pp. 294–299.

[27]   Zhou, Q., Aurelian, A., Chang, B., del Corral, C. & Koivo, H. N. 2004, 'Micro-assembly system with controlled environment', *Journal of Micromechatronics*, vol. 2, no. 3-4, pp. 227–248.

[28]   Sievers, T. & Fatikow, S. 2005, 'Visual servoing of a mobile microrobot inside a scanning electron microscope', *Proc. Int. Conf. on Intelligent Robots and Systems (IROS'05)*, Edmonton, Canada, pp. 1682–1686.

[29]   Jähnisch, M., Hülsen, H., Sievers, T. & Fatikow, S. 2005, 'Control system of a nanohandling cell within a scanning electron microscope', *Proc. Int. Symposium on Intelligent Control (ISIC'05) / Mediterranean Conference on Control and Automation (MED'05)*, Limassol, Cyprus, pp. 964–969.

[30]   PiezoMotor AB Sweden, 2007, 'Homepage', online: http://www.piezomotor.se/.

[31]   Nascatec GmbH Germany, 2007, 'Homepage', online: http://www.nascatec.de.

[32]   Kortschack, A., Hänßler, O. C., Rass, C. & Fatikow, S. 2003, 'Driving principles of mobile microrobots for the micro- and nanohandling', *Proc. Int. Conf. on Intelligent Robots and Systems (IROS'03)*, Las Vegas, USA., pp. 1895–1900.

[33]   Kortschack, A. & Fatikow, S. 2004, 'Development of a mobile nanohandling robot', *Journal of Micromechatronics*, vol. 2, no. 3-4, pp. 249–269.

[34]   Kortschack, A., Shirinov, A., Trüper, T. & Fatikow, S. 2005, 'Development of mobile versatile nanohandling microrobots: design, driving principles, haptic control', *Robotica*, vol. 23, no. 4, pp. 419–434.

[35]   Carl Zeiss SMT AG Germany, 2007, 'Homepage', online: http://www.smt.zeiss.com/.

[36]   Point Electronic GmbH Germany, 2007, 'Homepage', http://pointelectronic.de/.

[37]   Sievers, T. & Fatikow, S. 2005, 'Pose estimation of mobile microrobots in a scanning electron microscope', *Proc. Int. Conference on Informatics in Control, Automation and Robotics (ICINCO'05)*, Barcelona, Spain, pp. 193–198.

[38] Sievers, T. & Fatikow, S. 2006, 'Real-Time Object Tracking for the Robot-Based Nanohandling in a Scanning Electron Microscope', *Journal of Micromechatronics - Special Issue on Micro/Nanohandling*, vol. 3, no. 3-4, pp. 267–284(18).

[39] Sievers, T. 2006, 'Global sensor feedback for automatic nanohandling inside a scanning electron microscope', *Proc. Virtual Int. Conference on Intelligent Production Machines and Systems*, pp. 289–294, http://conference.iproms.org/.

[40] MicroE Systems MA USA, 2007, 'Homepage', online: http://www.microesys.com/.

3

Learning Controller for Microrobots

Helge Hülsen

Division of Microrobotics and Control Engineering,
Department of Computing Science,
University of Oldenburg, Germany

3.1 Introduction

3.1.1 Control of Mobile Microrobots

The **mobile microrobots** developed by AMiR and described in the previous chapter are controlled automatically or via teleoperation. Feedback during these processes is provided by a global pose sensor, which could be a scanning electron microscope (SEM), a light microscope or a video camera. As described in the next section, the mobile microrobot's pose controller contains several sequential subtasks that are performed before the signals that are applied to the microrobot's actuators are determined. These subtasks are either performed by a computer in automatic mode or by a human being in teleoperation mode (Figure 3.1), except for the actuator controller, which is typically implemented by an electronic device or a computer.

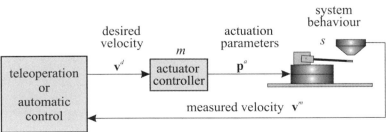

Figure 3.1. Actuator controller during teleoperation or automatic control

The **actuator controller** maps from a desired velocity vector $\mathbf{v}^d$ to a parameter vector $\mathbf{p}^a$, which defines the signals that drive the microrobot. An exact description of the actuator controller, however, is difficult to achieve because the mobile microrobot's motion behavior, *i.e.*, of the mapping s from actuation parameters $\mathbf{p}^a$ to a measured velocity $\mathbf{v}^m$, shows some unwanted properties. Generally, it requires a relatively high effort to exactly determine all parameters of a mobile microrobot's parameterized analytical model. The problems are variations of the parameters between different instances [1], complex measurements [2, 3], and variations of the parameters over time due to their dependency on environmental conditions like humidity and temperature [4, 5], and on the internal state of the microrobot like wear [6], piezo-drift [7], *etc.* Moreover, the microrobot's inverse behavior s^{-1}, and thus an actuator controller model m, is often not unambiguously defined, *i.e.*, different actuation parameters $\mathbf{p}^a$ lead to the same motion $\mathbf{v}^m$ [8, 9]. In the case of the considered mobile microrobots, there are six control channels for three degrees of freedom (DoF) of motion.

Different approaches are followed to determine the actuator controller model of currently existing mobile microrobots. For the **Abalone** platform, the parameters of an analytical model are found by measurements [2]. For the **MINIMAN** platform [10], the pose is either controlled by a neural network controller or a fuzzy logic controller [1]. The parameters for the approximate analytical model of the **MiCRoN** platform [11, 12] are estimated analytically, with help of neural networks [13] or with genetic programming [14]. The actuator controller model for the **Aoyama Labs platform** [15] can be found relatively easily due to the inch-worm locomotion principle and the actuator configuration. For the **NanoWalker** platform, a parameterized model of the motion behavior is briefly described in [16], but it is not used for the derivation of an actuator controller.

In all cases a transfer of the developed methods to other application domains, and in particular to other mobile microrobots, is quite difficult, because assumptions are made that rely on the specialties of each mobile microrobot. Moreover, an incremental adaptation to a changing system behavior has not been performed by any of the approaches.

3.1.2 Self-organizing Map as Inverse Model Controller

In a **control engineering context**, an inverse model maps from a desired system output $\mathbf{g}^d$ to a system input $\mathbf{p}^a$ such that the measured system output $\mathbf{g}^m$ is identical to the desired one (Figure 3.2a). In a closed-loop controller, an inverse model represents the final control element, sometimes referred to as *linearization*, with an inverse characteristic of the system (Figure 3.2b). A mobile microrobot is an example where an inverse model maps from desired velocities $\mathbf{g}^m = \mathbf{v}^d$ to actuation parameters $\mathbf{p}^a$, such that the measured velocities $\mathbf{g}^m = \mathbf{v}^m$ are as desired. A closed-loop controller then determines desired velocities $\mathbf{v}^d$ from the difference between desired poses $\mathbf{s}^d$ and measured poses $\mathbf{s}^m$.

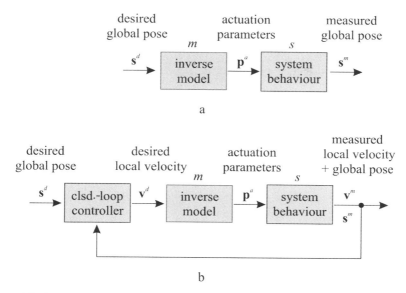

Figure 3.2. Inverse model in a. an open-loop control system and b. a closed-loop control system

Self-organizing maps (SOM) [17, 18] are a type of artificial neural network that learns unsupervised, *i.e.*, without an explicit teacher, and that are mainly used for data visualization and classification problems [19, 20]. There are some approaches, though, where SOM-based models are used in a control setup with supervised learning, resulting in so-called "self-supervised" models [21].

SOM consists of numbered nodes n_i, $i = 1...|N|$. The nodes are arranged in a topology that is defined by associated support vectors c_i in a topology space C. This topology can be one- or multidimensional and can have different shapes such as a rectangular grid (Figure 3.3) or a hexagonal grid. The neighborhood relation between two nodes n_i and n_k is expressed as a distance measure between their associated topology support vectors c_i and c_k. Each node n_i also has an input support vector g_i in an input space G and a corresponding output support vector p_i in an output space P associated to it, which is why this arrangement is called **associative memory** or **associative map**. SOM has the following two functions in the context of modeling and control:

1. Desired system outputs g^d are **mapped** to corresponding inputs $p^a = m(g^d)$ of the system to be controlled (Figure 3.3a).
2. System information, which is represented by pairs of training system inputs p^t and measured system outputs $g^m = s(p^t)$, is used to **learn** the mapping, such that the error $m(g^m) - p^t$ is minimized (Figure 3.3b).

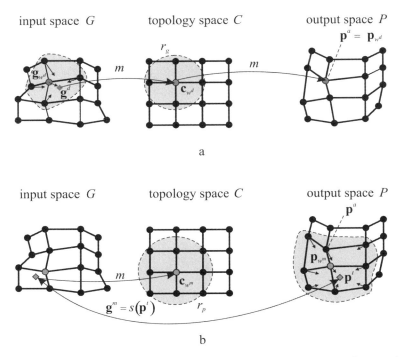

Figure 3.3. a. Mapping with a SOM and self-organization in input space. b. Learning of inverse system behavior by self-organization in output space.

The basic **SOM** approach shows some interesting properties for the implementation of a learning inverse model controller, as well as contrasting with classic artificial neural networks such as multilayer perceptron networks [22]. Firstly, since each node is exclusively responsible for a local area of the input space, the behavior of the input–output map is interpretable by human beings [18]. More important for the application to inverse model controllers, though, is the self-organization aspect of the SOM learning rule, where SOM's look-up-table-based data is ordered according to the pre-defined arrangement of topology support vectors [18]. Assuming that the map m is continuous, its inverse m^{-1}, which approximates a continuous system behavior s, becomes continuous, too [23, 24]. Moreover, the ordering process can be seen as an optimization process, which resolves possible ambiguities when learning the map m [25]. SOM provided a good starting point for the development of enhanced mapping and learning approaches for robotics and control. The most important are summarized in Table 3.1 The local linear map (**LLM**) [25] was the first complete SOM-based algorithm framework for robot control [21, 26-31]. Its local linear models allow for extrapolation outside the convex hull of the support vectors and provide a means for the estimation of output support vectors, such that the mapping error is minimized. Nevertheless, the local linear model's dependency on the inverse model's Jacobians makes an interpretation or an initialization with *a-priori* knowledge difficult.

Moreover, the sometimes global influence of the learning rule and the time-dependency of the learning parameters are not suitable for applications where the map is learned during operation and successive input vectors are correlated. Furthermore, LLM shows discontinuities at the borders of the nodes' influence range. By using the continuous SOM (**CSOM**) [31, 32] interpolation algorithm as an extension of LLM, these discontinuities can be avoided. The parameterized SOM (**PSOM**) [33, 34] does not model the inverse system behavior but models the system behavior itself and finds an optimal inverse during runtime. It provides a continuous and interpretable mapping, which is based on polynomials. Depending on the configuration of the map, these polynomials can lead to oscillations during interpolation and strong distortions during extrapolation [34]. PSOM inherits the learning properties of SOM, *i.e.*, time-dependence of locality, *etc.* The ambiguity resolution of PSOM is flexible and can even change its optimization criterion "on-the-fly", but for the price that the whole system behavior must be modeled. The last interesting approach is the continuously interpolating SOM (**CI-SOM**) [35–38], which mainly implements a continuous interpolation for SOM. The resulting mapping is continuous and local, only depends on the support vectors, and performs a meaningful extrapolation. The only drawback is the dependency on application- and configuration-specific interpolation coefficients. The focus of this development is the continuous mapping. The learning is probably performed by the SOM-learning rule, but detailed information on that topic was not available.

Table 3.1. Comparison of features of different SOM-based methods

Feature	SOM	LLM	CSOM	PSOM	CI-SOM	SOLIM
Mapping						
Continuity	−	−	×	×	×	×
Locality	×	×	×	×/−	×	×
Interpretability	×	−	−	×	×	×
Extrapolation	−	×	×	×/−	×	×
Learning						
Locality	×/−	×/−	×/−	×/−	[×/−]	×
Time-indep. params.	−	−	−	−	[−]	×
Ambiguity resolution	×	×	×	×/−	[×]	×

The analysis of existing approaches shows that there is no SOM-based approach that can directly implement a learning inverse model controller, which can learn incrementally. The self-organizing locally interpolating map (SOLIM) approach suggested in Section 3.3 tries to close this gap by providing a method for a continuous, local, and interpretable mapping that also performs a meaningful extrapolation, and to provide a new learning method that is local, does not use time-dependent learning coefficients, and that exploits the neighborhood relation for the resolution of the inverse model's ambiguities. Furthermore, the SOLIM algorithm framework is easy to use, *i.e.*, there are as few parameters as possible to be adjusted by the user. As shown in the following sections, this goal has been reached to a certain extent. However, the topology must still be provided by the user.

3.2 Closed-loop Pose Control

As shown in Chapter 2, the **control structure** of the automated microrobot-based nanohandling station (AMNS) consists of a control server, which is responsible for the execution of process primitives, and a master control program, which is responsible for the sequential calling and monitoring of the process primitives. The most basic process primitive of the AMNS is to change the measured pose $\mathbf{s}^m$ of a platform, microobject, or end-effector from a given starting pose $\mathbf{s}^s$ to a desired pose $\mathbf{s}^d$ along a linear trajectory. For every process primitive, the control server generates signals that drive the microrobots' actuators accordingly. The parameters of the signals are varied, depending on the feedback from a sensor or depending on the input from a teleoperation interface, where the user interprets the data from the sensors. In Figure 3.4 the control structure used is shown with a trajectory controller, a motion controller and an actuator controller, which together constitute the implemented low-level controller.

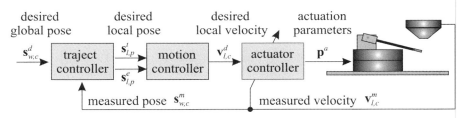

Figure 3.4. Low-level controller consisting of a trajectory controller, a motion controller, and an actuation controller

3.2.1 Pose and Velocity

The **pose vectors s** used in this chapter are either defined in Cartesian coordinates or in polar coordinates, indicated by the subscripts p and c, respectively. Furthermore, they are either given with respect to a world coordinate system or with respect to a local, moving coordinate system, indicated by additional subscripts w and l, respectively. The same notation applies for velocity vectors $\mathbf{v}$. A Cartesian pose vector $\mathbf{s}_c$ is described as (x, y, φ) with its components in the x-direction, in the y-direction and the angle φ around the z-axis. Correspondingly, a Cartesian velocity vector $\mathbf{v}_c$ is the first derivative $(\dot{x}, \dot{y}, \dot{\varphi})$ of a pose vector. A polar pose vector $\mathbf{s}_p$ is described by (r, ι, φ) with the distance $r = \text{sqrt}(x^2 + y^2)$, the direction $\iota = \text{atan2}(y/x)$ and the same angle φ. $\text{atan2}(y/x)$ yields the unambiguous angle of the vector (x, y) by modulating the sign of $\tan(y/x)$, depending on the sign of y and x. A polar **velocity vector** $\mathbf{v}_p$ is described accordingly by (v, ι, ω) with the velocity $v = \text{sqrt}\left(\dot{x}^2 + \dot{y}^2\right)$, the direction $\iota = \text{atan2}(\dot{y}/\dot{x})$ and the angular velocity $\omega = \dot{\varphi}$. Conversely, the components of Cartesian pose and velocity vectors can be easily determined as $x = r\cos(\iota)$ and $y = r\sin(\iota)$, and as $\dot{x} = v\cos(\iota)$ and $\dot{y} = v\sin(\iota)$, respectively.

3.2.2 Trajectory Controller

The trajectory controller (Figure 3.4 and Figure 3.5 – left side) determines a local deviation $\mathbf{s}_{l,p}^e$ from a linear trajectory and an orthogonal desired local pose $\mathbf{s}_{l,p}^t$ along that trajectory. These two poses correspond to the two distinct objectives of the subsequent motion controller. The first objective is to move the platform along the trajectory towards the desired pose and the second objective is to keep the platform on the trajectory.

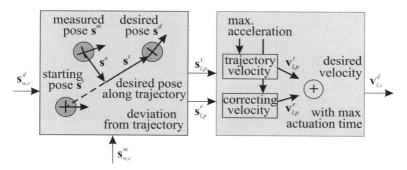

Figure 3.5. Left: The trajectory controller yields the local deviation from a given trajectory and a local pose along that trajectory. Right: The motion controller calculates corresponding velocities and combines them to a desired velocity.

First, an **intermediate pose** $\mathbf{s}_{w,c}^i$ on the line between the start pose $\mathbf{s}_{w,c}^s$ and the desired pose $\mathbf{s}_{w,c}^d$ is calculated, such that the intermediate pose has the shortest distance to the measured pose $\mathbf{s}_{w,c}^m$:

$$\mathbf{s}_{w,c}^i = \mathbf{s}_{w,c}^s + \mu^i \cdot \mathbf{s}_{w,c}^{sd}, \tag{3.1}$$

$$\mathbf{s}_{w,c}^{sd} = \mathbf{s}_{w,c}^d - \mathbf{s}_{w,c}^s. \tag{3.2}$$

μ^i describes the relative pose of the intermediate pose between the start pose $\mathbf{s}_{w,c}^s$, where $\mu^i = 0$, and the desired pose $\mathbf{s}_{w,c}^d$, where $\mu^i = 1$. The relative pose is calculated individually as $\mu^{i,t}$ for the measured position and as $\mu^{i,r}$ for the measured orientation by using only the first two components and the third component of the vectors in Equation 3.4, respectively. $\mu^{i,t}$ and $\mu^{i,r}$ are bounded to the range between 0 and 1, resulting in intermediate poses that are located between start pose and desired pose. μ^i is then composed as a weighted sum of $\mu^{i,t}$ and $\mu^{i,r}$, allowing for a dominance of translation or rotation during pose control. In the current implementation, when position and orientation are controlled simultaneously, the weights are $r^{\mu,t} = 0.7$ and $r^{\mu,r} = 0.3$, such that position control dominates.

$$\mu^i = r^{\mu,t} \cdot \mu^{i,t} + r^{\mu,r} \cdot \mu^{i,r} , \tag{3.3}$$

$$\mu^{i,t/r} = \left[\frac{\mathbf{s}^{sm} \cdot \mathbf{s}^{sd}}{\mathbf{s}^{sd} \cdot \mathbf{s}^{sd}} \right]_0^1 , \tag{3.4}$$

$$\mathbf{s}^{sm} = \mathbf{s}_{w,c}^m - \mathbf{s}_{w,c}^s . \tag{3.5}$$

The global **deviation** $\mathbf{s}_{w,c}^e$ **from the trajectory** is the difference between the measured pose and the intermediate pose, and the global **pose** $\mathbf{s}_{w,c}^t$ **along the trajectory** is the difference between the intermediate pose and the desired pose:

$$\mathbf{s}_{w,c}^e = \mathbf{s}_{w,c}^i - \mathbf{s}_{w,c}^m , \tag{3.6}$$

$$\mathbf{s}_{w,c}^t = \mathbf{s}_{w,c}^d - \mathbf{s}_{w,c}^i . \tag{3.7}$$

The global poses $\mathbf{s}_{w,c}^e$ and $\mathbf{s}_{w,c}^t$ are shifted by the negative measured pose $-\mathbf{s}_{w,c}^m$, and their position vectors are rotated by the negative measured angle $-\varphi_{w,c}^m$ to yield the corresponding local poses $\mathbf{s}_{l,c}^e$ and $\mathbf{s}_{l,c}^t$. These two steps, translation and rotation, can be comfortably performed with homogeneous matrices [39]. Finally, the local Cartesian poses are transformed into local polar poses $\mathbf{s}_{l,p}^e$ and $\mathbf{s}_{l,p}^t$ by applying the rules stated above.

3.2.3 Motion Controller

The motion controller (Figure 3.4 and Figure 3.5 – right side) calculates a desired velocity $\mathbf{v}_{l,c}^d$ such that the microrobot constantly decreases its velocity to zero at the desired pose.

The **correcting velocity** $\mathbf{v}_{l,p}^e$ and **trajectory velocity** $\mathbf{v}_{l,p}^t$ are calculated from the local deviation $\mathbf{s}_{l,p}^e$ of the mobile platform from the trajectory and from the local pose along the trajectory, respectively:

$$v_{l,p}^e = \sqrt{2 \cdot a^a \cdot s_{l,p}^e} , \tag{3.8}$$

$$\omega_{l,p}^e = \sqrt{2 \cdot \alpha^a \cdot \varphi_{l,p}^e} , \tag{3.9}$$

$$v^t_{l,p} = \sqrt{2 \cdot a^a \cdot s^t_{l,p}} \; ,$$
(3.10)

$$\omega^t_{l,p} = \sqrt{2 \cdot \alpha^a \cdot \varphi^t_{l,p}} \; .$$
(3.11)

The velocities are decreased with an acceleration a^a that is smaller than the lower bounds of the maximum acceleration a^{max}. Analogously, the angular velocities are decreased with an angular acceleration α^a that is smaller than the lower bounds of the maximum angular acceleration α^{max}. The values for a^{max} and α^{max} are estimated beforehand. There are two uncertainties during closed-loop control that motivate this artificial damping. Firstly, the measured velocity $\mathbf{v}^m_{l,c}$ can be higher than the desired velocity $\mathbf{v}^d_{l,c}$, mainly due to errors in the actuator controller model. Secondly, the sensor sampling rate and thus the update rate of the desired velocity, has some variance. With an artificially smaller acceleration, the maximum allowed desired velocity is smaller and a certain jitter in the sensor sampling rate is allowed before an emergency stop is required (Section 3.2.5).

The local velocity vectors $\mathbf{v}^e_{l,c}$ and $\mathbf{v}^t_{l,c}$ are then added to a **desired local velocity** vector $\mathbf{v}^d_{l,c}$, which is scaled down by the highest factor by which any of its components exceeds its maximum value. Depending on the implementation of the successive actuator controller, the output of the motion controller is given in polar coordinates $\mathbf{v}^d_{l,p}$ (for manually trained maps in Section 3.5.2) or in Cartesian coordinates $\mathbf{v}^d_{l,c}$ (for automatically trained maps in Section 3.5.3).

3.2.4 Actuator Controller

The actuator controller (Figure 3.4) finally tries to achieve the desired velocity $\mathbf{v}^d_{l,c}$ by manipulating the parameters $\mathbf{p}^a$ of the actuator signals correspondingly. The actuator controller is implemented as SOLIM as shown in Section 3.5.

3.2.5 Flexible Timing During Pose Control

For the control of microrobots in an SEM there are two **constraints** that have to be considered:

1. The sampling rates of the sensors differ and vary considerably. This property is due to the dependency of the acquisition and processing time on image properties like image size, model size, noise, *etc.* Especially when processing SEM images, the acquisition and processing times can vary between some milliseconds and some seconds.

2. The actuation time has a lower bound T^a_{min} and an upper bound T^a_{max} such that the robot does not move "blindly" when sensor sampling periods are too long.

Therefore, a controller update is triggered by the arrival of new sensor data. Then new actuation parameters are calculated by the low-level controller, the corresponding signal is copied to the ring buffer of a multichannel D/A board, and the output of the signal is started or continued.

In addition, two timers are set after the reception of a new sensor data to take into account the timing constraints. Firstly, a **stop timer** is set to T_{max}^a, which stops the robot after the maximum actuation time. The latest stop time T_{max}^a is calculated such that the microrobot does not move further than to the desired local pose along the trajectory. For a translation movement this means that after moving a time T_{max}^a with a velocity $(1+r^v)v_{l,p}^d$, there must be enough time to break with the maximum acceleration a^{max}:

$$T_{max}^a = \frac{s_{l,p}^t}{(1+r^v)v_{l,p}^d} - 0.5\frac{(1+r^v)v_{l,p}^d}{a^{max}}. \qquad (3.12)$$

The coefficient r^v estimates the velocity error, *i.e.*, if the desired velocity is $v_{l,p}^d$ the microrobot moves with $(1+r^v)v_{l,p}^d$ in the worst case. T_{max}^a is similarly calculated for rotation and the stop timer is started with the lower value. Secondly, a **request timer** is set to T_{min}^a, which requests the sensor server after the minimum actuation time to send new sensor data as soon as it has arrived. The minimum time T_{min}^a can be used to limit the communication traffic or the update rate of the controller. The two timers are implemented with help of the Windows multimedia timers, since they provide the necessary accuracy of about 1 ms, assuming that external disturbances from interrupts, *etc.* are prevented.

3.3 The SOLIM Approach

The main **structure** of self-organizing locally interpolating map is based on the standard SOM approach, where input support vectors are associated with output support vectors. The application of the SOLIM approach as ideal inverse model controller then defines the main working **principles**. SOLIM **mapping** is an extension of simplex interpolation as it is used in music synthesis control, and it also features extrapolation. SOLIM **learning** is divided into an approximation and a self-organization part. The approximation part locally learns the output support vector that has been most responsible for the measurement, and the new self-organization rule locally rearranges the neighboring vectors around this output support vector.

3.3.1 Structure and Principle

The **task** of the SOLIM network is to learn a continuous, smooth inverse model $m : G \mapsto P$ of a system behavior $s : P \mapsto G$, such that the difference between the desired system output $\mathbf{g}^d$ and the measured system output $\mathbf{g}^m$ is minimized (Figure 3.6a).

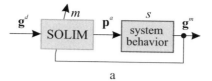

a

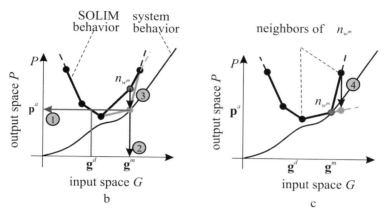

b   c

Figure 3.6. a. SOLIM learns an inverse model of the system behavior. b. Step 1: SOLIM mapping $m : \mathbf{g}^d \mapsto \mathbf{p}^a$, step 2: system response $s : \mathbf{p}^a \mapsto \mathbf{g}^m$, step 3: approximation by updating $\mathbf{p}_{w^m}$. c. Step 4: self-organization in output space by arranging all neighbors of n_{w^m}.

The **structure** of the SOLIM network is similar to the structure of extended SOM [18] (Figure 3.7). The network N consists of $|N|$ nodes $n_i = (\mathbf{g}_i, \mathbf{p}_i, \mathbf{c}_i, B_i)$. Each node n_i associates an input support vector $\mathbf{g}_i \in G$ with an output support vector $\mathbf{p}_i \in P$ and thus forms a kind of look-up table. In addition, each node n_i has an associated support vector $\mathbf{c}_i$ in topology space C and a set B_i of nodes $n_{i,k}^b$ that defines its neighborhood.

The following **steps** are performed during **operation** in the context of a learning controller, *i.e.*, during the learning of a model of an inverse system behavior (Figure 3.6b, Figure 3.6c, and Figure 3.7). These steps are explained in more detail in the following sections.

*   **Mapping.** A desired system output is mapped to actuation parameters $m : \mathbf{g}^d \mapsto \mathbf{p}^a$.
*   **System response.** The actuation parameters $\mathbf{p}^a$ are applied to the system and the system output is measured $s : \mathbf{p}^a \mapsto \mathbf{g}^m$.
*   **Approximation.** The pair $(\mathbf{g}^m, \mathbf{p}^a)$ describes the system behavior. Using $\mathbf{g}^m$ as map input yields a winning node n_{w^m} that has the highest influence with respect to $\mathbf{g}^m$. Therefore, the corresponding output support vector $\mathbf{p}_{w^m}$ is updated such that the mapped output of $\mathbf{g}^m$ is approximately $\mathbf{p}^a$.

- **Self-organization in output space.** The output support vectors around the winning node n_{w^m} are updated with the help of a self-organization rule to arrange the output support vectors according to the given topology.
- **Self-organization in input space.** The input support vectors can be pre-arranged according to the given topology or they can be learned with the help of a self-organization rule, where all input support vectors belonging to neighborhood of the winning node $\mathbf{g}^d$ are moved towards $\mathbf{g}^d$. The winning node n_{w^d} is the node that has the highest influence with respect to the input vector $\mathbf{g}^d$.

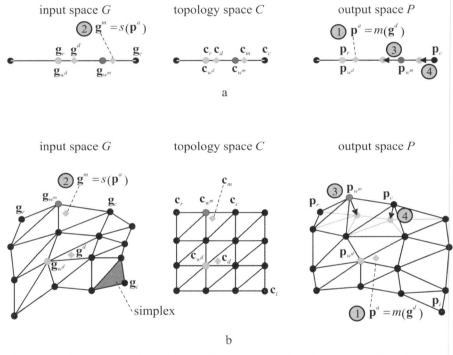

Figure 3.7. SOLIM operation with a. a 1D network and b. a 2D network. Step 1: SOLIM mapping $m : \mathbf{g}^d \mapsto \mathbf{p}^a$, step 2: system response $s : \mathbf{p}^a \mapsto \mathbf{g}^m$, step 3: approximation by updating $\mathbf{p}_{w^m}$, step 4: self-organization in output space by arranging all neighbors of n_{w^m}.

3.3.2 Mapping

The mapping part of the SOLIM algorithm framework finds a unique output vector $\mathbf{p}^a \in P$ for each input vector $\mathbf{g}^d \in G$ and is solely dependent on the network of nodes, *i.e.*, on the topology-supported association between input support vectors $\mathbf{g}_i \in G$ and output support vectors $\mathbf{p}_i \in P$. The input vector $\mathbf{g}^d$ may lie within the convex hull of neighboring input support vectors such that the output vector is determined as **interpolation** between the output support vectors. If the input vector lies outside the convex hull, the output vector is determined by **extrapolation** of the border nodes' behavior. In cases where the dimension d^C of the topology space is smaller than the dimension d^G of the input space (Figure 3.8), the separation is not that clear, and an input vector can be eligible for interpolation and extrapolation.

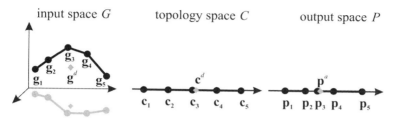

Figure 3.8. Mapping from a 2D input space to a 1D output space with a 1D network

The mapping is performed by first analyzing the input vector in input space and then synthesizing the output vector in output space.

- **Analysis.** The geometrical relation of the input vector $\mathbf{g}^d$ with respect to the arrangement of input support vectors $\mathbf{g}_i$, which is defined by their neighborhood B_i, is analyzed, yielding an interpolation value f_i and an extrapolation value x_i for each node n_i.
- **Synthesis.** The output vector $\mathbf{p}^a$ is synthesized, such that its geometrical relation with respect to the arrangement of output support vectors $\mathbf{p}_i$ corresponds to the interpolation values f_i and to the extrapolation values x_i.

By using the same characteristic influence values f_i and x_i during the analysis and during the synthesis, $\mathbf{p}^a$ has the same geometric relation to the arrangement of output support vectors as $\mathbf{g}^d$ has to the arrangement of input support vectors.

Before applying the output of the mapping to a real system, it is bounded by minimum and maximum values to prevent the system from being damaged by excessive voltages or currents. These values must be known beforehand and are parameters during the initialization phase.

3.3.2.1 Interpolation
SOLIM interpolation is based on **simplex interpolation** as it is used in music synthesis control [40]. A simplex is the d^C D analog of a line in 1D, a triangle in 2D and a tetrahedron in 3D. Generally, a d^C D simplex is defined as the convex hull of $d^C + 1$ nodes, often called *vertices*. In the scope of SOMs, the arrangement of topology support vectors defines simplices, which correspond to simplices in input and output space. During simplex interpolation, the relative pose of an input vector $\mathbf{g}^d$ within a simplex in input space is used to determine an output vector $\mathbf{p}^a$ with a corresponding relation to the associated simplex. Simplex interpolation has been chosen as the basis for development because it is a local technique, it is continuous, and it is relatively simple [41]. On the other hand, simplex

interpolation requires that any input vector appears in the convex hull of a simplex. This implies that topology and input space have the same dimension. In many robotic applications, however, it is more desirable to have a network with a topology space dimension d^C that is smaller than the input space dimension d^G, e.g., when a tool centre point is monitored in 3D by two cameras. 4D input support vectors that are arranged in a 3D topology would be the most efficient model.

SOLIM interpolation is based on simplex interpolation but changes the focus from simplices to nodes because with $d^C < d^G$ an input vector $\mathbf{g}^d$ will probably not be within a certain simplex defined by input support vectors, but close to the network of input support vectors e.g. a 2D input vector will probably not be exactly on one of the lines defined by neighboring input support vectors that are arranged in a 1D topology, but it will only be close to one or more lines (Figure 3.8). Therefore, instead of interpolating between the nodes of one simplex, each node has a continuous **influence** with respect to the current input vector. The influences are then normalized and used as weights for the corresponding output support vectors to generate the output vector.

During the **analysis** part of the mapping (see above), an influence f_i is calculated for each node n_i. This influence is maximal for input vectors $\mathbf{g}^d$ that are located at the corresponding input support vector $\mathbf{g}_i$ and decreases to 0 at limits that are solely defined by the associated set B_i of neighboring nodes (Figure 3.9). In the input space G, each limit $l_{i,j} \in L_i$ can be defined by a position vector $\mathbf{g}_{i,j}^l$ on the limit and a normal vector $\mathbf{g}_{i,j}^n$ perpendicular to the limit. The derivation of $\mathbf{g}_{i,j}^l$ and $\mathbf{g}_{i,j}^n$ from the set of neighboring nodes will be given later in this section.

The influence of a node n_i with respect to an input vector $\mathbf{g}^d$ depends on the relative distances $d_{i,j}$, $j = 1, 2, \ldots |L_i|$ of the input support vector $\mathbf{g}_i$ to the input vector $\mathbf{g}^d$ towards the limits $l_{i,j}$ (Equation 3.13). This relative distance measure is related to differences between support vectors and takes into account the direction $\mathbf{g}_{i,j}^n$ towards each limit, such that it implements a low-level normalization with respect to the network of support vectors. An explicit normalization is therefore not necessary.

$$d_{i,j} = \frac{\left| \mathbf{g}_{i,j}^n \cdot \left(\mathbf{g}^d - \mathbf{g}_i \right) \right|}{\left| \mathbf{g}_{i,j}^n \cdot \left(\mathbf{g}_{i,j}^l - \mathbf{g}_i \right) \right|}. \tag{3.13}$$

A blending function then maps the relative distance $d_{i,j}$ between input vector and input support vector to the influence $f_{i,j}$ of the node n_i with respect to the limit $l_{i,j}$:

$$f_{i,j} = b\left(d_{i,j} \right). \tag{3.14}$$

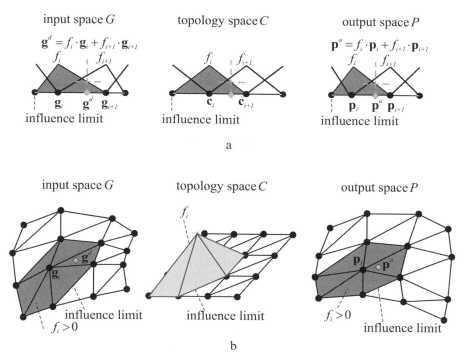

Figure 3.9. Influence f_i of a node n_i a. defined by two neighbors in a 1D network and b. defined by six neighbors in a 2D network. f_i is 100% for $\mathbf{g}^d = \mathbf{g}_i$ and decreases to 0% at the influence limits.

The blending function yields an influence of $f_{i,j} = 1$ for $d_{i,j} < 0$, an influence of $f_{i,j} = 0$ for $d_{i,j} > 1$, and a linear transition in-between:

$$b\left(d_{i,j}\right) = 1 - d_{i,j} \text{ for } 0 \le d_{i,j} \le 1. \tag{3.15}$$

When the input vector $\mathbf{g}^d$ lies within the simplex defined by the input support vector $\mathbf{g}_i$ and the neighboring input support vectors defining the influence limit $l_{i,j}$, each influence $f_{i,j}$ is identical to the barycentric coordinates of $\mathbf{g}^d$.

The influences $f_{i,j}$ of a node n_i with respect to each limit $l_{i,j}$ are then combined with the $\min(\cdot)$ operator to yield the influence f_i of the node (Equation 3.16). The $\min(\cdot)$ operator selects the highest of the limit-specific influences $f_{i,j}$ and thus selects the most relevant of the set of simplices that are defined by n_i and all of its neighbors (Figure 3.9). For the continuity of the whole mapping it is important that the $\min(\cdot)$ operator is a continuous operator:

$$f_i = \min\left(f_{i,j}\right) \tag{3.16}$$

In the **synthesis** part of the mapping, the influence values f_i, which describe the relative closeness of the input vector $\mathbf{g}^d$ to the input support vectors $\mathbf{g}_i$, are used as weights to combine the output support vectors $\mathbf{p}_i$ (Equation 3.17). Therefore, the output vector has the same relation to the arrangement of output support vectors as the input vector has to the arrangement of input support vectors. Both relations are expressed by the influence values f_i:

$$\mathbf{p}^a = \sum_{i=1}^{|N|} \tilde{f}_i \cdot \mathbf{p}_i \ .$$
(3.17)

$\tilde{f}_i$ are the $|N|$ influence values f_i, but normalized, such that their sum $\sum \tilde{f}_i$ becomes 1. In standard simplex interpolation, the sum of influences f_i is always 1. In SOLIM interpolation, however, the sum of influences f_i can have any positive value because the input vector does not necessarily fall within exactly one simplex and because nonlinear blending functions might be used. Therefore, downscaling is performed if the sum $\sum f_k$ of all $|N|$ neurons' influence values exceeds 1. If the sum $\sum f_k$ is smaller than or equal to 1, downscaling is not performed to avoid division-by-zero problems for the case when all $f_i = 0$. Instead, all influence values are increased uniformly such that the sum becomes 1:

$$\tilde{f}_i = \begin{cases} \dfrac{f_i}{\sum f_k} & \text{if} \sum f_k > 1, \\ f_i + \left(1 - \sum f_k\right) / |N| & \text{otherwise.} \end{cases}$$
(3.18)

3.3.2.2 Influence Limits

SOLIM interpolation relies on the definition of an influence f_i for each node that depends on the current input vector $\mathbf{g}^d$ and that defines the weight for the synthesis of the output vector $\mathbf{p}^a$. The influence for the node n_i is $f_i = 1$ for an input vector $\mathbf{g}^d = \mathbf{g}_i$ and decreases to $f_i = 0$ at its limits.

There are two groups of **nodes that influence the definition of a limit** $l_{i,j}$ for a node n_i (Figure 3.10). Firstly, the limit is defined by the neighboring nodes $n_{i,j,m}^l$, $m = 1, 2, \ldots d^C$, that, together with n_i, define the simplex $s_{i,j}$. This definition allows SOLIM interpolation to be similar to simplex interpolation because the influence of the node n_i is $f_i = 0$ for input vectors that are located at the facet that is opposite to $\mathbf{g}_i$ of the simplex in input space. This means that for an input vector $\mathbf{g}^d$ that is located at any input support vector $\mathbf{g}_i$ only the influence of the corresponding node n_i is $f_i \neq 0$ and the output vector is exactly the corresponding output support vector $\mathbf{p}^a = \mathbf{p}_i$. The definition of the limit $l_{i,j}$ with $n_{i,j,m}^l$ is sufficient for the case when the input space dimension d^G is equal to the topology space dimension d^C. In that case, d^C nodes $n_{i,j,m}^l$ define all of the limit's d^G degrees of freedom. When the input space dimension d^G is larger than the topology space dimensions d^C, the neighboring nodes $n_{i,j,m}^l$ only define d^C of the limit's d^G degrees of freedom. The remaining degrees of freedom are defined by the node n_i itself and the node $n_{i,j}^o$ that shares the facet given by $n_{i,j,m}^l$, $m = 1, 2, \ldots d^C$ of the simplex. This definition allows having the same limit definition for n_i and for $n_{i,j}^o$.

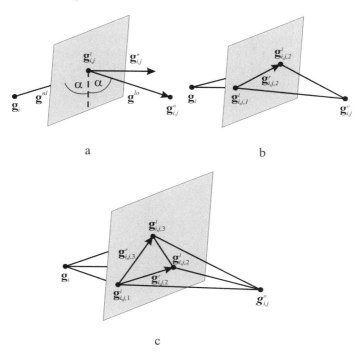

a

b

c

Figure 3.10. Limit of node n_i is defined by other nodes $n_{i,j,1}^l$, $n_{i,j,2}^l, \ldots, n_{i,j,d^C}^l$ of simplex $s_{i,j}$, by node n_i and by node $n_{i,j}^o$ on other side of simplex facet. a. 1D topology, line simplices. Limit normal defined by $\mathbf{g}_i$ and $\mathbf{g}_{i,j}^o$; b. 2D topology, triangle simplices. Limit normal defined by $\mathbf{g}_i$ and $\mathbf{g}_{i,j}^o$, and $\mathbf{g}_{i,j,2}^r$; c. 3D topology, tetrahedron simplices. Limit normal defined by $\mathbf{g}_i$ and $\mathbf{g}_{i,j}^o$, and $\mathbf{g}_{i,j,2}^r$ and $\mathbf{g}_{i,j,3}^r$.

During SOLIM interpolation the **limit** $l_{i,j}$ is defined by the position vector $\mathbf{g}_{i,j}^l$ on the limit and the normal vector $\mathbf{g}_{i,j}^n$, which is perpendicular to the limit (Figure 3.10). The **position vector** can be found as the input support vector corresponding to one of the neighboring nodes $n_{i,j,m}^l$ belonging to the simplex $s_{i,j}$.

$$\mathbf{g}_{i,j}^l = \mathbf{g}_{i,j,1}^l \tag{3.19}$$

The calculation of the **normal vector** is started with the sum of the normalized differences $\mathbf{g}^{nl} = \mathbf{g}_{i,j,1}^l - \mathbf{g}_i$ and $\mathbf{g}^{lo} = \mathbf{g}_{i,j}^o - \mathbf{g}_{i,j,1}^l$, which involve the node n_i itself and the node $n_{i,j}^o$ on the other side of the limit:

$$\mathbf{g}_{i,j}^n = \frac{\mathbf{g}^{nl}}{\left\|\mathbf{g}^{nl}\right\|} + \frac{\mathbf{g}^{lo}}{\left\|\mathbf{g}^{lo}\right\|}. \tag{3.20}$$

This definition leads to equal angles between the limit normal $\mathbf{g}_{i,j}^n$ and $\mathbf{g}^{nl}$ and between the limit normal $\mathbf{g}_{i,j}^n$ and $\mathbf{g}^{lo}$. As an example, assume that $\mathbf{g}_i$ is a 3D support vector belonging to a network with 1D topology space (Figure 3.10a). The position vector of one of its two limits $l_{i,1}$ would be defined by its neighbor $\mathbf{g}_{i+1}$ and the normal vector would be defined by the support vector $\mathbf{g}_i$ itself and the support vector $\mathbf{g}_{i+2}$ on the other side of the limit. There are cases when there is no node $\mathbf{g}_{i,j}^o$ on the other side of the facet which is opposite to n_i. Then the $\mathbf{g}^{lo}$-part of Equation 3.20 vanishes.

In the case of a 1D topology ($d^C = 1$), the definition in Equation 3.20 is sufficient. In all other cases ($d^C > 1$), the limit normal $\mathbf{g}_{i,j}^n$ must be perpendicular to the differences $\mathbf{g}_{i,j,k}^r$, $k = 2,3,\dots d^C$ between the support vectors defining the limit facet opposite to $\mathbf{g}_i$:

$$\mathbf{g}_{i,j,2}^r = \mathbf{g}_{i,j,2}^l - \mathbf{g}_{i,j,1}^l \tag{3.21}$$

$$\mathbf{g}_{i,j,3}^r = \mathbf{g}_{i,j,3}^l - \mathbf{g}_{i,j,1}^l \tag{3.22}$$

$$\dots = \dots \tag{3.23}$$

$$\mathbf{g}_{i,j,d^C}^r = \mathbf{g}_{i,j,d^C}^l - \mathbf{g}_{i,j,1}^l. \tag{3.24}$$

An iterative algorithm then makes the limit normal $\mathbf{g}_{i,j}^n$ as defined in Equation 3.20 perpendicular to all $\mathbf{g}_{i,j,k}^r$. In each step the difference vectors $\mathbf{g}_{i,j,k}^r$ that have not been applied on $\mathbf{g}_{i,j}^n$ yet are also made perpendicular to the current difference vector $\mathbf{g}_{i,j,k}^r$ to make the changes on $\mathbf{g}_{i,j}^n$, being linearly independent of each other.

- Make $\mathbf{g}_{i,j}^n$ and $\mathbf{g}_{i,j,3}^r, \dots, \mathbf{g}_{i,j,d^C-1}^r$ perpendicular to $\mathbf{g}_{i,j,2}^r$
- Make $\mathbf{g}_{i,j}^n$ and $\mathbf{g}_{i,j,4}^r, \dots, \mathbf{g}_{i,j,d^C-1}^r$ perpendicular to $\mathbf{g}_{i,j,3}^r$
- ...
- Make $\mathbf{g}_{i,j}^n$ and $\mathbf{g}_{i,j,d^C}^r$ perpendicular to $\mathbf{g}_{i,j,d^C-1}^r$
- Make $\mathbf{g}_{i,j}^n$ perpendicular to $\mathbf{g}_{i,j,d^C}^r$

3.3.2.3 Extrapolation

SOLIM extrapolation adds an **extrapolation component** $\mathbf{p}_i^x$ to each output support vector $\mathbf{p}_i$ before combining them into an output vector:

$$\mathbf{p}_i^x = \mathbf{x}_i \cdot \mathbf{p}_i^\eta, \tag{3.25}$$

$$\mathbf{p}^a = \sum_{i=1}^{|N|} \tilde{f}_i \cdot \left(\mathbf{p}_i + \mathbf{p}_i^x \right). \tag{3.26}$$

The weight x_i for the extrapolation component $\mathbf{p}_i^x$ defines the distance of the input vector $\mathbf{g}^d$ from the input support vector $\mathbf{g}_i$ in the direction of $\mathbf{g}_i^\eta$, as shown in Figure 3.11 and Figure 3.12:

$$x_i = \frac{\left(\mathbf{g}^d - \mathbf{g}_i \right) \mathbf{g}_i^\eta}{\mathbf{g}_i^{\eta\,2}}. \tag{3.27}$$

$\mathbf{g}_i^\eta$ is the mean difference vector between the input support vector $\mathbf{g}_i$ and all input support vectors $\mathbf{g}_{i,k}^\beta$ in an extrapolation neighborhood B_i^β. Analogously, $\mathbf{p}_i^\eta$ is the mean difference vector between the output support vector $\mathbf{p}_i$ and all output support vectors in the same extrapolation neighborhood B_i^β:

$$\mathbf{g}_i^\eta = \sum_{n_{i,k}^\beta \in B_i^\beta} \left(\mathbf{g}_i - \mathbf{g}_{i,k}^\beta \right), \tag{3.28}$$

$$\mathbf{p}_i^\eta = \sum_{n_{i,k}^\beta \in B_i^\beta} \left(\mathbf{p}_i - \mathbf{p}_{i,k}^\beta \right). \tag{3.29}$$

This neighborhood B_i^β is the set of the $3^{d^C} - 1$ direct neighbors in the d^C D topology space C, i.e. the two direct neighbors for $d^C = 1$, the eight direct neighbors for $d^C = 2$, etc.

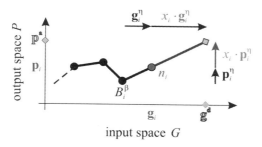

Figure 3.11. Extrapolation component $\mathbf{p}_i^x = \mathbf{x}_i \cdot \mathbf{p}_i^\eta$ corresponds to relative distance $\mathbf{g}_i^x = \mathbf{x}_i \cdot \mathbf{g}_i^\eta$ of input vector $\mathbf{g}^d$ from border input support vector $\mathbf{g}_i$

input space G topology space C output space P

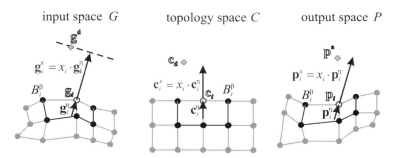

Figure 3.12. Extrapolation weight x_i: Distance between $\mathbf{g}^d$ and $\mathbf{g}_i$ in relation to mean distance $\mathbf{g}_i^\eta$ between $\mathbf{g}_i$ and its neighbors.

3.3.3 Learning

There are two **goals** for the **learning** task: SOLIM mapping approximates an inverse system behavior, and the input and output support vectors are both arranged according to a predefined topology, such that similar input vectors are mapped to similar output vectors and the mapping thus becomes smooth.

In standard SOM learning, approximation and self-organization in output space are performed in one step using the estimate for approximation as attractor for the self-organization of the output support vectors. In PSOM learning, the combined approximation and self-organization is even performed in a combined input-output space. In SOLIM learning, the single steps are separated to better control their influence on the map. The **approximation** step updates the output support vector of the node n_{w^m} that has the highest influence on the pair of system input $\mathbf{p}^a$ and measured system output $\mathbf{g}^m$. The **self-organization in output space** then locally arranges the output support vectors of the nodes around n_{w^m} according to the topology and thus ensures that the inverse of the mapping is continuous as well. The **self-organization in input space**, if activated at all, arranges the input support vectors of the nodes according to the topology.

3.3.3.1 Approximation
The **idea** behind SOLIM approximation (Figure 3.13a) is to change an output support vector $\mathbf{p}_{w^m}$ in the mapping $\mathbf{p}^m = \mathbf{m}\!\left(\mathbf{g}^m\right)$, such that $\mathbf{p}^a = \mathbf{p}^m$, *i.e.*, the SOLIM map passes through the point $\left(\mathbf{p}^a, \mathbf{g}^m\right)$ of the system behavior. There are several output support vectors that have an influence on $\mathbf{p}^m$ but since only one point of the system behavior is given, the output support vector $\mathbf{p}_{w^m}$ with the highest influence is used for updating.

First, the **synthesis equation** for $\mathbf{p}^m$ is **rearranged** to identify the influences of different output support vectors on the output:

$$\mathbf{p}^a = \mathbf{p}^m, \tag{3.30}$$

$$\mathbf{p}^a = \sum_{i=1}^{|N|} \tilde{f}_i^m \xi_i^m \mathbf{p}_i + \sum_{i=1}^{|N|} \tilde{f}_i^m x_i^m \mathbf{p}_i^h,$$ (3.31)

with $\tilde{f}_i^m$ and x_i^m being the interpolation and extrapolation weight, respectively, of each node n_i with respect to the input vector $\mathbf{g}^m$. ξ_i^m is just a substitution for $1 + x_i^m | B_i^\beta |$, where $| B_i^\beta |$ is the number of neighbors of node n_i contributing to the extrapolation component $\mathbf{p}_i^x$. $\mathbf{p}_i^h$ is the sum of the output support vectors of all these extrapolation-relevant neighbors. Equation 3.31 shows that the support vector $\mathbf{p}_{w^m}$ of the most relevant node n_{w^m} appears in the first sum of $\mathbf{p}^a$ and as part of $\mathbf{p}_i^h$. The second appearance can be neglected because $\mathbf{p}_{w^m}$ is only a neighbor of a few border nodes with a low influence $\tilde{f}_i^m < \tilde{f}_{w^m}^m$.

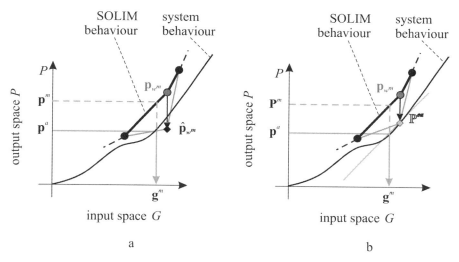

Figure 3.13. Approximation: a. Preliminary estimate $\hat{\mathbf{p}}_{w^m}$ by inversion of mapping. b. Better estimate $\mathbf{p}^{e,a}$ by scaling down support vector change $\hat{\mathbf{p}}_{w^m} - \mathbf{p}_{w^m}$.

Solving Equation 3.31 **for** $\mathbf{p}_{w^m}$ yields an expression for a **preliminary estimate** $\hat{\mathbf{p}}_{w^m}$ of $\mathbf{p}_{w^m}$, and calculating the difference between the estimate $\hat{\mathbf{p}}_{w^m}$ and the old value of $\mathbf{p}_{w^m}$, yields a relatively simple expression:

$$\hat{\mathbf{p}}_{w^m} - \mathbf{p}_{w^m} = \frac{1}{\tilde{f}_{w^m} \xi_{w^m}} \left(\mathbf{p}^a - \mathbf{p}^m \right).$$ (3.32)

The substitution ξ_{w^m} is given by the simpler notation $\xi_{w^m} = \xi_{i=w^m}^m$. The interpolation and extrapolation influences of the winning node n_{w^m} follow the same rule with $\tilde{f}_{w^m} = \tilde{f}_{i=w^m}^m$ and $x_{w^m} = x_{i=w^m}^m$, respectively.

If $\mathbf{p}_{w^m}$ were to be updated to $\hat{\mathbf{p}}_{w^m}$, $\mathbf{g}^m$ would be mapped to $\mathbf{p}^a$. But, assuming that the gradient of the SOLIM map is similar to the gradient of the system behavior (Figure 3.13b), **scaling down the change** $\hat{\mathbf{p}}_{w^m} - \mathbf{p}_{w^m}$ of $\mathbf{p}_{w^m}$ by $\tilde{f}_{w^m}$ yields a better **estimate** $\mathbf{p}^{e,a}$ for $\mathbf{p}_{w^m}$. The interpolation weight $\tilde{f}_{w^m}$ is chosen to scale down because it describes the ratio of $\mathbf{p}^{e,a} - \mathbf{p}_{w^m}$ to $\hat{\mathbf{p}}_{w^m} - \mathbf{p}_{w^m}$:

$$\mathbf{p}^{e,a} = \mathbf{p}_{w^m} + \tilde{f}_{w^m}\left(\hat{\mathbf{p}}_{w^m} - \mathbf{p}_{w^m}\right), \tag{3.33}$$

$$\mathbf{p}^{e,a} = \mathbf{p}_{w^m} + \frac{1}{\xi_{w^m}}\left(\mathbf{p}^a - \mathbf{p}^m\right). \tag{3.34}$$

Finally, $\mathbf{p}_{w^m}$ is moved towards $\mathbf{p}^{e,a}$ with a constant approximation learning rate $\varepsilon^{p,a}$:

$$\mathbf{p}_{w^m}^{(new)} = \mathbf{p}_{w^m}^{(old)} + \left(\mathbf{p}^{e,a} - \mathbf{p}_{w^m}^{(old)}\right) \cdot \varepsilon^{p,a}. \tag{3.35}$$

3.3.3.2 Self-organization in Output Space

During SOLIM **self-organization,** all neighbor pairs $(n_r, n_c) \in R$ of the winning node n_{w^m}, whose topology vectors are in-line with respect to $\mathbf{c}_{w^m}$ in topology space, are changed in the output space to be in-line with $\mathbf{p}_{w^m}$. Each of these neighbor pairs consists of a reference node $n_r = (\mathbf{g}_r, \mathbf{p}_r, \mathbf{c}_r, B_r)$, which has the smaller approximation error Δ_r and remains unchanged, and a candidate node $n_c = (\mathbf{g}_c, \mathbf{p}_c, c_c, B_c)$, which has the larger approximation error Δ_c and is updated with a self-organization learning rate $\varepsilon^{p,s}$ (Figure 3.14):

$$\mathbf{p}_c^{(new)} = \mathbf{p}_c^{(old)} + \varepsilon^{p,s} \cdot \left(\mathbf{p}^{e,s} - \mathbf{p}_c^{(old)}\right). \tag{3.36}$$

For each candidate node n_c an estimate for its output support vector $\mathbf{p}_c$ is found by aligning $\mathbf{p}_c$ to the difference $\left(\mathbf{p}_{w^m} - \mathbf{p}_r\right)$ and by adopting the relation between $\|\mathbf{g}_c - \mathbf{g}_{w^m}\|$ and $\|\mathbf{g}_{w^m} - \mathbf{g}_r\|$ in the output space (Figure 3.15)

$$\mathbf{p}^{e,s} = \mathbf{p}_{w^m} + \frac{\|\mathbf{g}_c - \mathbf{g}_{w^m}\|}{\|\mathbf{g}_{w^m} - \mathbf{g}_r\|} \cdot \left(\mathbf{p}_{w^m} - \mathbf{p}_r\right). \tag{3.37}$$

The **self-organization learning rate** is calculated for each candidate node and depends on the approximation learning rate $\varepsilon^{p,a}$, on a ratio r^p and on the candidate node's number of neighbor pairs $u_c = |B_c^\beta|/2$:

$$\varepsilon^{p,s} = 1 - \left(1 - r^p \cdot \varepsilon^{p,a}\right)^{1/u_c}. \tag{3.38}$$

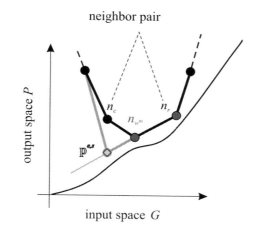

Figure 3.14. Self-organization in output space by aligning a candidate support vector $\mathbf{p}_c$ with respect to the winning support vector $\mathbf{p}_{w^m}$ and the opposite reference support vector $\mathbf{p}_r$

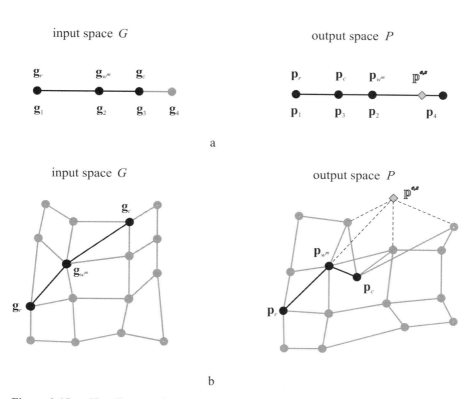

Figure 3.15. a. The alignment by a self-organization step in 1D. b. One of four alignments during a 2D self-organization step.

Averaged over time, an output support vector is moved u_c times in a self-organization step when it is moved once in an approximation step (Figure 3.16). Thus an output support vector is moved relatively with $d^{p,s} = 1 - \left(1 - \varepsilon^{p,s}\right)^{u_c}$ towards a self-organization estimate $\mathbf{p}^{e,s}$ when it is moved relatively with $d^{p,a} = \varepsilon^{p,a}$ towards an approximation estimate $\mathbf{p}^{e,a}$:

$$r^p = \frac{1 - \left(1 - \varepsilon^{p,s}\right)^{u_c}}{\varepsilon^{p,a}}. \tag{3.39}$$

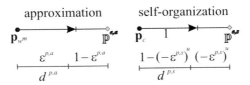

Figure 3.16. Left: Relative change $d^{p,a}$ of $\mathbf{p}_{w^m}$ towards its estimate $\mathbf{p}^{e,a}$ after one approximation step with update rate $\varepsilon^{p,a}$. Right: Relative change $d^{p,s}$ of candidate output support vector $\mathbf{p}_c$ towards its estimate $\mathbf{p}^{e,a}$ after u self-organization steps with update rate $\varepsilon^{p,s}$.

Solving Equation 3.39 for $\varepsilon^{p,s}$ then yields Equation 3.38. The ratio r^p is used as a parameter of the self-organization learning rate in Equation 3.38 and depends on the degree of disorder δ of the three nodes n_r, n_{w^m} and n_c, and on the approximation error Δ_c of the candidate node n_c :

$$r^p = \max\left(t\left(\Delta_c^l, \Delta_c^h, r^{p,\Delta,l}, r^{p,\Delta,h}, \Delta_c\right), t\left(\delta^l, \delta^h, r^{p,\delta,l}, r^{p,\delta,h}, \delta\right)\right), \tag{3.40}$$

where $y = t(x^l, x^h, y^l, y^h, x)$ describes a linear transition from y^l to y^h between x^l and x^h. Figure 3.17 shows the characteristic curve of r^p with typical values of $\Delta_c^l = 0.1$, $\Delta_c^h = 1.0$, $r^{p,\Delta,l} = 0$, $r^{p,\Delta,h} = 0.7$, $\delta^l = 30°$, $\delta^h = 120°$, $r^{p,\delta,l} = 0$, and $r^{p,\delta,h} = 1.0$. The idea behind Equation 3.40 is to decrease the self-organization learning rate smoothly from a high value for disordered SOLIM maps that approximate badly, to a low value for ordered SOLIM maps that approximate well (Figure 3.18).

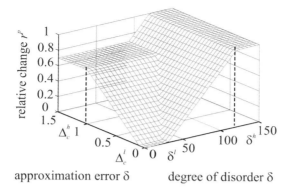

Figure 3.17. Characteristic of relative change r^p depends on candidate node's approximation error Δ_c and on degree of disorder δ of n_r, n_{w^m} and n_c

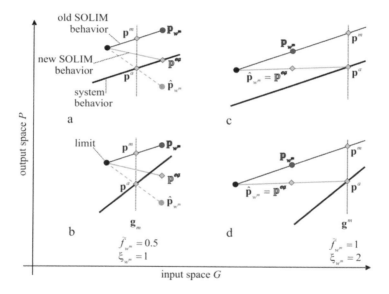

Figure 3.18. Different possible cases and possible result of self-organization (dashed arrows). a. System behavior $g^{d,1}, g^{d,2} \in [0.0, 1.0]$. Local disorder leads to locally high approximation error. b. System behavior $g^m = p_1^a + p_2^a$. Low approximation error but not well-ordered output support vectors. c. System behavior $g^{d,1}, g^{d,2} \in [-0.5, 1.5]$. Ordered output support vectors but locally high approximation error. d. In all cases, self-organization helps by decreasing the degree of disorder and the approximation error.

SOLIM self-organization minimizes the angle between $\mathbf{p}_c - \mathbf{p}_{w^m}$ and $\mathbf{p}_{w^m} - \mathbf{p}_r$, which is consequently chosen as a measure for the **degree of disorder** δ. A similar "degree of regularity" is defined in [42]:

$$\delta = \arccos\left(d_{cw} \cdot d_{wr}\right) \frac{180°}{\pi}, \tag{3.41}$$

$$d_{cw} = \frac{\mathbf{p}_c - \mathbf{p}_{w^m}}{\left\|\mathbf{p}_c - \mathbf{p}_{w^m}\right\|}, \; d_{wr} = \frac{\mathbf{p}_{w^m} - \mathbf{p}_r}{\left\|\mathbf{p}_{w^m} - \mathbf{p}_r\right\|}. \tag{3.42}$$

The **approximation error** is measured and stored for each node. The difference between $\mathbf{c}^m$ and $\mathbf{c}^d$, which are the corresponding mapping images of $\mathbf{g}^m$ and $\mathbf{g}^d$ in topology space, is stored as the approximation error Δ_{w^d} of the node n_{w^d}:

$$\Delta_{w^d} = \left\|\mathbf{c}^d - \mathbf{c}^m\right\|. \tag{3.43}$$

$\mathbf{c}^d$ and $\mathbf{c}^m$ are computed analogously to $\mathbf{p}^d$ and $\mathbf{p}^m$, respectively. $\mathbf{g}^d$ is computed analogously to d^G (compare Section 3.0):

$$\mathbf{c}^d = \sum_{i=1\ldots|N|} \tilde{f}_i^d \cdot \left(\mathbf{c}_i + \mathbf{c}_i^x\right), \tag{3.44}$$

$$\mathbf{c}^m = \sum_{i=1\ldots|N|} \tilde{f}_i^m \cdot \left(\mathbf{c}_i + \mathbf{c}_i^x\right). \tag{3.45}$$

Measuring the approximation error in topology space makes the measure relatively independent of scale differences between the dimensions and relates it to the smallest distance between topology support vectors.

3.3.3.3 Self-organization in Input Space
Self-organization in the input space G is often not necessary because in typical robotic applications the input support vectors can be predefined according to a topology, *e.g.*, to cover the desired velocity range of the mobile microrobot platform, the input support vectors can be placed on a 3D grid. If this is not the case, they can be learned with the Kohonen self-organization rule [18] with the input vectors $\mathbf{g}^d$ as attractors. Special attention must be given when succeeding input vectors $\mathbf{g}^d$ are strongly correlated. This is the case in many robotic applications if the tool centre point moves along a trajectory [43, 44]. The classic Kohonen self-organization rule will then let the input support vector network contract along the $\mathbf{g}^d$-trajectory if the learning rate is not small enough. In addition, when changing the input support vectors $\mathbf{g}_i$ the corresponding output support vectors $\mathbf{p}_i$ must be changed accordingly. Throughout this work, the input support vectors are predefined and kept constant.

3.3.4 Conclusions

The SOLIM approach has been designed to implement a learning inverse model controller.

The **SOLIM structure** is inherited from SOM, where input support vectors are associated with output support vectors and topology support vectors, together called nodes. In contrast to SOM, however, the neighborhood relation between two nodes must be given explicitly during a topology construction, and it also defines simplicial complexes in input space, in output space and topology space.

SOLIM mapping consists of an interpolation and an extrapolation component. Both components are based on transferring the position of a given input vector with respect to the input simplicial complex to an output vector having a similar position with respect to the output simplicial complex. The mapping is continuous because the dependency of the influence of each node on the input vector is continuous. Moreover, the influence of each node is limited by the input support vectors of its neighbors, such that the mapping is local and thus interpretable.

SOLIM learning also follows a different approach compared to existing SOM-based controllers. Firstly, the approximation and the self-organization phase are separated to better adjust their properties. For the derivation of the relatively simple approximation update rule, the mapping from a measured system output to the applied system input is solved for the output support vector with the highest influence on the measured system output (winning node). During self-organization, all neighbors of the winning node that are opposite in topology space are aligned to become opposite in output space, as well. The new learning approach is local because only the output support vector with the highest influence and its neighbors are updated during each iteration step. As a second prerequisite for incremental learning, all learning parameters are kept constant or are adapted automatically. Finally, the self-organization rule not only depends on the degree of disorder but also on the approximation error, such that inverse model ambiguities are resolved towards an ordered output support vector arrangement with a low approximation error.

3.4 SOLIM in Simulations

This section assesses the **mapping performance** and the **learning performance** with respect to systems with known behavior. The systems are modeled as functions $\mathbf{p}^a \mapsto \mathbf{g}^m$ with different characteristics.

3.4.1 Mapping

The first important feature of the SOLIM framework, the mapping performance, is evaluated by initializing the SOLIM map with known support vectors and by comparing the output with the output of the function the map approximates. A typical system for approximation tasks where numbers on approximation errors are available is the **2D Gaussian bell**, which is defined as

$$\mathbf{p}^a = \text{gauss}\left(\mathbf{g}^d\right) = \exp\left(-\frac{1}{2}\cdot\left(\left(\frac{g^{d,1}-\mu}{\sigma}\right)^2 + \left(\frac{g^{d,2}-\mu}{\sigma}\right)^2\right)\right). \tag{3.46}$$

In the following simulations, the mean and the standard deviation will be set to $\mu = 0.5$ and $\sigma = 0.1$, respectively (Figure 3.19a). The mapping is initialized with 5×5 nodes, whose input support vectors $\mathbf{g}_i$ are placed at the positions $g_{i,1}, g_{i,2} \in \{0.1, 0.3, 0.5, 0.7, 0.9\}$ and whose output support vectors $\mathbf{p}_i$ are placed at the corresponding positions $\mathbf{p}_i = \text{gauss}(\mathbf{g}_i)$.

To evaluate the **interpolation performance**, the RMS error for the SOLIM map is computed in the range $g^{d,1}, g^{d,2} \in [0,1]$ (Figure 3.19b). The RMS error of 0.043 is comparable to the RMS errors of other algorithms, which are listed in Table 3.2. To evaluate the **extrapolation performance**, the RMS-error is computed in a larger range $g^{d,1}, g^{d,2} \in [-0.5, 1.5]$ (Figure 3.19c). The extrapolation is reasonable, but the RMS error of 0.057 is relatively large because the falling slope at the borders is continued, although the Gaussian bell approaches the zero plane. This error can be decreased significantly by using more neurons that better represent the slopes, *e.g.*, by placing 7×7 neurons in the same area, yielding an RMS error of 0.024 (Figure 3.19d).

Table 3.2. RMS error for approximation of 2D Gaussian bell ($\mu = 0.5$, $\sigma = 0.1$) with 25 support vectors in the range $g^{d,1}, g^{d,2} \in [0,1]$. All values from [38], except for SOLIM, which is based on simulation of 51×51 points within same range (see also [45]).

Algorithm	RMS error
PSOM	0.79
local PSOM	0.049
SOLIM	0.043
RBF	0.041
I-SOM	0.040
C_R I-SOM	0.016

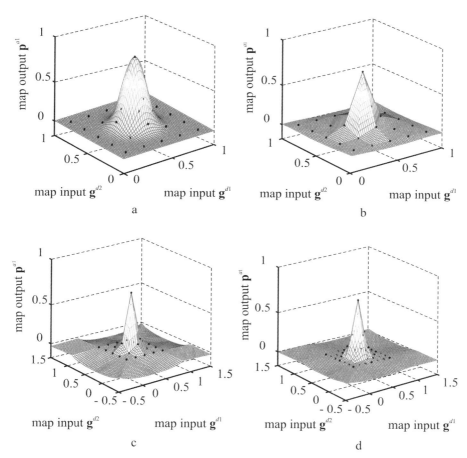

Figure 3.19. a. 2D Gaussian bell, $g^{d,1}, g^{d,2} \in [0.0,1.0]$. SOLIM approximation with b. 5 × 5 nodes, $g^{d,1}, g^{d,2} \in [0.0,1.0]$, c. 5 × 5 nodes, $g^{d,1}, g^{d,2} \in [-0.5,1.5]$ and d. 7 × 7 nodes, $g^{d,1}, g^{d,2}$ in $[-0.5,1.5]$.

3.4.2 Learning

The second important feature of the SOLIM framework, the learning performance, is evaluated by applying the SOLIM map output to a virtual system and by using the measured system output to learn the SOLIM map. For a virtual two-joint kinematic module with a changing behavior, an inverse map is learned incrementally at different similarity coefficients for successive desired system outputs.

3.4.2.1 Procedure
For the evaluation of the learning performance of the SOLIM approach, the same procedure is used for all simulations. It consists of an initialization phase, where

the network is set up, and the actual operation phase, where the network controls the system and is continuously updated.

For the **initialization phase** only three specifications are necessary:

1. The size of the network defines the number of nodes in each topology dimension and with it also the placement of the topology support vectors $\mathbf{c}_i$ in topology space C.
2. The input range defines the range for allowed desired input vectors $\mathbf{g}^d$ and, if the input space dimension d^G is equal to the topology space dimension d^C, it can also be used for initializing the input support vectors $\mathbf{g}_i$ in an ordered arrangement. If $d^G \neq d^C$, e.g., a 1D topology in a 2D input space, the input support vectors are prearranged manually or they are arranged during the operation phase by self-organization.
3. The output range defines the range for the initial random arrangement of the output support vectors $\mathbf{p}_i$ and the boundary for the output vectors $\mathbf{p}^a$. If knowledge on an inverse model already exists, it can be used to initialize the output support vectors and thus to provide a learning bias.

It is not necessary to initialize the learning parameters differently for each system behavior. During the simulations (and during all experiments of the next section) the parameters have been set to the values given in Table 3.3. These values have been found by alternately analyzing the influence of a quantity and selecting the best of the resulting new combinations. This procedure is similar to genetic algorithms, and its automation could reveal parameters that show a slightly better learning performance.

The operation phase consists of repeating the combination of mapping and learning, which is called incremental learning or online learning [33]. In more detail, the steps during operation are as follows (Figure 3.20):

*   A desired system output $\mathbf{g}^d$ is generated randomly.
*   SOLIM maps from the desired system output $\mathbf{g}^d$ to a system input $\mathbf{p}^a$.
*   The system input $\mathbf{p}^a$ is applied.
*   The system output $\mathbf{g}^m$ is measured.
*   SOLIM training is performed with the 3-tuple $\left(\mathbf{g}^d, \mathbf{p}^a, \mathbf{g}^m\right)$.

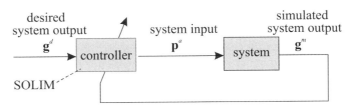

Figure 3.20. Simulation setup: The controller learns a direct inverse model $\mathbf{g}^d \mapsto \mathbf{p}^a$ of the system behavior $\mathbf{p}^a \mapsto \mathbf{g}^m$ by using the simulated system output $\mathbf{g}^m$.

Table 3.3. Learning parameters that are used during all simulations and experiments

Parameter	Value
learning rate for approximation in output space	$\varepsilon^{p,a} = 0.8$
lower bound for approximation error	$\Delta_c^l = 0.1$
upper bound for approximation error	$\Delta_c^h = 1.0$
lower bound for relative SO-change	$r^{p,\Delta,l} = 0.0$
upper bound for relative SO-change	$r^{p,\Delta,h} = 0.7$
lower bound for degree of disorder	$\delta^l = 30°$
upper bound for degree of disorder	$\delta^h = 120°$
lower bound for relative SO-change	$r^{p,\delta,l} = 0.0$
upper bound for relative SO-change	$r^{p,\delta,h} = 1.0$
threshold for deadlock counter	$\lambda^h = 10$
scaling of "allowed subspace" A for new random $\mathbf{p}_{w^d}$	$\rho = 2.0$

3.4.2.2 Inverse Kinematics

The SOLIM algorithm framework has been designed to be able to learn during operation and to adapt the inverse model to a system behavior that slowly changes over time. Furthermore, since all learning steps are performed locally, operation with correlated input vectors does not harm parts of the map that already have a low approximation error.

A model of a **two-joint serial kinematic** (Figure 3.21a) is used as system behavior. The model defines the mapping from the joint angle vector $\mathbf{p}^a$ to the global position vector $\mathbf{g}^m$;

$$g^{m1} = l_1 \cdot \cos\left(p^{a1}\right) + l_2 \cdot \cos\left(p^{a1} + p^{a2}\right), \tag{3.47}$$

$$g^{m2} = l_1 \cdot \sin\left(p^{a1}\right) + l_2 \cdot \sin\left(p^{a1} + p^{a2}\right). \tag{3.48}$$

The axes' lengths are $l_1 = 10$ and $l_2 = 5$ until iteration step 3000 and $l_1 = 12$ and $l_2 = 6$ after iteration step 5000. Between these two phases there is a linear transition of l_1 and l_2, *i.e.*, between iteration step 3000 and 5000 there is a slow **change of the system behavior**. The tool centre point (TCP) at the end of the second axis will reach positions in the range $g^{m1}, g^{m2} \in [4.5, 10.6]$, which are always reachable before, during, and after the system behavior's change (Figure 3.21b). Figure 3.21c

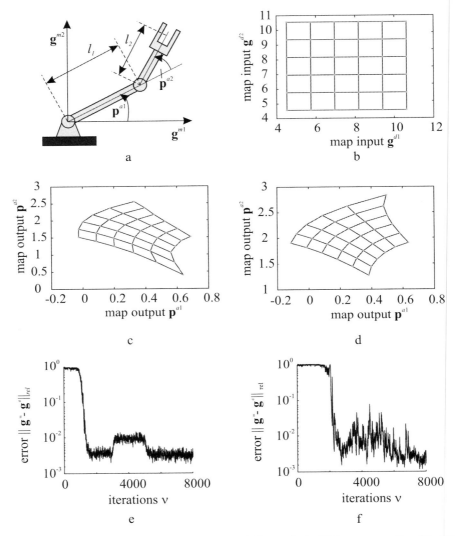

Figure 3.21. a. Configuration of a two-joint serial kinematic. b. 2D input space. Possible 2D output space c. before and d. after changing system behavior. Error development $\left| \mathbf{g}^m - \mathbf{g}^d \right|$ with similarity coefficient e. 0.0 and f. 0.9.

and Figure 3.21d show possible output support vector arrangements for the case $l_1 = 10$ and $l_2 = 5$ and for the case $l_1 = 12$ and $l_2 = 6$, respectively.

The SOLIM algorithm framework is able to adapt to changing system behavior during operation and even while the TCP is following given trajectories. "Following a trajectory" means that **successive input vectors $\mathbf{g}^d$** are **similar** to a certain extent c^s, e.g., a similarity coefficient of $c^s = 0.9$ means that an input vector is chosen randomly around the preceding input vector within a range of $\pm10\%$ of the whole range. For the first 2000 iterations, the similarity coefficient is $c^s = 0$ to allow the network to reach a stable state. This training phase is necessary since a map with random output support vectors cannot perform any useful operation. After this phase, the simulation is run with a similarity coefficient $c^s = 0.9$. The corresponding error developments for $c^s = 0.0$ and $c^s = 0.9$ are shown in Figure 3.21e and Figure 3.21f, respectively. In both cases a relative error of about $5 \cdot 10^{-3}$ is reached until iteration step $v = 3000$. In Figure 3.21f the training phase was not completed until $v = 2000$, but an error of about $5 \cdot 10^{-3}$ could also be reached until $v = 3000$, even with a similarity coefficient of $c^s = 0.9$. In both cases, the relative error is constantly at a higher mean level at about 10^{-2} between $v = 3000$ and $v = 5000$, but the error variation is higher for higher similarity coefficients. In both cases the relative error drops to about $5 \cdot 10^{-3}$ within a certain period of time, but this period increases with the similarity coefficient. The effect of the similarity coefficient on the error variation and the settling time can be explained by the time the input vectors $\mathbf{g}^d$ need before they have covered the whole input space. Because input vectors with a high similarity coefficient tend to "stay" in a limited region of the input space, the local learning rules only update the corresponding region of the map, such that relatively large parts of the map are untouched and keep their increased approximation error. The learning performance of the SOLIM approach thus depends on the relation between the speed of the system change and the speed of covering great parts of the input space by input vectors.

3.5 SOLIM as Actuator Controller

The SOLIM framework has been used to implement actuator controllers for different mobile microrobot platforms presented in Chapter 2. In this section, all measurements are performed with the mobile platform that is referred to as the effector platform. The same experiments with the stage platform have led to similar results.

3.5.1 Actuation Control

To make the robot move with a desired 3D velocity, typically the **amplitude** or the **frequency** of a sawtooth-shaped driving signal is **modulated**. In principle, the signal form can be modulated as well, but this approach would require some additional research. Amplitude modulation and frequency modulation have different application areas, as can be seen from their effect on the velocity of the effector platform in Figure 3.22a and c, and Figure 3.22b and d, respectively. At low voltage amplitudes, the velocity remains almost zero. Only above a certain threshold does the velocity increase fairly linearly with the amplitude. With increasing frequency, the velocity also increases linearly but the relation shows some nonlinearities and a lower repeatability at higher frequencies. The frequency can therefore be modulated to yield velocities in the lower range, while the amplitude can be modulated to yield velocities in the middle and higher range with a good repeatability. The threshold between the modulation methods can be found by relating a desired velocity $v_{l,p}^d$ and a desired angular velocity $\omega_{l,p}^d$ to the corresponding minimum values for amplitude modulation:

$$v_{l,p}^{d,rel} = \frac{v_{l,p}^d}{v_{l,p}^{d,min}},$$ (3.49)

$$\omega_{l,p}^{d,rel} = \frac{\omega_{l,p}^d}{\omega_{l,p}^{d,min}},$$ (3.50)

and by taking the RMS value of these desired relative velocities

$$r^f = \sqrt{\left(v_{l,p}^{d,rel}\right)^2 + \left(\omega_{l,p}^{d,rel}\right)^2}.$$ (3.51)

For values $r^f < 1$ frequency modulation is preferable, for values $r^f > 1$ amplitude modulation is preferable.

The frequencies of all actuation signals are set to be identical because they are generated simultaneously with the same multi channel DA board. With this constraint, the number of DoF that can be controlled by frequency modulation is limited to one such that the amplitudes must be modulated at small velocities to adjust the remaining two of three DoF. The current implementation of the motion controller uses the coefficient r^f defined above to scale down the step frequency and to keep the desired velocity and thus the resulting voltage amplitudes at a minimum level when the effective desired velocity becomes small ($r^f < 1$). Otherwise, amplitude modulation is used.

$$r^f \mapsto \min(r^f, 1),$$ (3.52)

$$f_{step} \mapsto f_{step} \cdot r^f,$$ (3.53)

$$v_{l,p}^d \mapsto v_{l,p}^d / r^f,$$ (3.54)

$$\omega_{l,p}^d \mapsto \omega_{l,p}^d / r^f.$$ (3.55)

The changes in Equations 3.52 to 3.55 are optional, but have shown to significantly decrease the achievable closed-loop position error. During training, however, only the amplitude is modulated because it is easier to implement and also yields satisfactory results. Frequency and signal form are kept constant.

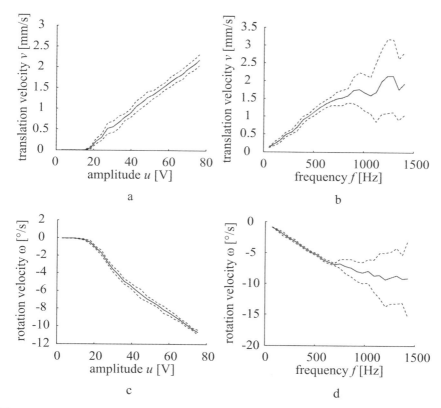

Figure 3.22. a. Influence of the signal amplitude on the platform's velocity ($u^3 = u^4 = 75$ V , $f_{step} = 700$ Hz). b. Influence of the signal frequency on the platform's velocity ($u^3 = u^4 = 50$ V , $f_{step} = 1400$ Hz). c. Influence of the signal amplitude on the platform's angular velocity ($u^2 = u^4 = u^6 = 75$ V , $f_{step} = 1400$ Hz). d. Influence of the signal frequency on the platform's angular velocity ($u^2 = u^4 = u^6 = 50$ V , $f_{step} = 1400$ Hz). Solid line: mean, dotted line: standard deviation of 10 cycles. 25 amplitude or frequency increments. 1000 motion steps per measurement.

3.5.2 Manual Training

As a first approach, knowledge from experiments is used to manually implement an actuator controller for the microrobot platform. The **actuator controller** implements a mapping from a local, polar 3D velocity vector $\mathbf{v}_{l,p}^d$ to a 6D amplitude vector $\mathbf{u}^a$:

$$\mathbf{v}_{l,p}^d = \begin{bmatrix} v_{l,p}^d & \iota_{l,p}^d & \omega_{l,p}^d \end{bmatrix}^T \mapsto \mathbf{u}^a = \begin{bmatrix} u^{a1} & \dots & u^{a6} \end{bmatrix}^T \qquad (3.56)$$

by mapping the translation velocity by a SOLIM map and the angular velocity by a linear map:

$$\left[v_{l,p}^{d} \quad t_{l,p}^{d} \right]^{T} \mapsto \mathbf{u}^{a,t}, \tag{3.57}$$

$$\omega_{l,p}^{d} \mapsto \mathbf{u}^{a,r}, \tag{3.58}$$

$$\mathbf{u}^{a} = \left(\mathbf{u}^{a,t} + \mathbf{u}^{a,r} \right) \cdot \tau. \tag{3.59}$$

The factor τ scales down the combined amplitude vector $\mathbf{u}^{a}$ such that none of its components is larger than the largest component of $\mathbf{u}^{a,t}$ or $\mathbf{u}^{a,r}$.

The manual training of the **SOLIM map for translation** is a tedious task where a person tries to find suitable amplitude vectors $\mathbf{u}^{a,t}$ for 36 desired velocity vectors. In the case of the effector platform, the desired velocity vectors are all combinations of the velocities $v_{l,p}^{d} \in \{0.0, 1.5, 2.5, 3.5\}$, given in mm/s, and the directions $t_{l,p}^{d} \in \{-180, -135, -90, -45, 0, 45, 90, 135, 180\}$, given in degrees. The desired velocity vectors are the input support vectors of the SOLIM map, and the amplitude vectors are the corresponding output support vectors. The SOLIM map is initialized as a 4×9 network, with the corresponding found input and output support vectors. In Figure 3.23a and b the input space and the resulting dependency of the first amplitude u^{a1} on the input vector is shown. Extrapolation is turned off in Figure 3.23b to avoid a problem, which appears when input space dimensions are very different. In this case, the scale difference between 3.5 mm/s in the first dimension and 360° in the second dimension is very high. The solution to this problem is either turning off extrapolation where it is not needed, changing the dimensions' scales to become similar, or finding a better extrapolation approach.

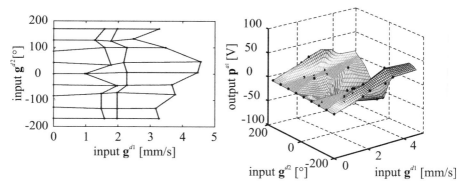

Figure 3.23. Effector platform: a. Input space covering absolute value and direction of translation velocity. b. Dependency of first actuation signal amplitude on desired translation velocity. Extrapolation is turned off.

The **linear map for rotation** is found by applying voltage signals to make the mobile platform rotate and by approximating the ratio of applied amplitude to measured angular velocity. The relation between voltage vector and angular velocity is approximated as

$$\mathbf{u}^{a,r} = \omega_{l,p}^{d} \cdot \begin{bmatrix} 1.8 & -5.5 & 1.8 & -5.5 & 1.8 & -5.5 \end{bmatrix}^{T}. \tag{3.60}$$

3.5.3 Automatic Training

The SOLIM framework is used to learn the mapping from desired local Cartesian velocities $\mathbf{v}_{l,c}^{d}$ in three DoF to the actuation signal amplitudes $\mathbf{u}^{a}$ in six DoF.

During the **initialization phase** of the 3D-network, the $5 \times 5 \times 5$ topology is associated with ordered input support vectors and random output support vectors within their predefined ranges. There is no more information required for the initialization and operation of SOLIM since all learning rates are fixed or adapted according to the network state. During all experiments the learning parameters as given in Table 3.2 are used.

Similar to the simulation procedure, the following steps are performed repeatedly (Figure 3.24) during the **operation phase**:

- A desired velocity vector $\mathbf{v}_{l,c}^{d}$ is generated randomly.
- SOLIM maps from the desired velocity $\mathbf{g}^{d} = \mathbf{v}_{l,c}^{d}$ to actuation parameters $\mathbf{p}^{a}$.
- The actuation parameters $\mathbf{p}^{a} = \mathbf{u}^{a}$ are applied to the microrobot for some time.
- The velocity $\mathbf{v}_{l,c}^{m}$ is measured with help of a camera and image processing.
- SOLIM is trained with the 3-tuple $\left(\mathbf{g}^{d}, \mathbf{p}^{a}, \mathbf{g}^{m} \right)$.

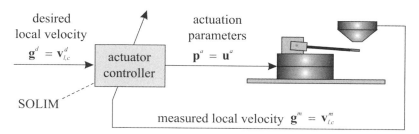

Figure 3.24. Experimental learning setup: The controller learns a direct inverse model $\mathbf{g}^{d} \mapsto \mathbf{p}^{a}$ of the system behavior $\mathbf{p}^{a} \mapsto \mathbf{g}^{m}$ by using the measured system output $\mathbf{g}^{m}$.

During the **initialization** of the SOLIM network, the input support vectors $\mathbf{g}_{i}$ of the 125 nodes are arranged equally spaced within the range of the velocity space, while the output support vectors $\mathbf{p}_{i}$ are randomly initialized within the allowed range for the amplitudes $u^{a1...6}$. With amplitudes in the range of [-80 V, 80 V] the effector platform can cover a velocity range of $\dot{x}^d_{l,c}, \dot{y}^d_{l,c} \in [-1\,\mathrm{mm/s}, 1\,\mathrm{mm/s}]$ and an angular velocity range of $\omega^d_{l,c} \in [-3°/\mathrm{s}, 3°/\mathrm{s}]$.

The **training of the effector platform** is performed with uncorrelated input vectors $\mathbf{g}^d$ that are uniformly chosen from the input range. The input space and the first three dimensions of the output space after 2000 iteration steps are shown in Figure 3.25a and Figure 3.25b, respectively. After about 1000 iteration steps, the norm of the relative velocity error $\left\| \mathbf{g}^m - \mathbf{g}^d \right\|_{rel}$ has fallen to about $7 \cdot 10^{-2}$ (Figure 3.25c), which is approximately 4% of the range of each input space dimension, or 0.08 mm/s for translation and 0.24°/s for rotation. The last 100 of the 2000 iteration steps are taken to evaluate the velocity error of the trained SOLIM map in more detail. Table 3.4 compares the velocity error of the manually trained combined SOLIM and linear map described above with the velocity error of the learning SOLIM map and with the velocity repeatability. The norm of the relative velocity error of the manually trained map is around 29%, while the norm of the relative velocity error of the learning SOLIM map is around 7%, which is a large improvement. The repeatability is computed as the mean standard deviation of 10 velocity measurements with the same voltage amplitudes. The norm of the relative repeatability is around 4% and thus in the same order of magnitude as the error of the learning SOLIM map.

Table 3.4. Effector platform: comparison of velocity error of trained SOLIM map, velocity error of static map and repeatability. Conditions: $f_{step} = 700\,\mathrm{Hz}$, 1000 motion steps per measurement.

	x [mm/s]	y [mm/s]	φ [°/s]	$\|\cdot\|_2$
velocity error manually trained map				
mean [std. dev.] of 100 iterat.	0.324 [0.201]	0.344 [0.229]	0.969 [0.726]	
...**relative to range**	$16.2\% \cdot 2$	$17.2\% \cdot 2$	$16.1\% \cdot 6$	28.6%
velocity error learning SOLIM map				
mean [std. dev.] of 100 iterat.	0.059 [0.047]	0.060 [0.048]	0.299 [0.227]	
...**relative to range**	$2.97\% \cdot 2$	$2.99\% \cdot 2$	$4.98\% \cdot 6$	6.53%
repeatability (10 measurements)				
mean [std. dev.] of 100 iterat.	0.040 [0.021]	0.038 [0.018]	0.151 [0.062]	
...**relative to range**	$1.99\% \cdot 2$	$1.90\% \cdot 2$	$2.51\% \cdot 6$	3.72%

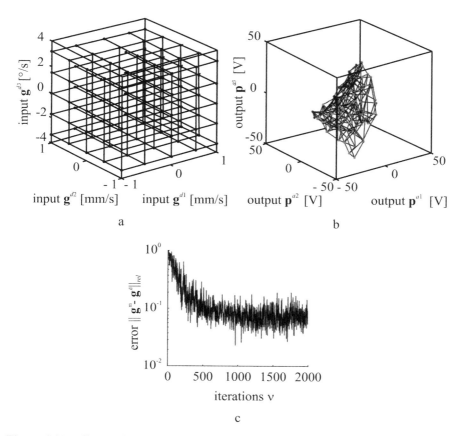

Figure 3.25. Effector platform: a. Input space, b. output space, c. error development after 2000 iteration steps. Conditions: $f_{step} = 700\text{Hz}$, 1000 motion steps per measurement.

As already indicated in the introduction of this chapter, the difference in the velocity error of an actuator controller influences the control performance when positioning the effector platform with feedback from a CCD camera. The mean positioning error of 100 positioning processes is about 0.0159 mm and 0.0179° when using the manually trained map. When using the automatically trained SOLIM map, the positioning error is reduced to 0.0041 mm and 0.0101°. Please note that all velocity and pose errors are given with respect to the vision system, which provides positions with a certain error, but which still allows comparison between the two differently trained maps.

3.6 Conclusions

3.6.1 Summary

The development of this work has been motivated by the problems that appear during the development of a **controller for mobile microrobots**. The main problem was the definition of a nonlinear, learning and unambiguous actuation control module that yields the necessary actuation parameters for each desired velocity. Controllers that are based on **self-organizing maps** show some interesting properties, which qualify them for the application to microrobots, mainly because they can learn a smooth controller model and because they are interpretable to a certain degree. On the other hand, existing SOM-based approaches still have problems when mapping from a desired system output to the corresponding system input and when learning an inverse model controller for a given system.

Details of a **closed-loop pose controller** for an automated microrobot-based nanohandling station are given with the components trajectory controller, motion controller and actuator controller. The latter will be replaced by the learning controller, which is the main contribution of this chapter.

The **SOLIM approach** inherits the basic structure of SOM by associating input support vectors, output support vectors and topology support vectors, whose combination is called a node. Important for any SOM-based approach is the definition of an order of the nodes, which in the case of SOLIM is done explicitly by assigning a set of neighboring nodes to each node. SOLIM maps exactly from any input support vector to its corresponding output support vector and otherwise inter- or extrapolates. SOLIM mapping is based on the similarity of the input vector with respect to the input support vector arrangement and the output vector with respect to the output support vector arrangement. The SOLIM framework has a separated update rule for approximation and for self-organization. The approximation part changes the output support vectors such that the cascaded SOLIM map and system behavior results in an identity mapping. The approximation update rule is based on solving the mapping formula for the value of the output support vector that would be most responsible for the measured system output. The self-organization rule aligns neighboring output support vectors whose topology vectors are in-line and determines the corresponding learning rate depending on their degree of disorder and their approximation error. Assuming an ordered arrangement of input support vectors, similar input vectors map to similar output vectors.

The **SOLIM** approach is validated in various **simulations**. SOLIM mapping is continuous and has a "meaningful" inter- and extrapolation. Its mapping error is comparable to other SOM-based interpolation approaches. When the SOLIM map is used as a learning controller for a virtual system, *e.g.*, an inverse kinematics system, it is able to learn a useful controller, even when its definition is ambiguous or when successive desired system output vectors are correlated.

Finally, the **SOLIM** map represents an **actuator controller**, which is trained and directly applied to map from desired velocities to amplitudes of the signals driving a microrobot. The training can be manual or automatic and successive desired velocities can even be correlated.

3.6.2 Outlook

Although some of the problems of current SOM-based controllers, which are identified in the introduction, are solved by the SOLIM approach, there is still a gap between the current state of SOLIM and the "ideal" inverse model controller. There are three main problems of the SOLIM approach that suggest directions for its enhancement. Firstly, the current **extrapolation** method can under certain circumstances lead to distortions. Secondly, SOLIM mapping and learning needs many **computing operations** for a high number of topology space dimensions, and finally, the **number of nodes** for each topology dimension must be known **beforehand**. Besides these extensions, it is also important to use the SOLIM approach for other **applications** and to gain information on possible future research directions in the field of SOM-based controllers.

3.6.2.1 Extrapolation

The current extrapolation method can be problematic if the input support vectors are not arranged on a regular grid, such that the interpolation border and the extrapolation border are misaligned. While the interpolation border is defined by the outer $(dC-1)$D faces of the d^CD simplicial complex, the extrapolation border is defined by the mean differences $\mathbf{g}_i^\eta$ between each border input support vector and its neighbors. There are thus areas outside the interpolation area where the extrapolation weight is $x_i = 0$, and there are areas inside the interpolation area where the extrapolation weight is $x_i > 0$. Since the sizes of these distorted areas are related to the angles between the extrapolation direction $\mathbf{g}_i^\eta$ of a border node and the normal $\mathbf{g}_{k,j}^n$ of a border facet, these distortions are larger when the input space dimensions have very different scales. The new extrapolation approach should therefore ensure that the extrapolation border is identical to the interpolation border and that the mapping result is identical with inter- and extrapolation at the border facets. Moreover, the extrapolation result must still be continuous, *i.e.*, the transition of the influence from one border node or border facet to another must be continuous.

3.6.2.2 Computational Load

The number of required computing operations rises faster than exponential with the number d^C of topology dimensions. There are two main ways of improvement: Optimize the calculation under certain constraints or introduce hierarchy and thus change the structure of the SOLIM approach.

The most promising approach for the optimization of the SOLIM computation is the **reduction of the number of limits** that have to be **evaluated** when mapping from an input vector to an output vector. A great reduction of computing time can be achieved, if the number d^G of input space dimensions is equal to the number d^C of topology space dimensions, such that an input vector is contained by only one d^CD simplex. In this case, classic simplex interpolation can be applied, where a search algorithm finds the responsible simplex with the influences of its $d^C + 1$ nodes.

Another approach for the reduction of the computational load is the **introduction of hierarchy**, which has also been proposed for LLM [25, 46] and SOM [18]. In hierarchical SOLIM the responsibility for different input dimensions could be divided into separate networks, which are arranged hierarchically, such that each node of a network has a subnetwork assigned. Subnetworks at the same level could be identical in input space and topology space, such that the input support vectors and topology support vectors would be arranged on a regular grid. The output support vectors, which are associated to each leaf of the hierarchy, must be different to implement a meaningful map. Approximation learning and self-organization learning could be performed according to the current SOLIM approach. In this case, the nodes with the highest influence on the measured system output $\mathbf{g}^m$ must be found by a hierarchical mapping. First approaches to the implementation of hierarchical SOLIM were undertaken in a student project [47] and in a diploma thesis [48]. The results show that the number of required computing operations increases linearly with topology dimension d^C when using a hierarchical topology with only 1D subnetworks, while it increased exponentially when using a d^CD network. The approximation error that has been achieved in that work is still high compared to current non-hierarchical SOLIM.

3.6.2.3 Predefined Network Size

Another big limitation of the SOLIM approach is the requirement to define the number of nodes in each topology dimension before operation. There are a number of methods for growing and shrinking SOM networks, and it would be a great enhancement to the SOLIM approach if a method for the dynamic topology construction could be integrated. Ideally, an inverse model controller should only need input and output range and optionally some kind of technical limitation for the SOM network, such as maximum computing time per iteration, maximum available memory or minimum approximation error. One of these practical criteria should then be used to direct the growth and shrinking of the network, *e.g.*, to ensure a uniform approximation error distribution. Special attention should be paid to the problem that approximation and self-organizing learning could interfere with the growing method.

3.6.2.4 Applications for SOLIM

For the further development of the SOLIM approach it is also important to know more about the demands of other applications where a learning inverse model controller can be used. In robotics, and especially in microrobotics, inverse kinematics and camera calibration are applications where SOLIM could be applied due to its general approach.

SOLIM could implement **inverse kinematics** $\mathbf{s}^d \mapsto \mathbf{u}$ for different manipulator kinematics $\mathbf{u} \mapsto \mathbf{s}^m$, like a voltage-controlled 3-DoF piezo-scanner, which positions micro tools. Applying SOLIM for this task is especially useful if the kinematics are nonlinear or if the parameter variance of different instances is very large.

Applying SOLIM is also practical for the control of different unknown kinematics, which are under development. Alternatively, the kinematics could also consist of axes with sensors and controllers for each axis, *i.e.*, a device that moves a tool centre point to a position $\mathbf{s}_w^m$ in world coordinates by moving its axes to local desired positions $\mathbf{s}_l^d$. SOLIM would then implement the inverse mapping from the TCP's desired world positions $\mathbf{s}_w^d$ to the axes' desired local positions $\mathbf{s}_l^d$. This application is practical if the kinematic behavior is position-dependent or nonlinear. In both cases, SOLIM could adapt to a time-variant kinematic behavior, assuming that data from a global sensor is available for incremental learning.

SOLIM could also be used for **camera calibration** by mapping from 3D camera coordinates $\mathbf{s}_c$ to 3D world coordinates $\mathbf{s}_w$. This would remove distortions of camera mappings $\mathbf{s}_w \mapsto \mathbf{s}_c$, which are mainly induced by lenses. Moreover, SOLIM could even perform this calibration if the camera was moved, assuming that this position could be measured. In this case, a hierarchical network would be the best solution, where one level covers the camera positions and a second level contains subnetworks taking care of the position-dependent camera calibration.

3.7 References

[1] Santa, K. & Fatikow, S. 1998, 'Development of a neural controller for motion control of a piezoelectric three-legged micromanipulation robot', *Proc. Int. Conf. on Intelligent Robots and Systems (IROS'98)*, Victoria, B.C., Canada, pp. 788–793.

[2] Zesch, W. 1997, 'Multi-degree-of-freedom micropositioning using stepping principles', Ph.D. thesis, Swiss Federal Institute of Technology ETH Zurich, Zurich, Switzerland.

[3] Bergander, A. 2003, 'Control, wear and integration of stick-slip micropositioning', thesis no 2843, École Polytechnique Fédérale de Lausanne (EPFL), Lausanne, Switzerland.

[4] Zhou, Q., Chang, B. & Koivo, H. 2004, 'Ambient environmental effects in micro/nano handling', *Proc. Int. Workshop on Microfactories*, Shanghai, China, pp. 146–151.

[5] Zhou, Q. 2006, 'More confident microhandling', *Proc. Int. Workshop on Microfactories (IWMF'06)*, Besancon, France.

[6] Bergander, A. & Breguet, J.-M. 2002, 'A testing mechanism and testing procedure for materials in inertial drives', *Proc. Int. Symposium on Micromechatronics and Human Science (MHS'02)*, Nagoya, Japan, pp. 213–218.

[7] Ronkanen, P., Kallio, P. & Koivo, H. N. 2002, 'Current control of piezoelectric actuators with power loss compensation', *Proc. Int. Conf. on Intelligent Robots and Systems (IROS'02)*, Lausanne, Switzerland, pp. 1948–1953.

[8] Martel, S. 2005, 'Fundamental principles and issues of high-speed piezoactuated three-legged motion for miniature robots designed for nanometer-scale operations', *Int. Journal of Robotics Research*, vol. 24, no. 7, pp. 575–588.

[9] Kortschack, A. & Fatikow, S. 2004, 'Development of a mobile nanohandling robot', *Journal of Micromechatronics*, vol. 2, no. 3–4, pp. 249–269.

[10] Fahlbusch, S., Doll, T., Kammrath, W., Weiss, K., Fatikow, S. & Seyfried, J. 2000, 'Development of a flexible piezoelectric microrobot for the handling of microobjects', *Proc. Int. Conf. on New Actuators (Actuator'00)*, Bremen, Germany.

[11]   Seyfried, J., Estaña, R., Schmoeckel, F., Thiel, M., Bürkle, A. & Woern, H. 2003, 'Controlling cm^3 sized autonomous micro robots operating in the micro and nano world', *Proc. Int. Conf. on Climbing and Walking Robots and their Supporting Technologies (CLAWAR'03)*, Catania, Italy, pp. 627–634.

[12]   Driesen, W., Varidel, T., Régnier, S. & Breguet, J.-M. 2005, 'Micro manipulation by adhesion with two collaborating mobile microrobots', *Journal of Micromechanics and Microengineering*, vol. 15, pp. 259–267.

[13]   Estaña, R. 2006, 'Public Report on EU Project MiCRoN <IST-2001-33567>', http://wwwipr.ira.uka.de/seyfried/MiCRoN/PublicReport_Final.pdf.

[14]   Vartholomeos, P., Loizou, S., Thiel, M., Kyriakopoulos, K. & Papadopoulos, E. 2006, 'Control of the multi agent micro-robotic platform MiCRoN', *Proc. Int. Conf. on Control Applications (CCA'06)*, Munich, Germany, pp. 1414–1419.

[15]   Misaki, D., Kayano, S., Wakikaido, Y., Fuchiwaki, O. & Aoyama, H. 2004, 'Precise automatic guiding and positioning of micro robots with a fine tool for microscopic operations', *Proc. Int. Conf. on Intelligent Robots and Systems (IROS'04)*, Sendai, Japan, pp. 218–223.

[16]   Martel, S., Sherwood, M., Helm, C., de Quevedo, W. G., Fofonoff, T., Dyer, R., Bevilacqua, J., Kaufman, J., Roushdy, O. & Hunter, I. 2001, 'Three-legged wireless minature robots for mass-scale operations at the sub-atomic scale', *Proc. Int. Conf. on Robotics and Automation (ICRA'01)*, vol. 4, Seoul, Korea, pp. 3423–3428.

[17]   Kohonen, T. 1982, 'Self-organizing formation of topologically correct feature maps', *Biol. Cyb.*, vol. 43, no. 1, pp. 59–69.

[18]   Kohonen, T. 2001, *Self-Organizing Maps*, 3 edn, Springer Verlag, Berlin.

[19]   Kaski, S., Kangas, J. & Kohonen, T. 1998, 'Bibliography of self-organizing map (SOM) Papers: 1981-1997', *Neural Computing Surveys*, vol. 1, no. 3&4, pp. 1–176. http://www.icsi.berkeley.edu/jagota/NCS/.

[20]   Oja, M., Kaski, S. & Kohonen, T. 2002, 'Bibliography of self-organizing map (SOM) Papers: 1998-2001 Addendum', *Neural Computing Surveys*, vol. 3, pp. 1–156. http://www.icsi.berkeley.edu/jagota/NCS/.

[21]   de A. Barreto, G., Araújo, A. F. R. & Ritter, H. J. 2003, 'Self-organizing feature maps for modeling and control of robotic manipulators', *Journal of Intelligent and Robotic Systems*, vol. 36, no. 4, pp. 407–450.

[22]   Nørgaard, M., Ravn, O., Poulsen, N. K. & Hansen, L. K. 2000, *Neural Networks for Modelling and Control of Dynamic Systems*, Advanced Textbooks in Control and Signal Processing, Springer Verlag, London.

[23]   Huggett, S. & Jordan, D. 2001, *A Topological Aperitif*, Springer Verlag, London.

[24]   Lawson, T. 2003, *Topology: A Geometric Approach*, Oxford University Press, New York.

[25]   Ritter, H., Martinetz, T. & Schulten, K. 1992, *Neural Computation and Self-Organizing Maps: An Introduction*, Addison-Wesley, Reading, MA.

[26]   Moshou, D. & Ramon, H. 1997, 'Extended self-organizing maps with local linear mappings for function approximation and system identification', *Proc. Workshop on Self-Organizing Maps (WSOM'97)*, Espoo, Finland, pp. 181–186.

[27]   Hülsen, H., Garnica, S. & Fatikow, S. 2003, 'Extended Kohonen networks for the pose control of microrobots in a nanohandling station', *Proc. Int. Symp. on Intelligent Control (ISIC'03)*, Houston, TX, U.S.A., pp. 116–121.

[28]   Garnica, S., Hülsen, H. & Fatikow, S. 2003, 'Development of a control system for a microrobot-based nanohandling station', *Proc. Int. Symp. on Robot Control (SYROCO'03)*, Wroclaw, Poland, pp. 631–636.

[29]   Ritter, H., Steil, J. J., Nölker, C., Röthling, F. & McGuire, P. C. 2003, 'Neural architectures for robot intelligence', *Reviews in the Neurosciences*, vol. 14, no. 1–2, pp. 121–143.

[30] de A. Barreto, G. & Araújo, A. F. R. 2004, 'Identification and control of dynamical systems using the self-organizing map', *IEEE Transactions on Neural Networks*, vol. 15, no. 5, pp. 1244–1259.

[31] Aupetit, M., Couturier, P. & Massotte, P. 1999, 'A continuous self-organizing map using spline technique for function approximation', *Proc. Artificial Intelligence and Control Systems (AICS'99)*, Cork, Ireland.

[32] Aupetit, M., Couturier, P. & Massotte, P. 2000, 'Function approximation with continuous self-organizing maps using neighboring influence interpolation', *Proc. Neural Computation (NC'2000)*, Berlin, Germany.

[33] Walter, J. 1997, *Rapid Learning in Robotics*, Cuvillier Verlag, Göttingen. http://www.techfak.uni-bielefeld.de/walter/.

[34] Walter, J., Nölker, C. & Ritter, H. 2000, 'The PSOM algorithm and applications', *Proc. of Int. Symp. on Neual Computation (NC'2000)*, Berlin, Germany, pp. 758–764.

[35] Göppert, J. & Rosenstiel, W. 1993, 'Topology-preserving interpolation in self-organizing maps', *Proc. NeuroNimes 93*, Nanterre, France, pp. 425–434.

[36] Göppert, J. & Rosenstiel, W. 1995, 'Toplogical interpolation in SOM by affine Transformations', *Proc. European Symposium on Artificial Neural Networks (ESANN'95)*, Brussels, Belgium, pp. 15–20.

[37] Göppert, J. & Rosenstiel, W. 1995, 'Interpolation in SOM: improved generalization by iterative methods', *Proc. Int. Conf. on Artificial Neural Networks (ICANN'95)*, Paris, France, pp. 425–434.

[38] Göppert, J. & Rosenstiel, W. 1997, 'The continuous interpolating self-organizing map', *Neural Processing Letters*, vol. 5, no. 3, pp. 185–192.

[39] Lay, D. C. 2000, *Linear Algebra and Its Applications*, 2 edn, Addison-Wesley.

[40] Nort, D. V., Wanderley, M. M. & Depalle, P. 2004, 'On the choice of mappings based on geometric properties', *Proc. Conf. on New Interfaces for Musical Expression (NIME'04)*, Hamamatsu, Japan, pp. 87–91.

[41] Amidror, I. 2002, 'Scattered data interpolation methods for electronic imaging systems: a survey', *Journal of Electronic Imaging*, vol. 11, no. 2, pp. 157–176.

[42] Göppert, J. & Rosenstiel, W. 1996, 'Regularized SOM-training: a solution to the topology-approximation dilemma?', *Proc. Int. Conf. on Neural Networks (ICNN'96)*, vol. 1, Washington, USA., pp. 38–43.

[43] Jockusch, J. & Ritter, H. 1999, 'An instantaneous topological mapping model for correlated stimuli', *Proc. Int. Joint Conference on Neural Networks (IJCNN'99)*, p. 445.

[44] Jockusch, J. 2000, 'Exploration based on neural networks with applications in manipulator control', Ph.D. thesis, University of Bielefeld, Bielefeld, Germany.

[45] Hülsen, H. & Fatikow, S. 2005, 'Extrapolation with a self-organising locally interpolating map', *Proc. Int. Conference on Informatics in Control, Automation and Robotics (ICINCO'05)*, Barcelona, Spain, pp. 173–178.

[46] Martinetz, T. M. & Schulten, K. J. 1990, 'Hierarchical neural net for learning control of a robot's arm and gripper', *Proc. Int. Joint Conf. on Neural Networks (IJCNN'90)*, Vol. 2, San Diego, CA, U.S.A., pp. 747–752.

[47] Schrader, C. 2005, 'Untersuchung eines Fuzzy-Kohonen-basierten Steuerungsalgorithmus und dessen Erweiterung um eine hierarchische Struktur', student project, University of Oldenburg, Oldenburg, Germany.

[48] Bruns, B. 2006, 'Untersuchung einer selbstorganisierenden, lokal interpolierenden Karte für die Online-Steuerung nichtlinearer, zeitvarianter Systeme', Diploma thesis, University of Oldenburg, Oldenburg, Germany.

Real-time Object Tracking Inside an SEM

Torsten Sievers

Division of Microrobotics and Control Engineering,
Department of Computing Science,
University of Oldenburg, Germany

4.1 Introduction

Within the last few years, a trend towards the automation of nanohandling processes has emerged. One key problem is the development of a global sensor to measure the position of handling tools and nanoobjects during the manipulation process. The sensor data is required as feedback to enable the closed-loop positioning of the tools and nanoobjects.

While the manufacturing of nanoobjects like carbon nanotubes (CNT) is mostly done automatically using batch processing [1, 2], the production of simple nanosystems is teleoperated, due to the incompatibilities of the different batch processes. A tool increasingly used to handle micro- and nanoobjects is the atomic force microscope (AFM) [3–5]. As mentioned in Chapter 1, one main drawback is that the AFM cannot be used simultaneously for manipulation and imaging. Thus, no sensor feedback is available, which is required for automation. In [6, 7] an AFM has been integrated into a scanning electron microscope (SEM) to image the handling process, but it was still carried out using teleoperation.

Other important tools for the handling of small objects are microrobots with piezodrive [8–10], which are equipped with application-specific tools [11, 12]. In [13–16], light microscopes are used to gain visual feedback for automated microhandling by microrobots. The limit of these systems is the resolving power of light microscopes, which does not allow reliable visualization of nanoscale objects.

The solution proposed in this chapter is to use an SEM, in combination with real-time image processing algorithms, as a **global pose sensor**. The calculated (x, y, φ)-pose of handling tools and nanoobjects is used as feedback for the control system to enable the automation of nanohandling tasks (Figure 4.1).

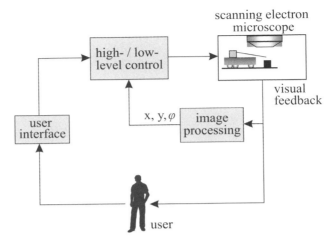

Figure 4.1. Automation of nanohandling tasks by image processing

The main benefits of an SEM are high resolution, short image acquisition time, and large depth of focus. Especially high frame rates are an advantage, as this enables high update rates of the estimated poses for fast closed-loop positioning. However, the drawback of fast scanning is strong additive noise. As averaging and filtering are too time-consuming, the tracking approaches must provide a high robustness against additive noise.

This chapter is structured as follows: in Section 4.2 relevant properties of the SEM are presented and the requirements for the image processing algorithms are defined. The software setup and a description of SEM integration in an image processing system to enable real-time image access and electron beam control is given in Section 4.3. The main part of this chapter is devoted to the implementation and validation of real-time tracking algorithms with high robustness against noise. In Section 4.4, a cross-correlation approach is described. An active contours algorithm with region-based minimization is presented in Section 4.5.

4.2 The SEM as Sensor

In addition to Section 2.2, where an SEM is introduced as a **sensor for nano-handling**, the relevant SEM properties for the image processing algorithms are described in this section. An SEM is, in terms of resolution (down to 3 nm), image acquisition time, and depth of focus, the most suitable sensor for the visualization of nanohandling processes. In addition, an SEM offers a large working distance and a continuous zoom. The application of scanning electron microscopy to different aspects of nanotechnology is described in [17]. The short image acquisition time is an especially important requirement for the automation of nanohandling processes, since high frame rates enable high update rates of the estimated poses for fast closed-loop positioning. Because image acquisition is done by pixel-wise scanning, the frame rate depends on the image size. The minimum acquisition time depends on the SEM type and varies for current devices between 125 and 250 ns for one pixel. Thus, a frame rate of 10 fps (5fps respectively) can be achieved with an image size of 1024 × 768 pixels. In Table 4.1 the minimum image acquisition times for four different image sizes are shown.

Table 4.1. Minimum image acquisition times for four different image sizes

Image size [pixels]	Acquisition time [ms]	Frame rate [fps]
1024 × 768	100	10
640 × 480	76.8	13
256 × 256	16.4	61
128 × 128	4.1	244

As described in the following, the real-time criterion is fulfilled if the pose of the target object can be calculated within the minimal image acquisition time. The drawback of fast scanning is **strong additive noise**, as only a few numbers of secondary electrons (SE) can be excited during such a short time period. Images with low noise are acquired by frame averaging or at lower frame rates (pixel averaging). An example of noise reduction by averaging is shown in Figure 4.2. To enable automation with high throughput, averaging and frame rate reduction are too time-consuming and therefore not applicable. Thus, the image processing approaches themselves must be very robust against additive noise.

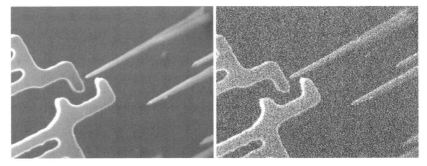

Figure 4.2. SEM image of a typical nanohandling scene with an image acquisition time of 40 seconds (left) and 90 milliseconds (right). The microgripper is described in [12] and [18].

Besides additive noise, two further sources of noise are important. The first one is gray-level fluctuation, which is a significant issue in SEM. Gray-level fluctuations occur due to electrostatic charge and due to variations of the alignment of target, electron beam, and secondary electron detector. Especially for pattern-matching approaches without invariance against amplitude variations, the fluctuations are problematic.

The third source of noise is clutter caused by objects in the image background and/or in the image foreground. Objects in the image foreground can occlude the target object, while background objects often mask edges. As occlusions of the target object or the nanotool are not unusual during a handling process, the image processing approaches to be developed must be robust against clutter as well.

Altogether, the following **requirements** must be fulfilled by the image processing algorithms:

- robustness against additive noise
- robustness against gray-level fluctuations
- robustness against occlusions
- real-time processing

As the frame rate of an SEM depends on the image size, real-time image processing is achieved if the processing time is smaller than or equal to the image acquisition time.

4.3 Integration of the SEM

A key problem to be solved is the integration of the SEM into the image-processing system. To enable **real-time tracking** it is necessary to access the acquired images directly and to control the electron beam. By controlling the electron beam a so-called **region of interest (ROI)** can be used. As the image acquisition time depends on the image size, a small ROI optimally adapted to the target object can be acquired much faster than a larger one. A direct beam access is necessary to control the size and the position of the ROI, which is calculated by the image-processing approach.

Today, most SEMs available on the market are not equipped with a software interface for real-time image access and beam control. In most cases, only a slow (low bandwidth) remote control interface is available. Therefore, the special image acquisition hardware DISS 5 is used. DISS 5 is equipped with a scan generator, an image acquisition unit, and a software interface based on a dynamic link library (dll). The minimum acquisition time is 250 ns per pixel. The device is connected to the external scan interface and to the SE detector of the SEM. Thus, the hard- and software to be developed can be used with virtually every SEM that is equipped with an external scan interface. The software setup of the image-processing system used is depicted in Figure 4.3. As described before, the image data and the beam control commands are transmitted to the SEM via the DISS 5 hardware.

In addition, the LEO 1450 is equipped with a remote control interface that is used to change SEM parameters like magnification, working distance, acceleration voltage, *etc.* As the variation of these parameters is not time-critical, the slow standard interface can be used. The real-time image processing used to calculate the pose of the target objects is the subject of the following sections.

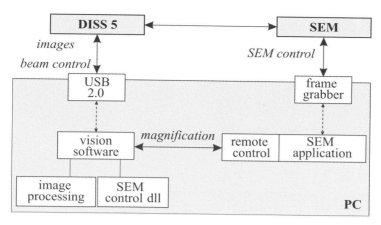

Figure 4.3. Software setup of the SEM image-processing system

4.4 Cross-correlation-based Tracking

Strictly speaking, correlation-based approaches do not belong to the group of object-tracking methods, but belong to the pattern-matching approaches. Tracking is achieved by correlating each input frame with an image of the target.

The cross-correlation of two two-dimensional functions f and g is equal to the folding of these functions [19]:

$$f \circ g \equiv \overline{f}(-x,-y) * g(x, y). \tag{4.1}$$

The correlation operation is denoted by $\circ$ and the folding operation by $*$. As a folding in the spatial domain is equal to a multiplication in the frequency domain (correlation theorem), Equation 4.1 can be rewritten as:

$$f \circ g = F\left[\overline{F}_{(v)} \cdot G_{(v)}\right]. \tag{4.2}$$

G denotes the Fourier transform of g, while $\overline{F}$ stands for the complex conjugated of the Fourier transform of f. In this case, f defines the input image, which contains an unknown scene, and g defines a filter mask containing the object to be found. The result of the cross-correlation operation is a measure for the similarity of the two functions. A high maximum in the result matrix indicates a high similarity. By definition of a threshold it can be decided whether the target object is located inside the image and at which position. Object recognition by cross-correlation is also called pattern matching. In the following, an example clarifies the approach. In Figure 4.4 a scene (left) is shown, in which an object (Figure 4.4, middle) is searched for. The cross-correlation matrix (Figure 4.4, right) is calculated using Equation 4.2. If the maximum value exceeds the defined threshold, the position of the maximum indicates the position of the target object in the input image.

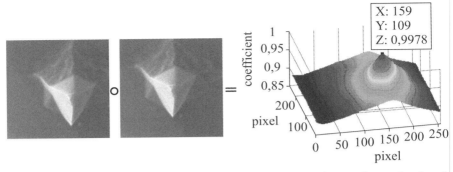

Figure 4.4. The cross-correlation matrix (right) can be used to estimate the (x, y)-displacement between an input image (left) and a pattern (middle)

Usually, the threshold value is determined empirically. The drawback is that, if the selected value is too low, false recognitions can occur, whereby a higher value will decrease the robustness against noise. Tracking is carried out by sequential object recognition in each input frame for an image sequence.

The advantages of cross-correlation-based tracking are a simple implementation and a high robustness against additive noise. In [20], several examples and variations of correlation-based approaches adapted to different physical noise are described. A disadvantage is the inaccurate modeling of the target object, since parts of the background also belong to the model. Thus, this approach is sensitive to clutter. A further restriction is that only translations of the target object can be estimated. If the target object rotates, deforms, and scales in comparison to the pattern (filter mask), the correlation decreases and a direct estimation of the transformation parameters is not possible. A quantitative determination of the rotation angle and invariance against scaling requires additional patterns that have to be folded with the input image. Thus, the computational cost increases significantly. Cross-correlation-based tracking in combination with an SEM as sensor is described in [21–22]. In [22], an integrated circuit mounted on a (x, y)-stage is positioned and inspected automatically. With the presented approach, the (x, y)-position and the orientation can be estimated. The drawback of both techniques is slow computation time. In [22] 20 s are needed to estimate the (x, y, φ)-pose. A further disadvantage is that no adaptive ROI is used. Instead, the SEM is connected to a PC via the analog-out interface.

In the following, an algorithm is described that enables real-time tracking for the continuous (x, y, φ)-pose estimation by using an adaptive ROI. To decrease the time-consumption caused by the orientation estimation with additional patterns, the maximum number of correlations per slope cycle is restricted. Therefore, a pattern vector with 360 components, each rotated by one degree, is generated. For every slope cycle only adjacent patterns $(\ldots, p_{j-1}, p_j, p_{j+1}, \ldots)$ of the vector are correlated, where j denotes the angle. If the starting orientation is known, the pattern p with the same orientation and the patterns with $+1°$ (p_{j-1}) and $-1°$ (p_{j+1}) are chosen. The pattern with the maximum correlation is estimated and set as p_j. After this, the new p_{j-1} and p_{j+1} are selected. The flow diagram of the algorithm is depicted in Figure 4.5. To detect rotations with more than 1 degree between two frames, the orientation estimation is repeated several times for the current frame. If the cross-correlation coefficient for p_{j+1} (respectively p_{j-1}) is smaller than p_j, the correct orientation is found.

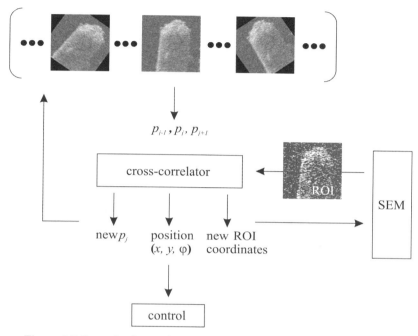

Figure 4.5. Setup for the continuous pose estimation using cross-correlation

The limit of this approach is that only continuous orientation changes can be measured efficiently. In the present application, this prerequisite is fulfilled.

The cross-correlation approach is validated in consideration of the requirement for robustness against additive noise and real-time processing. For both experiments, the image acquisition time is set to the lowest value, which means that it takes 250 ns to acquire one pixel. To measure the robustness against additive noise, the standard deviation of x, y, and φ is measured for a sequence of 100 frames, while the target object is not moved. The result is compared with the theoretical accuracy of the approach, which is 1 pixel and 1°. As target objects, a CNT and a gripper tip were chosen (Figure 4.6).

For both objects the accuracy of the pose estimation is 1 pixel for the (x, y)-position and 1° for the orientation. Thus, the additive noise does not affect the theoretical accuracy, which indicates high robustness for this experiment.

Real-time processing means that the pose can be calculated within the image acquisition time. As the image acquisition time varies with the size of the ROI, the processing time is measured for different sizes. The parameters on which the processing time primarily depends are the size of the ROI and the number of cross-correlations per frame (for the orientation estimation). In addition, the processing time depends on the hardware and on the implementation of the FFT. For all measurements, an Intel Pentium 4 standard PC with 2.6 GHz processor was used. The algorithm was implemented in C++, and the FFTW3 algorithm was used as implementation for the fast Fourier transform [23]. An advantage of the FFTW3 is the capability to efficiently process images with sizes not equal to 2^N. Therefore, the user does not have to use patterns with, *e.g.*, 128×128 or 256×256 only or to expand smaller patterns, which would decrease speed and robustness.

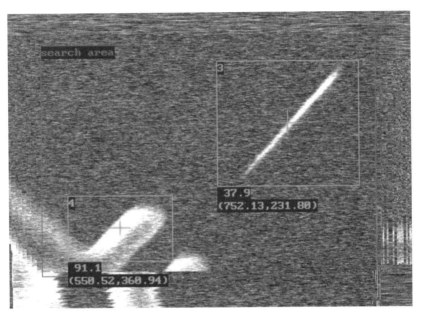

Figure 4.6. Tracking of a gripper tip and a CNT using cross-correlation and an adaptive ROI. The length of the CNT is 6 μm

In Table 4.2, the image acquisition and processing time is shown for three different ROI sizes. The processing time is calculated for one cross-correlation step and will increase depending on the number of calculations needed for the orientation estimation. Real-time processing is achieved if the average number of cross-correlations per frame is between three and four. Thus, the maximum rotation speed that can be tracked in real-time is approximately 1° per image acquisition time, if the highest frame rate is used.

Table 4.2. Frame rates and correlation calculation time for different ROIs. The correlation time is measured for one correlation operation.

ROI [pixels]	Image acquisition time [ms]	Processing time [ms]
256×256	16.4	4.2
140×192	6.7	1.5
128×128	4.1	0.8

4.5 Region-based Object Tracking

The basic idea of this approach is to use an **active contour** that is minimized to the contour of the target object. Active contours have been introduced by Kass, Witkin, and Terzopoulos [24] to segment images, *e.g.*, objects from the image background. The contour can be implemented by a polygon or spline curve and is minimized using an energy function, which considers image features (external energy) and the shape of contour (internal energy). Edges, color, gray-level statistics, *etc.* can be used as image features.

In [25, 26], active contours for visual tracking with edge-based minimization are described. The drawback of this approach is a low robustness against additive noise, as strong edges are needed for the feature detector. In [27], the application of this tracking approach to SEM images is presented. A minimization depending on the gray-level statistic leads to higher robustness against noise [28, 29]. This approach is also called region-based active contours. In Figure 4.7, two examples are shown, where two regions are segmented depending on statistical parameters.

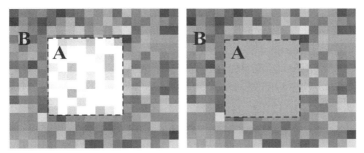

Figure 4.7. Two examples for segmenting an image in two regions A and B. In the left image, the segmentation criterion is the mean gray level. A different standard deviation of the gray levels is used as segmentation parameter in the right image.

In the left image, the active contour segments pixels with higher gray levels from the darker background, whereas pixels with equal gray level are segmented in the right image. In the following, the energy functions and the minimization algorithm needed to use region-based active contours for visual tracking in SEM images are described.

4.5.1 The Energy Functions

The internal energy function depends on the shape of the contour, whereas the external energy function changes with the gray levels of the input image. Two assumptions are used for the definition of the external energy function.

1. The probability of all gray levels in the image is defined by one probability function.
2. The probability parameters (*e.g.*, mean and/or standard deviation) are different for the regions to be segmented.

This image model is also known as **statistical independent region** (SIR) model [28]. The first assumption ensures that the image noise does not depend on the image content, or, in other words, only uncorrelated noise is considered. As this is the case for SEM images, the assumption can be made. The second assumption ensures that a criterion can be defined for the segmentation. Which probability distribution function has to be chosen depends on the sensor and is not significant for the SIR model.

If an SEM is used as sensor, the Bernoulli and the Poisson probability distribution functions have to be considered (Table 4.3).

Table 4.3. Probability distributions that describe the gray-level distribution in SEM images

Probability distribution	$P(x)$	Parameter
Bernoulli	$p\delta(x)+(1-p)\delta(1-x)$	P
Poisson	$\sum_{n\in N}\delta(x-n)e^{-p}\dfrac{p^{n}}{n!}$	P

In an SEM image, the distribution of the gray levels is described by the Poisson distribution [30]. In addition, the Bernoulli distribution is needed to describe binary images, which are acquired if only single electrons are detected for each pixel.

A maximum likelihood estimate $\hat{\mu}$ is defined for the parameter set μ of the given probability distributions that describes a gray level set (image) s:

$$\hat{\mu}(s) = \arg\max_{\mu} L(s\mid \mu) = \arg\max_{\mu} \prod_{x\in s} P(x\mid \mu). \tag{4.3}$$

In Equation 4.3, assumption 1 is considered and L is the likelihood of s if μ is given. If an image s is segmented into two regions a and b by an active contour θ, the estimated probability $L_{pseudo}(s,\theta)$ of s can be calculated by substituting the maximum likelihood estimates $\hat{\mu}_a$ and $\hat{\mu}_b$.

$$L_{pseudo}(s,\theta) = L(a\mid \hat{\mu}_a,\theta)L(b\mid \hat{\mu}_b,\theta), \tag{4.4}$$

with $\int L_{pseudo}(s,\theta)ds \neq 1$. From this, it is the aim to find the $\hat{\theta}(s)$ that maximizes $L_{pseudo}(s,\theta)$:

$$\hat{\theta}(s) = \arg\max_{\theta} L_{pseudo}(s,\theta) = \arg\max_{\theta}\left(L(a\mid \hat{\mu}_a,\theta)L(b\mid \hat{\mu}_b,\theta)\right). \tag{4.5}$$

By using Equation 4.3 and Table 4.3, the pseudo-likelihood for a Poisson distribution is defined as

$$L_{pseudo}(s,\theta) = L(a \mid \hat{\mu}_a, \theta) L(b \mid \hat{\mu}_b, \theta)$$

$$= \prod_{x \in a} P_{Poisson}(x, \hat{\mu}_a) \prod_{x \in b} P_{Poisson}(x, \hat{\mu}_b) \ . \tag{4.6}$$

After substituting Equation 4.6 in Equation 4.5 it follows that

$$\hat{\theta}(s) = \arg\max_{\theta} L_{pseudo}(s, \theta)$$

$$= \arg\max_{\theta} \left(-N_a f \left(\frac{1}{N_a} \sum_{x \in a} x \right) - N_b f \left(\frac{1}{N_b} \sum_{x \in b} x \right) \right), \tag{4.7}$$

with $f(z) = -z \ln z$. From this, the energy function to minimize is defined as

$$J_{Poisson}(s, \theta) = N_a f \left(\frac{1}{N_a} \sum_{x \in a} x \right) + N_b f \left(\frac{1}{N_b} \sum_{x \in b} x \right). \tag{4.8}$$

The energy function that considers the Bernoulli probability distribution is defined accordingly as

$$J_{Bernoulli}(s, \theta) = N_a f \left(\frac{1}{N_a} \sum_{x \in a} x \right) + N_b f \left(\frac{1}{N_b} \sum_{x \in b} x \right), \tag{4.9}$$

with $f(z) = -z \ln z - (1-z) \ln(1-z)$.

The energy minimization segments an image into two regions a and b with N_a and N_b pixels. Both energy functions only depend on the sum over the gray levels $x \in a, b$. An efficient algorithm to minimize the energy function is described in Section 4.5.3.

While the external energy function is mainly used to segment an image, an additional internal energy function is needed to smooth the contour. The internal energy J_{int} does not depend on the image but on the shape of the contour. A low internal energy leads to a smooth contour, while a complex shape contains a high energy. By the weighted summation of both energy functions, the total energy is given by

$$J(s, \theta) = J_{ext}(s, \theta) + \alpha J_{int}(\theta), \tag{4.10}$$

where α is a parameter to weigh the internal energy function in relation to the external. In the following, four different internal energy functions are described. If a polygon is used to implement the active contour, the internal energy function can be described as

$$J_{int1}(\theta) = \sum_{i=0}^{N-1} d_i^2 \ . \tag{4.11}$$

The length of the polygon segment i is denoted by d_i. The minimization of Equation 4.8 leads to a shrinking polygon with equal point distances [29].

In [20], an internal energy function is presented, including the curvature of the polygon:

$$J_{int2}(\theta) = \sum_{i=0}^{N-1} d_i'^2 \ , \tag{4.12}$$

with

$$d_i' = \left| P_i - \frac{1}{2}(P_{i-1} + P_{i+1}) \right| . \tag{4.13}$$

A further definition of the internal energy can be given by the compactness of the polygon [31]:

$$J_{int3}(\theta) = \frac{L^2}{A} \ . \tag{4.14}$$

The length of the polygon is denoted by L and the area by A. This energy function supports circular polygons. An advantage of J_{int3} is the low computational cost, as A is calculated with the external energy.

4.5.2 Fast Implementation

The external energy is calculated by the summation of the gray levels (Equations 4.8 and 4.9) located inside and outside the active contour:

$$\sum_{(x,y)\in\Delta} s(x,y)^k \ , \text{ for } k = 1,2 \ . \tag{4.15}$$

To reduce the computational cost, an ROI Δ that contains the target object was used, as described in Section 4.5.3. Nevertheless, the computing time is still very costly due to the number of operations to be carried out for an iterative minimization. Thus, an efficient implementation that does not use Equations 4.15 is required to enable the real-time capability of this approach.

The application of the Green–Ostrogradsky theorem

$$\iint_{\Delta} \mathrm{div}\left(\vec{V}\right) dxdy = \int_{\delta} \vec{V} \cdot \vec{n} dt \tag{4.16}$$

allows a significant improvement of the computing speed. $\vec{n}$ denotes the vector normal to the surface, and div is the divergence operator. The theorem states that the surface integral of a scalar function is equivalent to the line integral over the contour δ of the surface. Rewritten for the present case, Equation 4.16 becomes

$$\iint_{(x,y)\in\Delta} s^k (x,y) dxdy = \int_{(x,y)\in\delta} V_k (x,y)\vec{e}_x \cdot \vec{n} dt , \tag{4.17}$$

with

$$V_{k(x,y)} = \int_0^x s^k (u,y) du \tag{4.18}$$

for the continuous case and with

$$F_k (x,y) = \sum_{t=0}^{x} s(t,y)^k \tag{4.19}$$

for discrete images [32]. $F_k(x, y)$ is the sum over the k-th power of the gray levels in row y from the left border to column x. Thus, Equation 4.15 can be replaced by a summation over the contour δ:

$$\sum_{(x,y)\in\Delta} s(x,y)^k = \sum_{(x,y)\in\delta} c(x,y)F_k (x,y) \tag{4.20}$$

The coefficient $c(x, y)$ can be –1, 0, or 1 and is defined for each pixel on the discrete contour. The lines of the image are divided into sections that belong to Δ and sections that do not belong to Δ. A section belonging to Δ is limited by the contour δ on both sides. If the pixels between the limits do not belong to δ, $c = -1$ on left and $c = 1$ on the right boundary. All other pixels of δ are weighted with 0. The sum over c is always 0 for a single row. In Figure 4.8, the coding of the contour is clarified with an example.

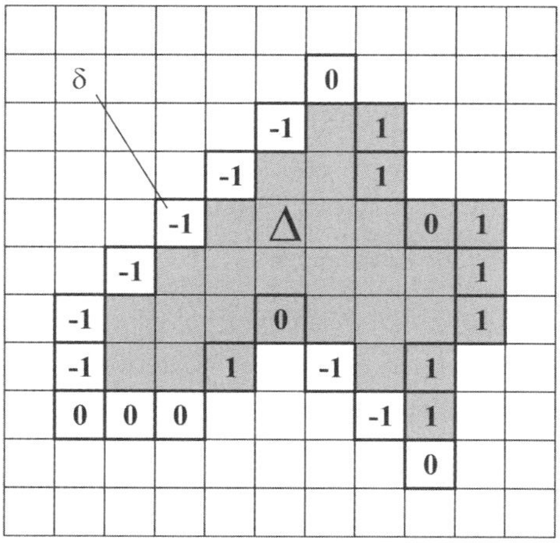

Figure 4.8. Coding of the contour δ

By using Equation 4.20, the summation of the gray levels inside the contour is replaced by a summation along the contour. The precondition is that the contour does not contain intersections. A drawback is the additional calculation of F. But F has to be calculated only once per acquired image, whereas the energy has to be calculated several times per frame as the minimization is done iteratively (see next section). Thus, the computational cost is significantly reduced.

4.5.3 Minimization

The minimization of Equation 4.10 is carried out iteratively in consideration of the transformation space of the target object. In the following, the Euclidean similarities are used as transformation space, which contains translations, rotations, and scale variations. Thus, the degrees of freedom required for nanohandling are covered.

The minimization is carried out consecutively for all degrees of freedom. First the direction (*e.g.*, left or right for the x-position) in which the energy decreases is evaluated. Then an interval is estimated wherein the global minimum is located. From this, the minimization is carried out stepwise, whereby the interval is downsized to increase the accuracy. The minimum interval is three pixels for the (x, y)-position. Thus, the maximum accuracy is one pixel.

Frame-by-frame tracking is achieved by continuous minimization, whereby the minimized contour of the current frame is used as the start contour for the following frame. In addition to the high robustness against additive noise and due to the consideration of the gray level probability distribution, a transformation space enables a high robustness against clutter, since only certain shapes of the contour are allowed.

As for the cross-correlation approach, an ROI is used to decrease the image acquisition time. The size of the ROI changes with the dimension of the contour (Figure 4.9), and the position is shifted with the center of gravity. Thus, the ROI is automatically defined for each frame.

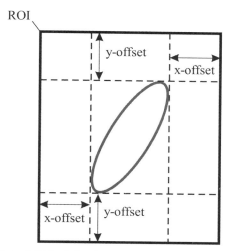

Figure 4.9. The size of the ROI is defined depending on the size of the contour. 100 pixels are added to the maximum dimension

Because region-based tracking is a frame-by-frame minimization, a first initialization step is needed to estimate the first ROI. Therefore, the target object has to be recognized in the first input image, which can be done by cross-correlation as described in Section 4.4. The initial shape of the contour can be defined manually or, if available, from a CAD model. In addition, the contour can be initialized by a rectangular polygon, if detail shape information is not available. In this case, the contour is minimized without a restriction to the transformation space (free segmentation) using an approach described in [20].

For the (x, y, φ)-pose calculation, the parameterization of the contour is used. The translation (x, y) is determined by the centroid, which can be defined as [26]:

$$\bar{r} = \frac{\int_0^L r(s)|r'(s)|\,ds}{\int_0^L |r'(s)|\,ds},$$ (4.21)

with $r(s) = (x(s), y(s))^T$ and s defined on $[0\ldots L]$. $|r'(s)|$ is the "throughput speed" and

$$\int_0^L |r'(s)|\,ds$$ (4.22)

defines the length of the contour. For a polygon, Equation 4.21 is rewritten for the discrete case

$$\begin{pmatrix} \bar{x} \\ \bar{y} \end{pmatrix} = \frac{\sum_{i=0}^{N-1}\left(d_i\begin{pmatrix} x_i \\ y_i \end{pmatrix}\right)}{\sum_{i=0}^{N-1}d_i}. \tag{4.23}$$

$(x_i, y_i)^T$ is the center of polygon segment i and d_i is its length. The orientation of the contour is defined by the principal axes, which are calculated by the moments of the contour [31]

$$m_{pq} = \sum_x \sum_y x^p y^q s(x,y). \tag{4.24}$$

m_{pq} denotes the moment of order $p+q$. The centered moments are defined as

$$\mu_{pq} = \sum_x \sum_y (x-x_c)^p (y-y_c)^q s(x,y), \tag{4.25}$$

with

$$x_c = \frac{\sum_x \sum_y xs(x,y)}{\sum_x \sum_y s(x,y)} = \frac{m_{10}}{m_{00}}, \tag{4.26}$$

$$y_c = \frac{\sum_x \sum_y ys(x,y)}{\sum_x \sum_y s(x,y)} = \frac{m_{01}}{m_{00}}. \tag{4.27}$$

The point $(x_c, y_c)^T$ denotes the center of the contour and is used as the reference point in the following. Centered moments are translation invariant. In addition, invariance against scaling is needed, which is obtained by normalizing Equation 4.25:

$$\eta_{pq} = \frac{\mu_{pq}}{\mu_{00}^{\frac{p+q}{2}+1}}. \tag{4.28}$$

From this, the orientation φ of the contour can be calculated by

$$\varphi = \frac{1}{2}\arctan\frac{2\eta_{11}}{\eta_{20}-\eta_{02}}. \tag{4.29}$$

4.5.4 Evaluation and Results

In the following, the region-based tracking approach is validated in consideration of the requirements defined in Section 4.2. The performance and accuracy measurements were performed with a test chessboard as target object, which is shown in Figure 4.11. The robustness against clutter and gray level fluctuations is validated by a nanohandling task. In every experiment, the image acquisition time is set to the minimum value (no averaging, 250 ns per pixel) to test the approach with the strongest additive noise.

4.5.4.1 Performance
The intention of this section is to demonstrate that the region-based approach enables real-time tracking, which means that the pose can be calculated within the image acquisition time. As the image acquisition time varies with the size of the ROI, the processing time is measured for different sizes. The parameters that the processing time primarily depends on are the size of the ROI and the length of the contour. A large ROI requires more calculation steps for the estimation of F_k, and increasing length of the polygon makes the calculation of Equation 4.8 more complex. For the performance measurement, it can be considered that the length of the contour and the size of the ROI are correlated. Because the shape of the contour is restricted by the transformation space, only scale variations can change the length. And as described before, the size of the ROI is set depending on the contour. Thus, both parameters can be analyzed in parallel.

The measured image processing time is compared to the image acquisition time, because the time changes with the ROI size as well. Within the test sequence, the scale of the target object was changed due to a variation of the magnification. For all measurements, an Intel Pentium Core 2 Duo standard PC with 2.4 GHz was used. The result of the performance test is shown in Figure 4.10.

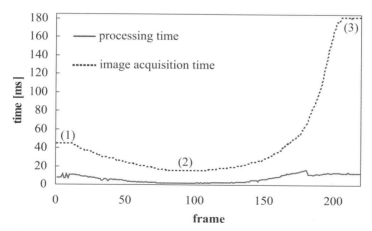

Figure 4.10. Processing in comparison to the image acquisition time. During the sequence the scale of the contour and the ROI is changed due to a variation of the SEM magnification. At the beginning of the contour the ROI size is 397 × 451 pixels (1), the minimum size is 243 × 254 pixels (2), and the final size is 766 × 922 pixels (3).

Figure 4.10 shows that the processing time is below the image acquisition time during the whole sequence. On average, the pose estimation is carried out seven times faster than the image acquisition. The difference between processing and acquisition increases for larger ROIs. In Table 4.4, the time needed for the calculation of F_k and for the minimization is shown for four different ROI sizes.

Table 4.4. Processing time for four different ROI sizes scanned at highest frame rate. The processing time is the sum of F_k and minimization.

ROI [pixel]	Image acquisition [ms]	Processing [ms]	F_k [ms]	Minimization [ms]
243 × 254	15	2	0.2	1.8
397 × 451	45	9	4	5
511 × 596	76	11	6	5
766 × 922	177	13	7	6

Except for very small ROIs, the minimization and the calculation of F_k take the same amount of time. In addition, the extra performance enables multiple-object tracking even in real-time.

4.5.4.2 Robustness Against Additive Noise

The experimental validation of the robustness against additive noise was also carried out with the chessboard test sample. Due to the minimization algorithm, the position can be calculated with pixel accuracy and the orientation angle with an accuracy of one degree. The objective of this experiment is to verify that the pose of the target can be calculated with the theoretically possible accuracy even in the presence of strong additive noise. To measure the robustness, the standard deviation of x, y, and φ is measured for a sequence of 100 frames, while the target object is not moved. The measurement is repeated for magnifications between 200× and 50,000×, which is necessary as the noise and blurring increase with higher magnifications. To enable a measurement over this large magnification range, the target object is changed at 2000× (Figure 4.11). Every measurement is done with the shortest image acquisition time.

The results of the accuracy measurement are shown in Table 4.5. Besides the accuracy in pixels, the corresponding sensor resolution in microns is calculated using the calibration of the SEM. The accuracy decreases marginally, but persists below one pixel up to a magnification of 20,000×. At 50,000×, the accuracy of the x position is still two pixels or 4.6 nm in world coordinates. For the y position, an accuracy of 2.3 nm is achieved. Thus, no significant impact of the additive noise on the accuracy can be measured.

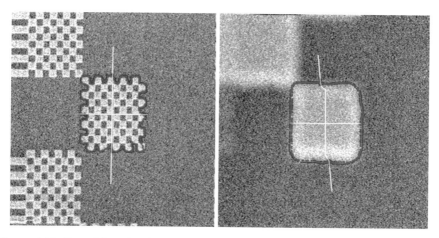

Figure 4.11. Test object for the accuracy measurement to verify the robustness against additive noise. Measurements in the range 200× – 2000× are carried out with a 10 by 10 chessboard (left image). In the range 5000× – 50,000× a single quadrate with dimension 1 μm × 1 μm is used.

Table 4.5. Results of the accuracy measurement

Magnification	σ_x [pixel]	σ_x [μm]	σ_y [pixel]	σ_y [μm]	σ_φ [°]
200	0.0158	0.5718	0.2095	0.5718	1.198
500	0.0219	0.2287	0.0186	0.2287	0.3621
1000	0.063	0.1144	0	0.1144	0
2000	0.42	0.0572	0.118	0.0572	0
5000	0.4984	0.0229	0.374	0.0229	1.017
10,000	0.5086	0.0114	0.5351	0.0114	0.6
20,000	0.938	0.0057	0.776	0.0057	0.79
50,000	1.9719	0.0046	1.184	0.0023	0.5276

The accuracy of the orientation angle oscillates for different magnifications. In general, the accuracy is lower, if the contour contains only a few pixels. However, an accuracy of 1° is suitable for most nanohandling tasks.

4.5.4.3 Robustness Against Clutter

The validation of the robustness against clutter is done on the basis of a typical nanohandling experiment, where a microgripper and a CNT are positioned for a gripping task. During the positioning process, the CNT and the gripper mask each other. First, the CNT is tracked, while 1/3 is occluded by the gripper. In Figure 4.12 three images of the sequence are shown.

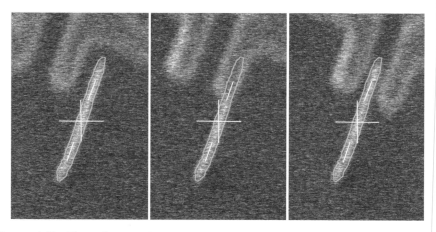

Figure 4.12. Three frames of a sequence wherein a tracked CNT is occluded by a microgripper. During the sequence the microgripper is moved over the CNT, while the CNT is fixed. The length of the CNT is 8 μm.

In spite of the occlusion by the gripper, the CNT is tracked reliably. In Figure 4.13, the (x, y)-position of the CNT is shown for 37 frames.

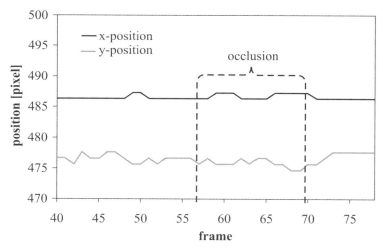

Figure 4.13. (x, y)-position of the CNT while it is occluded by the gripper tip

In a second experiment, one gripper tip is tracked, while being moved over the CNT. With this experiment, the robustness against background clutter is tested. To increase the difficulty, the magnification is changed while the gripper tip is above the CNT. In Figure 4.14, four frames of the test sequence are shown.

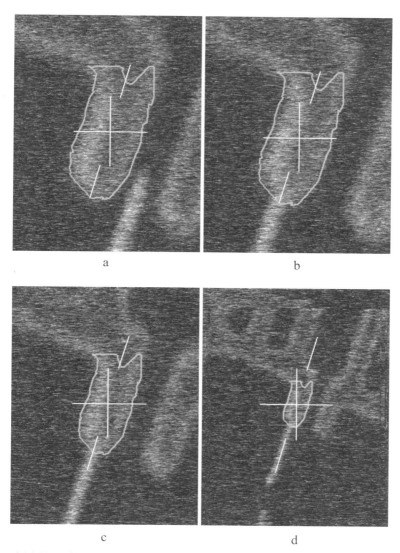

a   b

c   d

Figure 4.14. Four frames of a sequence wherein a tracked gripper tip is masked by a CNT. In images a and b, the gripper tip is moved over the CNT. The magnification is changed in image c, and d.

As in the experiment before, the target object is tracked reliably even while the magnification is changed.

4.5.4.4 Robustness Against Gray-level Fluctuations

During nanohandling processes, gray-level fluctuation often occurs when the nano-tool touches a surface. The reason is a charge transport between tool and surface. In Figure 4.15, an example is shown where a gripper touches a surface. It can be seen that the average gray level increases significantly. To validate the robustness against gray-level fluctuations, the position of a CNT is measured while the wafer surface is contacted by a gripper. The experiment is repeated three times.

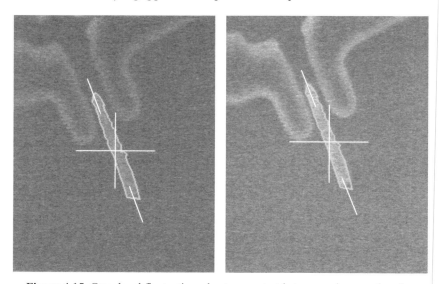

Figure 4.15. Gray-level fluctuations due to a contact between gripper and surface

In Figure 4.16 (top), the mean gray-level of the each frame is shown during the contacts. Three abrupt jumps can be seen. Directly after the contact, the gripper is lifted over the surface and the additional charges drain off slowly. The (x, y)-position of the CNT is shown in Figure 4.16 (bottom). In spite of a jump of approximately 24 gray levels, the CNT is not lost during the whole sequence. While the y-position is not affected by the fluctuation, a displacement of the x-position is measured for all three contacts. During the second contact, the y-position increases about 11 pixels. The reason is global image drift caused by the charge transport, which can be measured by correlation.

4.6 Conclusions

4.6.1 Summary

So far, the handling of nanoobjects is mostly teleoperated, which is costly due to time consumption and the required experience. To decrease the costs, automation is proposed by combining tools for nanohandling like microrobots with a global sensor system. Up to now, the SEM is considered the most suitable imaging sensor for nanohandling considering resolution, frame rate, and depth of focus. The main drawback of an SEM is the lower image quality due to strong additive noise, if the frame rate is high. Thus, the requirements for an image-processing approach are,

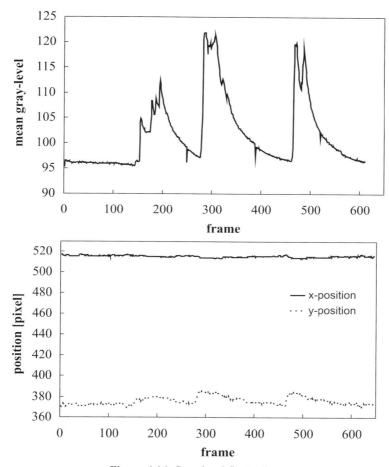

Figure 4.16. Gray-level fluctuations

especially, high robustness against additive noise and, in consideration of the nanohandling task, robustness against clutter and gray-level fluctuations. In addition, invariance against scaling is needed as the SEM is a zooming sensor.

An important precondition for the proposed sensor system is the integration of the SEM into the image-processing system. The proposed solution enables direct access to the acquired images and the control of the electron beam, which can be performed by a ROI, whereby the tracking efficiency is significantly increased.

For continuous pose estimation, two image-processing approaches have been described. The first one, correlation-based tracking, is very robust against additive noise and easy to implement. One drawback is that additional patterns are needed to estimate the orientation angle and no invariance against scaling is provided. The main drawback is the low robustness against clutter due to the fixed filter mask.

To overcome these drawbacks and to finally fulfill all requirements, a new active-contour-tracking approach has been introduced. The contour is implemented as polygon and minimized depending on the gray-level probability distribution. By taking into account the Poisson distribution, a high robustness against additive noise has been achieved for SEM images. It has been shown that an efficient implementation of the statistical parameter calculation is possible to meet the real-time requirement. A high robustness against clutter is achieved by restricting the minimization to the transformation space of the target object. The Euclidean similarities are an example for a transformation space, which is suitable for most nanohandling tasks.

Subsequently, the region-based tracking approach has been validated in consideration of the requirements. Real-time processing has been proven with different ROI sizes. The processing time depends on the ROI size and is 7 to 13 times faster than the image acquisition time. To validate the robustness against additive noise, accuracy measurements using the shortest image acquisition time (highest noise) have been carried out for different magnifications. A significant influence of the additive noise on the theoretically achievable pixel accuracy could not be measured. With an SEM magnification of 50,000×, the accuracy is up to 2.3 nm. The robustness against clutter and gray-level fluctuations has been validated by a typical nanohandling task, wherein a microgripper and a CNT are positioned. Despite a 30% occlusion of the CNT, the measured pose is not influenced. Gray-level fluctuation has been induced by a contact between microgripper and wafer. It has been shown that only the global image drift is measured.

All requirements defined in Section 4.2 have been fulfilled by the region-based tracking approach. In combination with the SEM as image sensor, a high-resolution sensor for several applications in the field of nanohandling is available.

4.6.2 Outlook

The results described in this chapter are very promising and will trigger more research work in this direction. Beside the continuous pose estimation of nanotools and -objects, additional information, estimated by image-processing algorithms, could increase the efficiency of automated nanohandling processes. In addition to the automation of the handling tasks, it is important to automate the quality control as well. Therefore, an automated analysis of the contour parameter is proposed. The calculated data can be used to verify the alignment of a target object in the gripper or at the target position after the place operation. Future work will also deal with the integration of other existing SEM detectors. In addition to the SE detector, a backscattered electron detector (BSE detector) and an energy dispersive X-ray detector (EDX detector) can be used to analyze the material distribution of samples. For example, an automated analysis of the detector signals by image processing can be used for an automatic quality control of bonding processes.

4.7 References

[1]   Dai, H. 2001, *Nanotube Growth and Characterization*, Springer, pp. 29–51.
[2]   Ziaie, B., Baldi, A. & Atashbar, M. 2004, *Introduction to Micro and Nanofabrication*, Springer, chapter 5, pp. 147–184.

[3]   Hansen, L., Kühle, A., Sørensen, A., Bohr, J. & Lindelof, P. 1998, 'A technique for positioning nanoparticles using an atomic force microscope', *Nanotechnology*, vol. 9, pp. 337–342.

[4]   Sitti, M. & Hashimoto, H. 2000, 'Controlled pushing of nanoparticles: modeling and experiments', *IEEE/ASME Transactions on Mechatronics*, vol. 5, no. 2, pp. 199–211.

[5]   Williams, P., Papadakis, S., Falvo, M., Patel, A., Sinclair, M., Seeger, A., Helser, A., Taylor, R., Washburn, S. & Superfine, R. 2002, 'Controlled placement of an individual carbon nanotube onto a microelectromechanical structure', *Applied Physics Letters*, vol. 80, no. 14, pp. 2574–2576.

[6]   Troyon, M., Lei, H., Wang, Z. & Shang, G. 1998, 'A scanning force microscope combined with a scanning electron microscope for multidimensional data analysis', *Scanning Microscopy*, vol. 12, no. 1, pp. 139–148.

[7]   Joachimsthaler, I., Heiderhoff, R. & Balk, L. 2003, 'A universal scanning-probe-microscope based hybrid system', *Measurement Science and Technology*, vol. 14, pp. 87–96.

[8]   Fatikow, S. & Rembold, U. 1997, *Microsystems Technology and Microrobotics*, Springer.

[9]   Breguet, J.-M., Pernette, E. & Clavel, R. 1996, 'Stick and slip actuators and parallel architectures dedicated to microrobotics', *Microrobotics: Components and Applications, SPIE's Photonics East, Boston*, Vol. 2906, pp. 13–24.

[10] Fahlbusch, S., Fatikow, S., Seyfried, J. & Buerkle, A. 1999, 'Flexible microrobotic system MINIMAN: design, actuation principle and control', *Proc. of IEEE/ASME Int. Conference on Advanced Intelligent Mechatronics*, pp. 156–161.

[11] Kim, P. & Lieber, C. 1999, 'Nanotube nanotweezers', *Science*, vol. 286, pp. 2148–2150.

[12] Mølhave, K. & Hansen, O. 2005, 'Electro-thermally actuated microgrippers with integrated force-feedback', *Journal of Micromechanics and Microengineering*, vol. 15, pp. 1265–1270.

[13] Sulzmann, A., Breguet, J. & Jacot, J. 1997, 'Micromotor assembly using high accurate optical vision feedback for microrobot relative 3D displacement in submicron range', *Proceeding of IEEE Int. Conf. on Solid State Sensors and Actuators*, vol. 1, pp. 279–282.

[14] Vikramaditya, B. & Nelson, B. J. 1997, 'Visually guided microassembly using optical microscopes and active vision techniques', *Proceeding of IEEE Int. Conf. on Robotics and Automation (ICRA)*, pp. 3172–3177.

[15] Yang, G., Gaines, J. A. & Nelson, B. J. 2003, 'A supervisory wafer-level 3D microassembly system for hybrid MEMS fabrications', *Journal of Intelligent and Robotic Systems*, vol. 37, pp. 43–68.

[16] Sun, Y., Greminger, M. A. & Nelson, B. J. 2004, 'Nanopositioning of a multi-axis microactuator using visual servoing', *Journal of Micromechatronics*, vol. 2, no. 2, pp. 141–155.

[17] Liu, J. 2005, *High Resolution Scanning Electron Microscopy*, Kluwer Academic Publishers, Chapter 11, pp. 325–359.

[18] Mølhave, K., Wich, T., Kortschack, A. & Bøggild, P. 2006, 'Pick-and-place nanomanipulation using microfabricated grippers', *Nanotechnology*, vol. 17, no. 10, pp. 2434–2441.

[19] Papoulis, A. 1962, *The FourierIntegral and Its Applications*, McGraw-Hill.

[20] Goudail, F. & Réfrégier, P. 2004, *Statistical Image Processing Techniques for Noisy Images*, Kluwer Academic/Plenum Publishers.

[21] Rosolen, G. & King, W. 1998, 'An automated image alignment system for the scanning electron microscope', *Scanning*, vol. 20, pp. 495–500.

[22] Tan, H., Phang, J. & Thong, J. 2002, 'Automatic integrated circuit die positioning in the scanning electron microscope', *Scanning*, vol. 24, pp. 86–91.

[23] Frigo, M. & Johnson, S. 2005, 'The design and implementation of FFTW3', *Proc. of IEEE*, Vol. 93 of *2*, pp. 216–231.

[24] Kass, M., Witkin, A. & Terzopoulos, D. 1988, 'Snakes: active contour models', *International Journal of Computer Vision*, vol. 1, pp. 321–331.

[25] Isard, M. & Blake, A. 1998, 'Condensation - conditional density propagation for visual tracking', *International Journal of Computer Vision*, vol. 29, pp. 5–28.

[26] Blake, A. & Isard, M. 2000, *Active Contours*, Springer.

[27] Sievers, T. & Fatikow, S. 2006, 'Real-time object tracking for the robot-based nanohandling in a scanning electron microscope', *Journal of Micromechatronics*, vol. 18, pp. 267–284.

[28] Réfrégier, P., Goudail, F. & Chesnaud, C. 1999, 'Statistically independent region models applied to correlation and sementation techniques', *Proc. of Euro-American Workshop on Optoelectronic Information Processing*, pp. 193–224.

[29] Cremers, D., Tischhäuser, F., Weickert, J. & Schnörr, C. 2002, 'Diffusion snakes: introducing statistical shape knowledge into the Mumford-Shah functional', *International Journal of Computer Vision*, vol. 50, no. 3, pp. 295–313.

[30] Reimer, L. 1998, *Scanning Electron Microscopy – Physics of Image Formation and Microanalysis*, Springer.

[31] Nischwitz, A. & Haberäcker, P. 2004, *Masterkurs Computer Graphics and Image Processing*, 1 edn, Friedr. Vieweg & Sohn Verlag/GWV Fachverlage GmbH. In German.

[32] Chesnaud, C., Réfrégier, P. & Boulet, V. 1999, 'Statistical region snake-based segmentation adapted to different physical noise models', *IEEE Transactions on Pattern Analysis and Machine Intelligence*, vol. 21, pp. 1145–1157.

5

3D Imaging System for SEM

Marco Jähnisch

Division of Microsystems Technology and Nanohandling,
Oldenburg Institute of Information Technology (OFFIS), Germany

5.1 Introduction

Handling processes with the aim of changing the position of objects relative to each other can be carried out by a microrobot-based nanohandling station (Chapters 1 and 2), which makes the teleoperated or automated handling of nanoobjects possible. Because the object size is in the range of µm, sub-µm, and even down to a few nm, scanning electron microscopes (SEM) are increasingly employed to observe these processes. They allow enormous magnifications to be achieved, so that manipulation processes on the nanometer scale can be observed.

Through observation, information such as the position of the objects and of the tools can be determined. This information is needed by the user or by a robot control system. It is therefore important to know the relative position of the objects and tools in all three dimensions of space. In addition, the determination has to be carried out precisely, robustly, fast, and without interfering with the handling process.

Commercial standard SEMs deliver 2D images without depth information. However, for precise handling and working without disturbing the object, 3D information is needed. This chapter considers various possibilities to obtain 3D information from SEMs with respect to nanohandling actions. Section 5.2 describes some of the basic concepts from mathematics, biology, and microscopy, which are necessary for the understanding of the following chapter. In Section 5.3, current systems used to obtain 3D information using SEMs are presented. Finally, a 3D imaging system for nanohandling is described in detail in Section 5.4, and Section 5.5 presents its use in the manipulation of nanoobjects, *e.g.*, nanotubes.

5.2 Basic Concepts

In this chapter, the most important basics of this topic will be described. In Section 5.2.1, an overview of the principles of stereoscopic image approaches is provided. In Section 5.2.2, the principles of stereoscopic imaging for SEMs are shown and in 5.2.3, the mathematical basics are presented. Section 5.2.4 gives an introduction to the subject of biological vision systems.

5.2.1 General Stereoscopic Image Approach

The **stereoscopic principle** is described here with the help of the human visual system. This enables depth perception by using two eyes, which observe the surroundings from slightly different perspectives. This shift allows the brain to reconstruct the depth data.

Figure 5.1 shows a schematic representation of the principle function. The eyes focus on a fixed point (point F), which thus impacts on the middle of the retinas of both eyes. The optical axes intersect at point F. If the retinal projections of point P are considered, which is at the same distance from the observer, it can be seen that relative to F, point P appears to be equally shifted in both the left and the right eye. On the other hand, point Q is placed in front of F and therefore results in dissimilar shifts in the left and right eyes. In this example, the magnitudes of the shifts are equal, but not the directions. The difference in the shifts between the left and right eyes is known as the **disparity**.

For P, the disparity $(F_l - P_l) - (F_r - P_r) = 0$. Points with a disparity of 0 have the same distance from the observer as the fixed point.

For Q: $(F_l - Q_l) - (F_r - Q_r) \neq 0$. Depending on whether the disparity is positive or negative, the point is located either in front of or behind the **fixed point**.

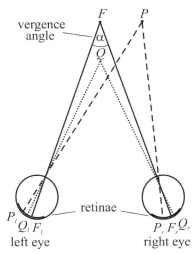

Figure 5.1. Schematic diagram of the stereoscopic effect: F – fixed point; P – a point with equal disparity to F; Q – a point with a different disparity

The **disparity** is a relative measure of the distance. With this alone, it is not possible to determine the absolute distance, because the absolute distance is dependent on other parameters such as the distance of the eyes from each other and the angle of vergence (see below). For the calculation of absolute object coordinates from measured disparities and for a detailed consideration of geometrical relationships, information can be found in [1]. In the context of this chapter, only the determination of the relative distance or depth is considered.

The problem of the (relative) determination of depth is therefore reduced to the determination of the disparities of all points in an image. The objective then is to find the corresponding disparity values for as many points as possible, in order to include the relative depth relationships in the resulting image.

5.2.1.1 The Cyclopean View

Although humans possess two eyes, they perceive only one image of the surroundings. The brain combines the two images with their slightly differing perspectives to form a complete image (fusion). The resulting image has a perspective corresponding to a single eye positioned between left and right. This is known as the **cyclopean view**.

Considering disparities, the question arises, which view should these relate to? If a point in the left image is observed, it appears shifted to the left in the right image (positive disparity). On the other hand, observed from the right image, it is shifted to the right in the left image (negative disparity). It is therefore important to differentiate between left, right, and cyclopean view and the corresponding disparity maps. Some algorithms, *e.g.*, in [2], use bidirectional matching, in which two disparity maps can be generated and where the results relate to the left or right view. Here, one of the input images is used as the main image and the other as reference image. In [3], on the other hand, the input images undergo a cyclopean analysis, that is, the determined disparities are based on the cyclopean view. This makes the construction of the image with a cyclopean view possible.

It is assumed in this chapter that the disparities are based on the cyclopean view. The calculation of the disparities at an image location therefore describes a pixel coordinate from a cyclopean view.

5.2.1.2 Disparity Space

Disparity is a local attribute of an image, meaning that at different parts of a stereoscopic image pair, different disparities can arise. These can be both positive and negative. If we take d_{min} and d_{max} to represent the minimum and maximum disparities in a stereoscopic image pair, then the value range of the disparities [d_{min}; d_{max}] is called the disparity space.

In the calculation of the disparities using algorithms, all procedures have in common that they can only look for a limited number of disparities. In most cases, the search area is limited, so that the calculation time is kept to a reasonable level. The area of values for the disparities, which is searched for, is known as the 'disparity search area'. If a disparity arising from an image lies outside of this search area, it cannot be detected. The disparity search area should therefore completely cover the disparity space.

According to the research literature, negative disparities relate to points which are located in front of the fixed point and positive disparities to points behind the fixed point [4–6]. Without prejudice to the general approach, this chapter takes the opposite position, as the points needed for the disparity map should be brighter the closer they are to the observer (brighter points are represented with higher pixel values and darker ones with lower values). The disparity region of a stereoscopic image pair can be inverted, so that left and right sides of the image exchange positions.

5.2.1.3 Vergence and Version

The angle at which the optical axes of a visual system intersect is called the **vergence** angle. It is dependent on the distance from the fixation point. The closer this is, the bigger is the angle of vergence. For example, if the eyes in Figure 5.2 focus on point Q, the vergence angle is larger than for a focusing on point F. Depending on the vergence angle, the eyes are focused more in either an inward or outward direction. Although the eyes move with a change in the vergence angle (that is, the fixation distance), this has no effect on the cyclopean line of sight, known as **version**. Similarly, the line of sight has no effect on vergence. The following diagram illustrates this.

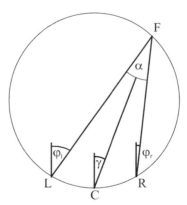

Figure 5.2. The Vieth–Müller circle

Figure 5.2 shows the Vieth–Müller circle arranged for the left eye, right eye, and fixation point. In the figure, α represents the vergence angle and γ the version angle, which corresponds to the cyclopean line of sight. The angles of the optical axes for each eye relative to the direction of the head are indicated by φ_l and φ_r. According to Thales' theorem, α is equal for all points F on the Vieth–Müller circle. A change in version has therefore no effect on vergence. This can also be seen by considering φ_l and φ_r (that is, the position of the axes). The angles are related as follows:

$$\alpha = \varphi_l - \varphi_r, \tag{5.1}$$

$$\gamma = \frac{1}{2}(\varphi_l + \varphi_r).$$

(5.2)

If φ_l and φ_r change in harmony (that is, the eyes turn equally to the left or right), then only version γ, the line of sight, changes; the vergence α remains constant. If φ_l and φ_r change inversely (the eyes turn inwards or outwards – convergence or divergence) only the vergence α, the fixation distance, changes. The version γ remains constant. If only one eye turns, both version and vergence change.

To enable the reconstruction of depth data, vergence is of particular importance as it affects the fixation distance. The cyclopean image cannot be created with a simple overlaying of two partial images. If both partial images are being overlaid, only limited areas of the images merge (local merging).

The control of the eye muscles is closely linked to disparity detection. If the eyes have to focus on a new **fixation point**, the required eye movement can be calculated (or at least estimated fairly accurately) based on the actual measured disparities. For example, if a point is fixed, which lies further away (but in the same cyclopean line of sight), then the **vergence signals**, which control the movement of the eyes, are determined by the measured disparities of the new fixation point relative to the actual fixation point. Figure 5.3 shows an example of this. In Figure 5.3a, the eyes are focused on the level F_1. Points on this level have disparity 0.

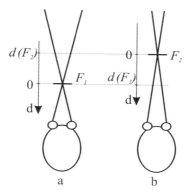

Figure 5.3. Focusing on two different depths: a. eyes are focused on level F_1; b. eyes are focused on level F_2

The points on the more distant level F_2 have a negative disparity $d(F_2)$. The greater the extent of this disparity, the further do the eyes have to turn outwards in order to focus on the level F_2 (Figure 5.3b). Through the changes in the fixation level, the disparities in the new level become 0. The disparity region is shifted; it is always relative to the fixation point. In Figure 5.3a, F_2 always has a negative disparity to F_1. In Figure 5.3b, F_1 always has a positive disparity to F_2.

5.2.1.4 Vergence System

As referred to above, the disparity search area for the algorithmical processing of stereoscopic images has a limited extent. Disparities outside of this search area cannot be detected. Because the value range of the arising disparities in a stereoscopic image pair is not known in advance, the question arises as to how the search area can be selected. One possible answer is to make the search area sufficiently large such that the probability of disparities lying outside the area is reduced. However, as the calculation time is strongly influenced by the size of the search area, this is not a good solution.

The use of a **vergence system** is therefore sensible, *i.e.*, a system that deals with the orientation of an image pair to each other. The idea here is to analyze a stereo image pair with a coarse resolution, which reduces the disparities arising in the image data so that the area of detectable disparities is effectively extended. Naturally, the disparity calculation for coarse resolutions is less precise, but it serves as the rough orientation of the images, so that a finer resolution can be fixed for the **disparity search area**. Figure 5.4 shows this schematically.

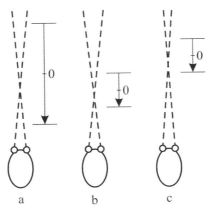

a   b   c

Figure 5.4. Coarse selection of a fixation point prior to a more exact calculation: a. larger depth range can be included; b. selected disparity search area; c. selection of a additional fixation point

Through the analysis of the images with a coarse resolution, the effective disparity search area is increased, that is, a larger depth range can be included (Figure 5.4a). After this, the disparity search area can be selected for fine resolutions so that the disparities arising in the image data are included in the search area (Figure 5.4b). If there are still disparities outside the search area, these can be relocated through the selection of a further fixation point so that the missing disparities can be recognized (Figure 5.4c).

5.2.2 Principle of Stereoscopic Image Approaches in the SEM

5.2.2.1 Structure of the SEM

Because of the "long" wavelength of light, it is not suitable for the observation of structures on a nanometer scale (Chapter 2). In this subsection, a short overview of the structure of the SEM is given. This is needed for stereoscopic image generation, which is clearly described below. More details can be found in Chapter 10.

In the SEM, the sample to be examined is scanned with electrons instead of photons, and the resulting reaction is measured. The whole of the sample is not scanned with electrons at once, but individual points are scanned one after the other in a grid pattern before (in modern equipment) being composed into a digital image. Figure 5.5 shows a schematic representation of a scanning electron microscope. The complete setup is under vacuum during operation, mainly to allow the electrons to move freely and not be obstructed or diverted by gas molecules [7].

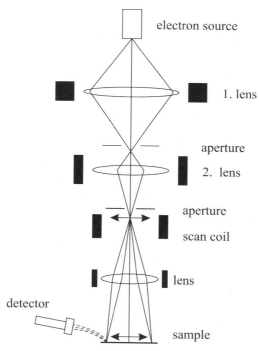

Figure 5.5. Schematic representation of a scanning electron microscope

The operating mode of an SEM can be described as follows:

- **Creation of an electron beam**. An electron cannon is the source of the beam. The free electrons generated at the cathode (such as a tungsten filament) are accelerated through an electric field and shot at the sample at high speed.

- **Electron optics.** In order to control the beam, different lens systems are used. These normally consist of charged coils to generate the magnetic field through which the electrons can be deflected. (Note – the actual lenses are not solid, that is, they consist of magnetic fields. The lenses shown in Figure 5.5 indicate only that the function is similar to that of an optical lens). The condenser lenses and apertures have the function of bundling the beam, and the quality is frequently improved by the use of several lenses. Finally, the beam is deflected through the scanning coils and focused via the objective onto a sharply defined point on the sample. In this way, different parts of the sample are scanned, one after the other.
- **Interaction with the probe.** When the beam of electrons strikes the sample, several types of interaction of result. The electrons which are set free allow inferences to be drawn about material qualities. The beamed electrons are backscattered, as are so-called secondary electrons which are knocked out of the sample. These give information about the topography of the material.
- **Detector.** Different types of detector are used depending on the information required to be collected. The most frequent type used (also in the context of this work) for topographical data is the secondary electron detector. The number of electrons knocked out at each contact point determines the brightness of the corresponding point on the image, which is also why only gray-toned images are produced.

SEMs are used mainly when the resolution capability of a light microscope is not adequate. Light microscopes achieve magnifications up to 1500 times. On the other hand, SEMs achieve magnifications of up to 400,000 times and more. As a result of the bundling of the electron beam and the use of very fine apertures, images arise that have a very high depth of focus compared to light microscopes. Because the sample is sequentially scanned, there is a time delay in the generation of the complete image. Thus, there is always a compromise between image quality and the speed of image creation. Images are generated in less than a second (low scan speed), for example, are usually particularly noisy. Images of very high quality (high scan speed) may require 30 seconds or more.

5.2.2.2 Generation of Stereoscopic Images in the SEM

For the generation of an image, the SEM scans individual points of the sample sequentially. This makes the generation of a simultaneously synchronized stereo image pair impossible. Even if it were technically feasible to apply two electron beams alternately activated during the scanning process, this would be out of proportion to the technical and financial expenditure required. Therefore, the images of a stereoscopic image pair must be sequentially generated in a process in which the angle of observation of the sample is changed between both scans. This can be done in two basic ways: by tilting the sample table or by deflecting the electron beam (Figure 5.6).

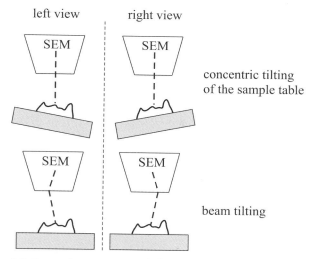

Figure 5.6. Generating a stereoscopic image in an SEM

- **Sample table tilting:** it is possible to generate an image pair by tilting the sample table along with the sample between the scans. The effect of this is that the electron beam hits the sample from two different angles, resulting in two different perspectives of the sample. The problem with this method is with respect to the concentric tilting. It cannot generally be assumed that the actual tilting axis of the sample table is at the same height as the sample itself. In the case of the SEM used for this work (LEO 1450), the actual tilting axis is clearly under the carrying medium. This results in the sample disappearing from the area of impact of the electron beam during tilting (Figure 5.7). Because of this, for concentric tilting to work, the sample table has to be translatively corrected around the physical axis after tilting [8]. Tilting the sample table results in the sample moving to or away from the detector, so that more or fewer electrons, respectively, are measured. Images produced in this way therefore normally have a contrast difference [9].
- **Beam deflection:** another method of generating an image pair is to deflect the electron beam. This means that the position of the sample is not changed, but instead, the electron beam is deflected by an extended lens system causing it to hit the sample at a different angle. The differences in angle allow for two different perspectives of the sample. The advantage of this method is that it is not necessary to make mechanical changes and so the angle of view can be very quickly changed. Of course, the lens system must be very precise to ensure that the electron beam hits the same place on the sample in each case. In [10–12], a specially developed electron column with integrated lens system, which allows for tilting at several angles, is used.

5.2.2.3 Influences on the Disparity Space

The greater the disparity space in a stereoscopic image pair is, the greater the **disparity search area** that has to be selected so that all disparities can be found. Because the size of the disparity search area has a direct influence on the processing time for an image pair, the disparity region should be kept to a reasonable size. This is dependent on three factors. It is limited by the highest and lowest points of the sample the tilt angle, and the set magnification. This is exemplified in Figure 5.7a, showing a side view of the sample table on which a sample is placed.

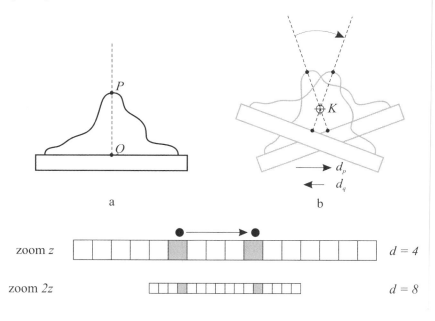

Figure 5.7. How the disparities are affected by the height of the point, the tilt angle, and the set magnification: a. side view of the sample table; b. tilting the sample; c. altering the points on a scanline by changing the magnification

The highest point of the sample P and the lowest point Q are observed. Tilting the sample means that both points (as shown in Figure 5.7b), because of their different heights, move different distances, and this is observable in stereoscopic images as disparities. In this example, P will have a positive disparity d_p, because it lies above the tilt axis K, and Q will have a corresponding negative disparity d_q. As P and Q represent the highest and lowest points of the sample, the disparity region is limited to d_p and d_q. If the **tilt angle** is increased, the disparity region increases as well (and decreases in the case of a reduction). The same happens in the case of the magnification factor. Figure 5.7c shows schematically the places on a scanning line where a point shifted by tilting is imaged, when the magnification is changed. Therefore the change of magnification also affects the disparity region.

5.2.3 Mathematical Basics

In this section, the most important mathematical basics are described. The relationship between these basics and the stereoscopic image processing will be explained later in the work.

5.2.3.1 Convolution

A convolution is a mathematical integral defined by two functions f(x) and g(x) which produce a result function h(x). In a continuous case, h(x) is the integral over the product of f(x) and a mirror-image shifted version of g(x). With discrete functions, the integral is a sum, of

$$h(x) = (f \otimes g)(x) = \sum_n f(n) \cdot g(x-n). \tag{5.3}$$

Convolutions are frequently used as a filter operation in image processing. The image is presented as a two-dimensional function *f(x,y)* and filtered through a convolution mask *g(x,y)*. Different convolution masks are used for different filtrations. For example, the convolution mask:

$$\begin{pmatrix} \dfrac{1}{9} & \dfrac{1}{9} & \dfrac{1}{9} \\ \dfrac{1}{9} & \dfrac{1}{9} & \dfrac{1}{9} \\ \dfrac{1}{9} & \dfrac{1}{9} & \dfrac{1}{9} \end{pmatrix}$$

has a smoothing effect, which suppresses image noise but also makes the image less sharp. Such masks are known as smoothing masks.

Calculations involving convolutions are normally time-consuming and require a large number of steps, because for every value of *h(x)*, it is necessary to calculate a sum of products, even though function values of a **convolution mask** frequently have value 0, meaning that it is not necessary to carry out all the operations. However, the two-dimensional functions arising in the image processing increase the calculation effort considerably.

5.2.3.2 Frequency Analysis

According to the French mathematician Jean B.J. Fourier, any periodical signal in a continuous sequence of harmonic (sinusoidal) signals is composed of different frequencies, amplitudes, and phases. In order to describe the signal over time, either the value of the amplitude at that point in time can be given, or the amplitudes of frequencies occurring in the signal can be given with the corresponding phases.

For the usual system of representing images in the so-called spatial domain, each image coordinate is assigned a brightness intensity. In this way, each image can be clearly described. In image processing, images are also frequently composed as (discrete) two-dimensional spatially moving signals. The image data can then be represented through a **Fourier transform** as a so-called frequency domain, in which an image is described only in terms of the occurring frequencies and their associated phases.

For the data content of an image, it is the same whether this is given in the spatial or frequency domain. The data does not change; it is simply presented in a different way. However, different ways of presenting the data may be more or less suitable to show particular contents or to allow for more efficient processing.

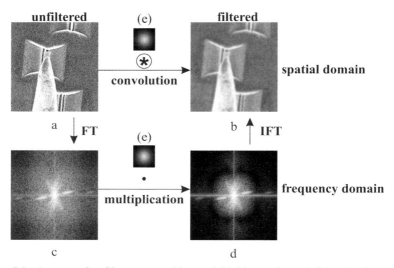

Figure 5.8. An example of image smoothing: a. initial image in spatial domain; b. a filtered image in spatial domain; c. frequency domain; d. a filtered image in frequency domain; e. convolution mask in spatial and frequency domain

Figure 5.8 shows an example of a smoothing operation to illustrate the point. Normally, the initial image (a) would be convoluted with, *e.g.*, a Gaussian smoothing mask (e) in order to achieve the desired results (b). A smoothing operation needs, however, a lot of effort, so it is helpful to avoid it. Therefore, the image is first transformed into the frequency domain, as shown in (c). (The frequency images (c) and (d) are brightened here for simplification.) In this representation, the brightness of a point represents the amplitude of an occurring frequency in the initial image. The lower frequencies are located in the middle region, and the higher frequencies in the outer region. Because of the high level of noise in the initial image, a lot of high frequencies are present. Smoothing the image corresponds to a **low-pass filter** in which high frequencies are suppressed. This is simple to achieve in the frequency domain as the frequency spectrum can, for example, be multiplied by a Gaussian function (d). Through a subsequent inverse transformation into the spatial domain, the same result arises as for the convolution of the initial image in the spatial domain. To achieve an equivalence of both operations, the convolution mask must be transformed likewise. The particular point about the Gaussian smoothing mask used here is that in the frequency domain, a Gaussian function is also represented, as shown in (e).

The conversion of the image data from spatial to frequency domain is done using the Fourier transform, in which a corresponding frequency and phase spectrum results. The following representation relates to the discrete two-dimensional Fourier transform, as used in image processing:

$$F(u,v) = \frac{1}{MN} \sum_{x=0}^{M-1} \sum_{y=0}^{N-1} f(x,y) e^{-i2\pi(\frac{ux}{M}+\frac{vy}{N})}. \qquad (5.4)$$

$f(x,y)$ represents the image in the spatial domain, $F(u,v)$ in the frequency domain. Here, u and v are frequencies occurring along the x or y axes. M and N are the dimensions of the image in pixels. An important feature of this function is that it is reversable. The inverse Fourier transform is calculated by:

$$f(x,y) = \sum_{u=0}^{M-1} \sum_{v=0}^{N-1} F(u,v) e^{i2\pi(\frac{ux}{M}+\frac{vy}{N})}. \qquad (5.5)$$

It should be noted that $F(u,v)$ is a complex value and contains frequency as well as phase data. For the example shown in Figure 5.8, the phase data is not interesting, which is why only the frequency spectrum of the transformed image is shown. The phase data can, however, be used in the context of disparity detection.

5.2.3.3 Gabor Function

Gabor functions, named after the physician D. Gabor, are often used in the frequency analysis of image data. Basically, a Gabor function is a (co-)sine wave under a Gaussian bell. In a one-dimensional case, the Gabor function is described by

$$g(x; \omega, \varphi, \sigma) = e^{-x^2/2\sigma^2} \cos(\omega x + \varphi), \qquad (5.6)$$

where ω is the frequency and φ is the phase of the cosine function. σ indicates the width of the **Gaussian function**. Figure 5.9a gives an example of a one-dimensional Gabor function.

In the two-dimensional case, the additional parameter θ describes the orientation of the function:

$$g(x,y,\omega,\theta,\varphi,\sigma) = e^{-\frac{x'^2+y'^2}{2\sigma^2}} \cdot \cos(\omega x' + \varphi),$$
$$x' = x \cdot \cos(\theta) + y \cdot \sin(\theta), \qquad (5.7)$$
$$y' = -x \cdot \sin(\theta) + y \cdot \cos(\theta).$$

Figure 5.9b and c show a two-dimensional Gabor function in two different orientations.

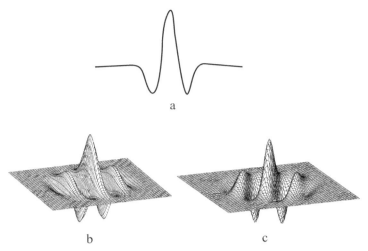

b   c

Figure 5.9. Examples of Gabor functions: a. one-dimensional; b. and c. two-dimensional in different orientations

The main advantage of these functions is that they are limited in both the spatial domain and the frequency domain (Figure 5.10a and b). They are therefore suitable for carrying out a **spatial band-pass filtering** of image data. In this process, σ influences the bandwidth and ω the middle frequencies of the filter. If the frequency spectrum of an image was multiplied by, for example, the function shown in Figure 5.10b, certain frequencies with a particular orientation at a local region of the image would be filtered out.

For digital image processing, these functions are made discrete, *i.e.*, scanned at discrete locations. In this way, there is a finite number of filter coefficients which can be used as convolution masks for filtering image data.

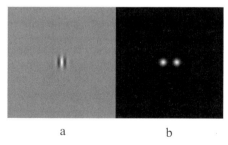

a   b

Figure 5.10. A two-dimensional Gabor function in a. spatial domain and b. frequency domain with image size 256×256 pixels; $\lambda = \dfrac{2\pi}{\omega} = 16$ pixels, $\theta = 0°, \varphi = 0°, \sigma \approx 9$

5.2.4 Biological Vision Systems

For many problems known to science, there are analogies in nature. Humans have increasingly made use of natural systems and processes which have been perfected by nature over millions of years. To determine, how disparity information can be subtracted from stereoscopic image pairs, it is apposite to examine biological vision systems. The following section deals in short with the functioning of the neurons which make up biological vision systems.

5.2.4.1 Neuron Models

The neurons in a neuronal network are connected to each other in a complicated way. If they receive sufficient stimulation, they are able to fire off so-called **action potentials** (APs) which are transmitted to the next neuron, where they may cause another fire-off. A neuron normally receives APs from several other (precursor) neurons and then transmits its APs to several other (successor) neurons. A neuron is schematically represented in Figure 5.11. The number of precursor neurons from which the neuron receives APs, and which therefore influence its response, is known as the "receptive field".

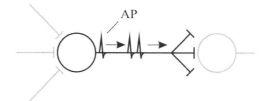

Figure 5.11. A schematic representation of a neuron

The APs arriving from the neurons of the receptive field have varying impacts on the behavior of the neuron; they are weighted. This weighting can also be negative; that is, having an inhibiting effect on the neuron. Whether a neuron fires off an AP to its successors or not depends on the weighted sum of the incoming potential.

In fact, the exact firing behavior of a neuron is not just dependent on the weightings, but in addition on the timing at which the APs arrive. The behavior of a group of neurons is therefore particularly complex. It is still not completely understood how neurons code and process information.

The **neuron model** which is used here is a simplified version. The firing rate, that is, the (mean) number of action potentials which are set off per unit of time, is the measure for the firing of a neuron. The firing rate f of a neuron and the incoming firing rates of the precursors $f_1, ..., f_n$, with the weights $g_1, ..., g_n$ are calculated by:

$$f = \sum_{i=1}^{n} g_i f_i.$$

(5.8)

Because f can have a negative value but neurons cannot code a "negative" firing, this is a simplification of the situation in which a value is described by two neurons, one for positive values and one for negative values, whereby only one of the two is able to fire [4].

5.2.4.2 Depth Perception in Biological Vision Systems

During the earlier period of research into **stereoscopy**, it was assumed that the impression of depth that humans received from the surroundings based on disparities had mainly a geometrical cause and arose from the different spatial projections of lines or object edges. In [13], experiments were carried out in the 1960s with so-called random dot stereograms and it was shown that even for structureless images, a perception of depth arose. Observed individually, they consist of randomly arranged points with no order and indicating no structure. Seen stereoscopically, however, depth structures can be recognized.

The fact that the **human visual system** is able to abstract disparity information from such images leads to the conclusion that this is not only a result of the analysis of shapes or colors. In addition, the experiments showed that the recognition of disparities takes place relatively early in the visual data path of biological perception systems and not only after a time-consuming monocular analysis.

In [14] experiments with cats showed that there are neurons in the visual cortex which are sensitive to light stimulation of particular types. In the main, a differentiation is made between two groups of neurons: simple cells and complex cells. The reactions of individual neurons have been measured experimentally in [4]. They suggested a model which would describe the behavior of these neurons and which can be used for the investigation of disparities. It was found from this that the **receptive fields of simple cells** could be modeled using Gabor functions. This leads to the suspicion that the brain carries out a frequency analysis to determine disparities. "[…] the brain has at its disposal the two-dimensional Fourier transform of the presented brightness distribution" [15].

The model developed in [4] is described as an "energy model". Because it has a central importance in this work, it will be described more fully in the following section.

5.2.4.3 Energy Models

For the calculation of signal energies, **energy models** use so-called energy neurons. Figure 5.12 shows a schematic outline of such a neuron. It is made from two simple cells and one complex cell. The receptive field of a simple cell stretches over a limited region of the retina. The input weights are described through Gabor functions, that is, the input stimulus (a limited local area of the retina) is weighted through Gabor coefficients. The resulting sum represents the response of a simple cell. Such a neuron is described as a linear neuron.

An individual linear neuron cannot be used as a measure for the signal energy as its response is dependent upon which phase of the stimulus from the receptive field arrives. The example of a sinusoidal stimulus is considered, which has the same frequency and phase as the Gabor function which is used. In this case, the response of the linear neuron is maximized. If the phase of the stimulus shifts, the response is reduced.

Through the use of two simple cells, the construction of a response which is independent of the stimulus phase can be construed. Both simple cells possess the same receptive field but filter the stimulus with different Gabor functions, the phases of the functions differing by 90°. Such a pair of filter functions is described as a **quadratic pair**. The response of the two **linear neurons** is squared by a complex cell and summed in order to maintain a constant measure of signal energy. This construction is helped by the fact that $\sin^2(x) + \cos^2(x) = 1$ is valid.

Energy neurons can be used for the detection of disparities, since they can be extended to binocular energy neurons (Figure 5.12b).

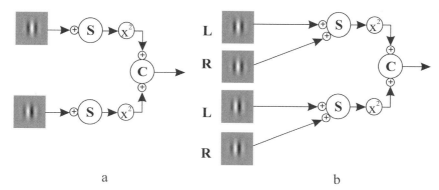

Figure 5.12. a. Structure of an energy neuron; b. structure of a binocular energy neuron

Here, a simple cell receives signals from both the left and right image, which are filtered with separate filter functions. If we observe the phase at which a stimulus arrives at the retinas in both the left and right eye, we see that the response of a simple cell is maximized when the phases are the same [16]. This corresponds to a disparity of 0. If there are differences in the phases, the disparity is not equal to 0 and the response is reduced. Because the response of a simple cell is not just dependent on the **relative phase difference**, but also additionally on the absolute phase as in a monocular case, a quadratic pair is also used here. A complex cell squares and sums the response of two simple cells which have a filter function differing in the phase by 90°.

5.3 Systems for Depth Detection in the SEM

Various methods are used to determine depth of objects in micro- and nano-handling operations in SEMs. In addition, in the field of materials research, surfaces with different properties are characterized. The following section presents an overview of the available procedures described in the literature.

5.3.1 Non-stereoscopic Image Approaches

In [17], the use of a so-called **touchdown sensor** is described. It consists of two flexible piezos, whereby the first serves as an actuator and the second as a sensor. A current is applied across the actuator which causes this to oscillate at its own frequency. The oscillations and their amplitude can be measured by the second piezo through a small phase shift. At the tip of the sensor, a nanogripper is attached. If the gripper now comes into contact with another object, its shape deforms. This leads to a sudden change in the measured amplitude at the sensor. This change can be measured and thus serves as a contact detector. One difficulty in using this procedure is that there is contact between the gripper tip and the object to be gripped. This can lead to damage of the tool or object because of the oscillations.

In [18], another approach is described. Here, the SEM is equipped with an additional light microscope for automatic object manipulation. They use a simplified experimental design in which spherical objects with a diameter of 30 µm can be positioned according to a set pattern. The objects are lifted and placed using a needle-shaped micromanipulator. The electron microscope is used to deliver the top view of the objects. By means of an edge detection algorithm, the 2D position of the objects is determined. The additional light microscope, which is mounted in the sidewall of the SEM, delivers a side view of the objects. Therefore, it is possible to detect whether the object is lifted up or not. Due to limited resolution of the light microscope and the fixed magnification of 1000×, only relatively large objects can be observed.

In [19] a triangulation approach for the determination of the third dimension is used and two methods are proposed. In the first method, a light microscope is used to project a line from a laser onto the object surface at an angle. The relative depth can be calculated through the relative shift between the base line and the projection line. The higher the shift, the higher is the object. The surface of the object must first be treated with a luminescent material. When the beam hits the object surface, a luminous spot appears, which is detected by a miniaturized light microscope that is mounted inside the vacuum chamber. Due to the resolution limits of the light microscope (CCD camera), the accuracy of this approach is 1-5 µm.

In [20], a **four-quadrant backscattered electron (BSE) detector system** for the three-dimensional surface characterization of microstructures is used. The four-quadrant BSE detector system consists of four separate backscatter electron detectors that are placed symmetrically to the sample in the sample chamber of the SEM. If the sample is scanned with an electron beam, different images of the surface structure arise at each of the detectors. This results in four different gray-toned images produced from the information provided by the respective detectors.

In these images, an apparent direction of illumination can be determined, arising from the different shadow areas in the images. This property can be used to produce a 3D model of the sample from the four images. The accuracy of the measurements in the procedure is in the single-digit µm range.

The above approaches are not suitable for the support of nanohandling inside the SEM, because the resolution is not high enough, the systems are too bulky, or they have an impact on the handling process.

5.3.2 Stereoscopic Image Approaches

A frequent area of application for the stereoscopy in the SEM is the analysis of **surface topographies**, for example in researching crack formation in materials.

In [21], surfaces with pyramidal structures, which are used in solar technology, are analyzed. They use an edge detection algorithm to create a wire frame model of the pyramidal structures for each of the left and right views of a stereo image pair. Using the disparities at the lattice points, a height model can then be created. In the analysis of regular structures, in which there are many edges, a **feature-based stereo algorithm** for the processing of the images is used. The stereo image pair is created with a tilt angle of 30°.

In [22], a combination of several procedures is used. In the first calculation phase, features from the images are extracted. They serve to provide the rough alignment and the geometrical equalization of the images. After this, the disparities of some of the reference points are calculated using a **correlation-based approach**. From this, a not very dense but, on the other hand, precise (accurate at sub-pixel level) disparity map arises. Finally, the density of the disparity map is increased by means of the reference points and the application of an interpolation procedure.

The standard software package MeX (Version 4.1) is based on the work in [23]. It calculates a digital surface model (DSM), a three-dimensional height map from a stereoscopic image pair following a correlation-based procedure. Following this, the reconstructed surface can be analyzed and measured. The software has a graphical user interface for data input and the analysis of the results. As data is fed in, the mapping of the input image is given, as well as the relative tilt angle, the operating distance, and the pixel size of the images. The latter is used primarily for the calculation of absolute object coordinates. The other information enables the overall displacement of the images to align them against each other. One function allows the images to be automatically aligned in this way.

Once all the data has been fedin, the calculation of the DSM can begin. This is carried out interactively in several stages. To start with, the user establishes a region of interest (ROI), that is, an image extract for which the calculation is to be carried out. Then, an "input DSM" is calculated and displayed. This is a rough-textured DSM in which few points of the image data can be used for calculating. With this snapshot, the user can work out whether the image data is usable or whether operation parameters need to be modified. At this point, the calculation can either be stopped or continued. With a continuation, a detailed DSM will be calculated. This can then be used for comprehensive volumetric measurements.

With stereoscopic calculations, a correlation-based procedure is used in which the size of the correlation window is dynamically calculated. The image data is processed hierarchically using a coarse-to-fine process. With an input image size of 1024 × 768 pixels, the disparities are determined from about 15,000 sampling points. From this, a wire frame structure is derived, which is used for the interpolation of the DSM. The reference image is superimposed as a polygon texture over the height map when the DSM is displayed, in order to complete the three-dimensional view. As an alternative, the height information can be observed in the form of a depth map, corresponding to a top view of the DSM.

The described stereo approaches have not been developed for the support of nanohandling. They use a standard SEM that is only equipped with a tiltable sample table. One disadvantage of this approach is that the handling station, which consists of the handling object, robot, and tools, has to be tilted as well. This can take several minutes and there is a high risk that the object changes its position. Therefore, this approach is not suitable for the generation of stereo images for the observation of handling processes.

Also, images of technical handling processes are different from images of surfaces. The major difference in technical process images is the **high number of low-texture** regions. This leads to heavy noise in these regions as a result (disparity map) of the stereo algorithm. Thus, misinterpretation of the object and tool positions may result.

The existing approaches do not fulfill the requirement for nanohandling inside the SEM.

5.4 3D Imaging System for Nanohandling in an SEM

5.4.1 Structure of the 3D Imaging System for SEM

In this section, the structure of the 3D imaging system, which fulfills the requirements of micro- and nanohandling, is described [24–26]. The system is schematically represented in Figure 5.13 and consists of five main components.

Because the 3D imaging system has to be suitable for observing manipulation processes, the technique of beam tilting is used. To overcome the known problems of handling in the beam-tilting system, the system described below was developed.

This system does not need a special electron column and can be flexibly and universally applied in each standard SEM. The stereo images are acquired through the beam-tilting system which is controlled by the beam-control unit. The images are then suitable for the following processing work. Before the images are transferred into the **3D module**, they are pre-processed and filtered by standard methods such as the median filter. Noise levels in the images are therefore reduced and a contrast enhancement is carried out.

After the pre-processing, the images are sent to the 3D module. This consists of two sub-modules: the **vergence system and the stereo system**. The vergence system is necessary to line up the images against each other and to compensate for unwanted shifts, rotations, and different zoom scales, where it is necessary. In this way, prepared images are sent to the stereo system which processes the image data and provides the 3D information (disparity maps). These results are then processed in the last module (Figure 5.13) which provides the data for the graphical user interface (GUI) or a robot control system.

The image acquisition system and the 3D module are described in more detail in the following sections.

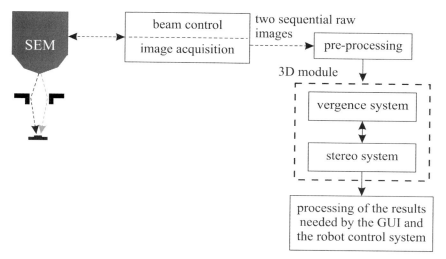

Figure 5.13. Structure of the 3D imaging system

5.4.2 Image Acquisition and Beam Control

Beam-tilting is a suitable stereo image generation technique for nanohandling tasks in the SEM. Until now, however, this technique has only been available in SEMs that have a special electron beam column [10–12], and it cannot be used in standard SEMs. Therefore, the described system provides an external and flexible lens system which can be installed in every SEM [24].

For the **beam-tilting system** a magnetic lens is used. This is fixed under the electron gun with the help of a metal arm. The arm can be flexibly mounted on the side panel of the SEM and the position of the lens can be adjusted. Additionally, the lens is supplied with power via a vacuum-suited cable entry.

The task of the **external lens system** is to deflect and tilt the beam twice. With this procedure, two images from different perspectives of a sample are generated by deflected electron beams. By using the integrated beam-shift unit of the SEM, the beam is deflected in one direction, and with the additional external lens system the beam is then deflected in the opposite direction. In Figure 5.14, the principle is shown schematically. With this approach a tilt angle of max. 3° can be achieved.

In order to generate two different perspectives, a beam control unit is needed. This unit is schematically represented in Figure 5.15 and consists of four main components.

With the integrated SEM **beam-shift unit** the beam entry position can be changed with reference to the center position of the lens. The image acquisition unit acquires two images, which have to be generated from two different perspectives. This means that the view direction has to be changed. Thus, the beam has to be shifted in the right position and the polarity of the lens must be changed as well. This is carried out by the control system, which sends the necessary control commands to the electronic polarity unit and to the beam shift unit.

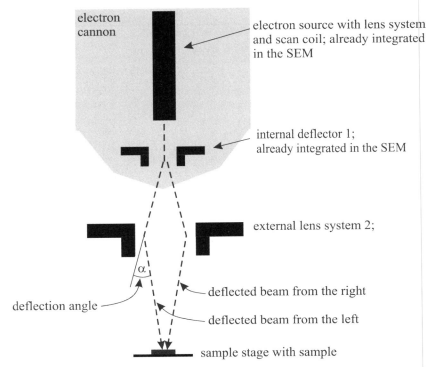

Figure 5.14. Principle of the beam deflection

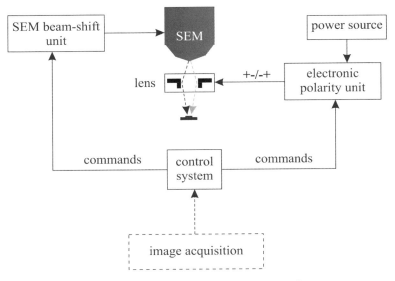

Figure 5.15. Structure of the beam control

5.4.3 The 3D Module

The 3D module is a very important part of the 3D imaging system. Its purpose is to extract 3D information from stereoscopic images. It consists of two components, the vergence system and the stereo system (Figure 5.16). The functionality and the algorithms of these systems are described below.

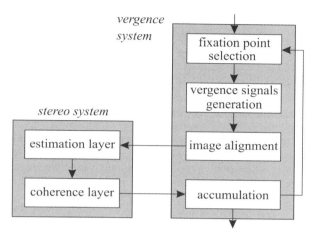

Figure 5.16. Flowchart of the 3D module

The disparity calculations and therefore the acquisition of the 3D information are carried out in the **stereo system**. This receives a stereoscopic image pair which is processed over two layers. The estimation layer calculates disparity estimates, and the coherence layer determines stable disparity values and presents them in the form of a disparity map. Apart from this, other results, which are described in the next section, can also be calculated. Because of the additional special image acquisition in an SEM, the stereo images are potentially distorted. Vertical shifts, counter-rotation, or different zoom scales could result between the two stereo images. This makes it difficult for the stereo system to extract the image disparities. However, it is possible to compensate these distortions within limits by a vergence system, so that the images can be properly processed by the stereo system.

The stereo system then receives its input from the **vergence system**, which aligns the stereo image pair. The vergence system can be passed through a number of times. In each pass, a **fixation point** is chosen and the image pair is aligned according to the vergence signals (shift parameters). The aligned images are sent to the stereo system to calculate the disparities relative to the chosen fixation point. The results of each pass are accumulated, which provides robust 3D information.

In the next section the stereo system is described first, because the results of this system are needed for the vergence system, which will be described afterwards.

5.4.3.1 Stereo System
The presented stereo algorithm consists of two processing layers, the estimation layer and the coherence layer (Figure 5.17).

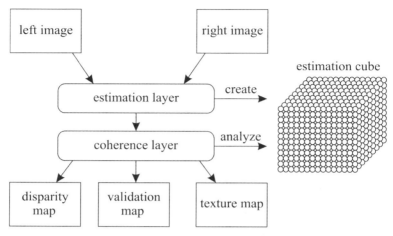

Figure 5.17. Structure of the stereo algorithm

In the estimation layer, a number of disparity estimations for every pixel are calculated. This leads to approximately correct disparity values but also to random and incorrect values. Therefore, the coherence layer has the task of separating the correct from the incorrect estimations and calculating a stable result. The central data structure of the algorithm is the estimation cube. It is created by the estimation layer and analyzed by the coherence layer for disparity detection. Both the estimation and the coherence layers are described in more detail in the following sections. The results of the stereo algorithm are a sharp and high-density disparity map, which gives the relative depth, a validation map, which gives the reliability of the disparity values, and a texture map showing the regions with low or high texture in the images. The validation and texture values are needed for the vergence system which is explained below.

The task of the **estimation layer** is to estimate several disparities for all image pixels. The base unit of this approach is a **disparity estimation unit (DEU)**. This unit is capable of estimating the disparity down to one pixel position. The estimation cube consists of lots of such units.

Figure 5.18 schematically shows a DEU (bottom of the figure). Such a DEU is built up of several neuronal layers, which are based on the concept of the energy model described in [3, 27, 28]. The two major layers are the simple cell layer and the complex cell layer. The receptive field of a simple cell is modeled by Gabor [3, 28, 29].

The discrete filter coefficients are normalized to eliminate the DC component. The response of one simple cell is given by:

$$s(x,y) = \sum_{i=-n}^{n} \tilde{g}(i;0) \cdot I_l(x+i,y) + \tilde{g}(i;0) \cdot I_r(x+i,y). \tag{5.9}$$

This means that a convolution between an image scan line and the coefficients of the **Gabor function** is carried out. The **complex cells** are modeled as a **quadratic filter** and the receptive fields are calculated by the squared input from at least two simple cells.

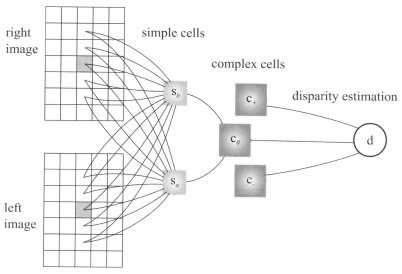

Figure 5.18. Structure of a disparity estimation unit (DEU): the simple cells of c_- and c_+ are not shown

With these complex cells, disparity estimations can be carried out. However, it is known that the response of complex cells depends on the local contrast. To avoid this undesirable effect, a solution described in [27] is used. For the disparity estimation, three complex cells are used and the responses are compared. The complex cells differ in their parameters. The first complex cell (c_+) detects slightly positive disparities, the second detects slightly negative ones (c_-) and the third detects zero disparities (c_0). The difference between c_+ and c_- is normalized by c_0. This approach covers only a small disparity range.

Since the detection of larger disparities is desirable, a combination of the phase shift and the position shift model [4, 5, 16, 27, 30] is used. This leads to the response of one DEU by

$$d(x,y;s) = \frac{c_+(x,y;s,+\phi) - c_-(x,y;s,-\phi)}{c_0(x,y;s,0)} - s, \tag{5.10}$$

where

$$c(x,y;s,\phi) = \sqrt{s_a(x,y;s,\phi)^2 + s_b(x,y;s,\phi)^2},$$
(5.11)

$$s_a(x,y;s,\phi) = \sum_{i=-n}^{n} \tilde{g}\left(i - \frac{s}{2}; +\phi\right) \cdot I_l(x+i,y)$$
$$+ \tilde{g}\left(i + \frac{s}{2}; -\phi\right) \cdot I_r(x+i,y),$$
(5.12)

$$s_b(x,y;s,\phi) = \sum_{i=-n}^{n} \tilde{g}\left(i - \frac{s}{2}; \frac{\pi}{2} + \phi\right) \cdot I_l(x+i,y)$$
$$+ \tilde{g}\left(i + \frac{s}{2}; \frac{\pi}{2} - \phi\right) \cdot I_r(x+i,y).$$
(5.13)

The parameter s determines the tuned disparity of one DEU and ϕ is fixed to $\pi/4$. Because of the fact that $(c_+\text{-}c_-/c_n)$ gives the relative disparity to s, s must be subtracted from the result. A group of DEUs with different s parameters can cover a bigger disparity space than a single DEU. If the number of DEUs increases, then the detectable disparities will increase as well. Values for s from -4 to 4 at the step size of $1/3$ are suitable for most cases.

The estimation cube (Figure 5.17 right) is a three-dimensional structure which consists of numerous DEUs. The size of the cube in the x-y-direction is the same size as the input images. This is because the disparities for all pixels of the images are estimated. The size of the third dimension depends on the disparity search space, that is, how many DEUs are necessary to cover a desired disparity search space.

The **coherence layer** analyzes the results of the estimation layer which are stored in the estimation cube. The result of one DEU can either approximately represent the correct disparity or can have a more or less random value. The latter case will occur if the true disparity lies outside the working range of the DEU. However, the estimation cube consists of several DEUs with **overlapping working ranges** for one image position. So, the disparity search space can be increased and a robust estimation can be carried out.

To find the correct disparity for one image position (x,y), all relevant DEUs are grouped to a disparity stack, according to [3, 28]. Using coherence detection, it is possible to estimate the correct disparity, so the biggest cluster of DEUs which has similar disparity estimations is calculated. The final disparity of a position (x,y) is calculated from the average of the estimated results of the coherence cluster:

$$Dis(x,y) = \frac{1}{|C_{x,y}|} \sum_{i \in C_{x,y}} d(x,y;S(i)),$$
(5.14)

where $C_{x,y}$ is the set of DEUs in the coherence cluster of the disparity stack at position (x, y) and $S(i)$ is a mapping function mapping i to the tuned disparities s of the DEU.

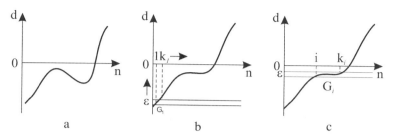

a   b   c

Figure 5.19. Search for the coherence cluster: a. typical curve in a disparity stack; b. first group of sorted DEUs; c. biggest cluster

In every disparity stack, **the biggest DEU group** with similar disparity values is searched (Figure 5.19). In Figure 5.19a, a typical curve in a disparity stack is shown. The DEUs are shown on the abscissa and the calculated disparities are shown on the ordinate.

In the first step, the DEUs are sorted by their disparity values in ascending order. Let the disparity values of the sorted DEUs be $D(i)$ $(i \in 1,...,N)$, with N being the size of the disparity stack. Then N groups of DEUs $(G_1,...,G_N)$ are built up and determined by:

$$G_i = \{D(i),...,D(k_i)\}, \tag{5.15}$$

$$k_i = \max(\{k \mid i < k \leq N \wedge D(k) - D(i) \leq \varepsilon\}). \tag{5.16}$$

Figure 5.19b shows the first group of sorted DEUs. Let $\overline{d}(G_i) = D(k_i) - \underline{D}(i)$, then the group consists of as many consecutive DEUs as possible, where $\overline{d}(G_i)$ is maximal and is not greater than ε. The biggest cluster found in this way is the coherence cluster $C_{x,y}$ of the **disparity stack** at the position (x, y). Figure 5.19c shows the biggest cluster for the example. In the worst case, several groups with the same size are found. In this case, the group is chosen in which $\overline{d}(G_i)$ is minimal, as a large group with a small range of disparity values indicates a stable estimation.

The parameter ε should be carefully chosen. If ε is too low, only small groups are found, and the likelihood for clusters of the same size is increased, which in turn leads to ambiguities. If ε is too high, the average of the disparity result is less precise, because noise coding DEUs start to appear in the coherence cluster. Additionally, the calculation time increases. It was found experimentally that a value of $\varepsilon = 0.5$ is a good compromise for SEM images.

One major problem of the generated images is that in most cases the probability of regions having low or no texture is high. In these regions (*e.g.*, a black background) only small changes in the intensity exist. Therefore, no disparity can be detected by simple cells and the disparity map is noisy in these regions. In this chapter, a solution using a **dynamic texture-based filter** for overcoming this problem is described. One important part of the filter is the texture map. This shows regions with low and high texture of the images. Using this information, a dynamic texture-based filter can be built up, which reduces the noise of the disparity map. To create the texture map, the complex cells $c_0(x, y; s, 0)$ of all DEUs have to be used, since they detect the local texture in the stereo image pair. The final result is determined by averaging all complex cell c_0 responses of a disparity stack. Therefore, the texture of an image position (x, y) is given by:

$$T(x, y) = \frac{1}{N} \sum_{i=1}^{N} c_0(x, y; S(i), 0),$$ (5.17)

where $S(i)$ as above is a mapping function for mapping the index i to the tuned disparity s of the DEU and N is the size of the disparity stack. If the texture map is calculated for every position (x,y), then each disparity value will have a corresponding texture value. In order to reduce the noise in the disparity map, all regions with a texture value which is smaller than a threshold are taken out.

Because the range of texture values depends on the image data, a fixed threshold is not suitable. A **dynamic adaptation** of the threshold is therefore proposed. A solution can be found by analyzing the histogram of the texture map. Figure 5.20a shows a typical histogram of an image pair with several textureless regions. It can be seen that most of the values are found in the left part of the texture domain which is limited by the smallest $\min(\{T(x, y)\})$ and the biggest $\max(\{T(x, y)\})$ value. In contrast, Figure 5.20b shows a histogram of an image pair with few low-texture regions. Here, most texture values are located in the middle. These histograms lead to the conclusion that a threshold close to the dotted line is preferable. With such a threshold, large regions of the disparity map can be filtered if the noise level is high, and on the other hand it is possible to filter only small regions of the map if the noise level is low. The threshold can be calculated by:

$$t_{thres} = (\max(\{T(x, y)\}) - \min(\{T(x, y)\})) \cdot p$$
$$+ \min(\{T(x, y)\}).$$ (5.18)

where all disparity values $T(x, y) < t_{thres}$ are ignored and filtered. The parameter p determines the portion of the filtered texture values. A value of $p = 0.05$ yields good results in most cases for images showing handling processes.

5.4.3.2 Vergence System

Because of the special acquisition of the stereo images in the SEM, the images might differ in how well they can be used for processing. With a high-quality stereo image pair, only horizontal disparities occur, meaning that image regions are shifted either to the left or to the right (or not shifted at all in the case of zero disparities). However, inaccuracies during image acquisition could lead to additional vertical disparities, which allow image regions also to be shifted up or down. Figure 5.21 shows three examples.

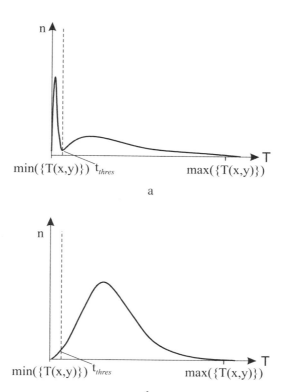

a

b

Figure 5.20. Typical texture histograms: a. many low-texture regions; b. few low-texture regions

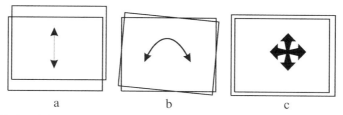

a   b   c

Figure 5.21. Possible distortions in the images: a. vertical shifts; b. rotation; c. different zoom

If the SEM beam does not hit the specimen on exactly the same spot after changing the viewing angle, the resulting images will have a global vertical shift (Figure 5.21a). A counter rotation of the images will lead to **vertical disparities**, particularly in the left and right areas of the images (Figure 5.21b). If the images differ in a slight zoom scale, vertical disparities will occur in the top and bottom areas of the images (Figure 5.21c).

The first case is the easiest to compensate, because the vertical disparities are constant over all image regions. The other two cases are more difficult because the vertical disparities differ from region to region.

The occurrence of **vertical shifts** will strongly affect the results of the stereo system, as it can only detect horizontal disparities. During image processing, it is not known if and where vertical disparities are present. Therefore, a pre-processing of the images, which compensates, for example, for rotational differences, is not possible.

The concept of a vergence system as described in [24, 31] can be used to reduce these problems by aligning the images not only horizontally, but also vertically. So, vertical disparities are taken out. Images with different vertical disparities in different image regions can be processed in multiple passes using appropriate vertical shifts in each pass. The **validation values** of the coherence layer indicate the image regions where the disparity calculations have been successful. Therefore, after each pass, the disparities that have higher validation values than the disparities calculated in prior passes are accumulated.

The vergence system is used as an enhancement of the stereo system. The stereo system is not altered in its functionality, but used multiple times with different input parameters. Before each call, a fixation point is chosen. The **fixation point selection** is carried out using the validation and **texture map**. Vergence signals are generated accordingly, that is, horizontal and vertical shift parameters to fuse the input images at the fixation point. The aligned images are sent to the stereo system for calculation and the results are accumulated.

5.5 Application of the 3D Imaging System

In this section, the results of the 3D imaging system are first presented and discussed. Afterwards, two applications of the imaging system, the handling of CNTs and the handling of crystals, are shown and discussed.

5.5.1 Results of the 3D Imaging System

Figure 5.22a,b shows the input images and Figure 5.22c–i the results of the stereo algorithm. The tilt angle is small (about 2°), so only small disparities are detectable. The images show a gripper [32] with a jaw size of about 2 μm and a nanotube (produced by Engineering Department at Cambridge University) of about 300 nm diameter and 12 μm length seen from the side with a magnification of 3000 times. The texture around the objects is low and there is a contrast difference between the images as well. The calculated disparity map (Figure 5.22f) shows dark regions for large distances and bright regions for small distances with reference to the observer. In Figure 5.22g a 3D plot of the disparity map is shown.

The stereo algorithm calculates a high-density disparity map in sub-pixel accuracy. The smallest step size of the gripping process in the direction of the nanotube was 1 μm which could be detected by the approach.

It is also noteworthy that the nanotube in the disparity map is displayed wider than in the input images. This is due to the size of the receptive fields of the DEUs.

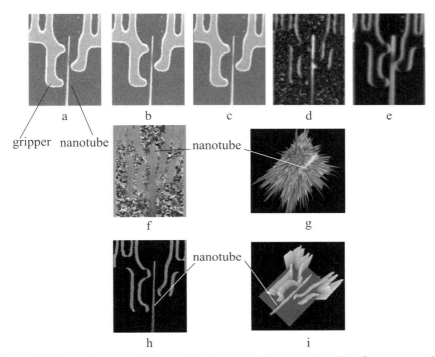

Figure 5.22. Input images and the resulting images of the stereo algorithm for one nanotube and a gripper: a. and b. input images; c. cyclopean view; d. validation map; e. texture map; f. unfiltered disparity map; g. 3D plot of the unfiltered disparity map; h. filtered disparity map; i. 3D plot of the filtered disparity map

This effect can be reduced by increasing the image resolution (and possibly by defining a region of interest) or reducing the size of receptive fields.

It should be noted that, because of the low-texture regions around the gripper and the nanotube, noise arises around it in the disparity map (Figure 5.22f) and the tool and the tube are not clearly seen in the 3D plot in Figure 5.22g. This is a common problem for all stereo algorithms, since the corresponding problem cannot be solved in textureless regions. However, the described stereo algorithm provides a texture map (Figure 5.22e) which shows **regions of high and low texture** in the input images. Therefore, with this map, a dynamic threshold can be calculated and used to filter the noisy disparity map. The result of the filtering is shown in Figure 5.22h and the 3D plot is shown in Figure 5.22i. In the result image and the 3D plot, the noisy regions of the disparity map are removed and the contours are thinned by means of the determined cyclopean view (Figure 5.22c). Therefore the objects are clearly shown and only the important regions are available. This result is more suitable for the further processing steps than the noisy disparity map. Because of the textureless region inside the gripper, only the contour is shown in the disparity map.

5.5.2 Application for the Handling of CNTs

As previously referred to, the manipulation of carbon nanotubes has great importance in nanotechnology. There is a big advantage in the use of an SEM for observing manipulation processes. The manipulation task consists of gripping the nanotube in a controlled way and placing it on a TEM (transmission electron microscope) grid (Chapter 2). The procedure is **teleoperated** or carried out **automatically**.

For this purpose, multiwalled carbon nanotubes (MWCNT) are used. These are produced on a silicon substrate through the chemical vapor deposition (CVD) procedure. The availability of these multiwalled tubes was facilitated through the Engineering Department of Cambridge University, UK [33].

The main difficulty here is to bring the gripping tool and the nanotubes to the same level. Only in this way is it possible to achieve a precise gripping of the tube.

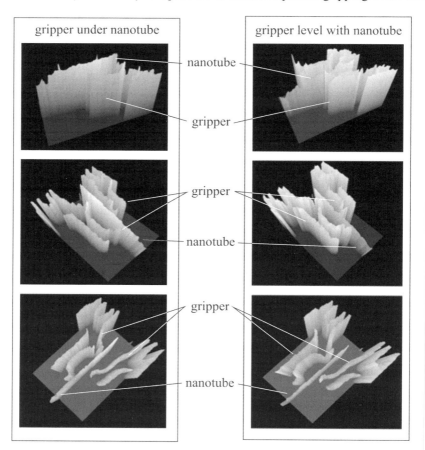

Figure 5.23. Left: different views from the gripper, which is under the nanotube: Right: different views from the gripper at the same level as the nanotube.

Through the 3D imaging system, continuous stereo images of the manipulation process are produced. These are displayed on the GUI as an **anaglyph image** or they can be shown as a **3D plot**. The anaglyph image can be used by the operator, with the help of suitable red–green glasses, to produce a stereoscopic impression of the manipulation procedure. The 3D plot displays the results of the stereo algorithm. These results can also be used by the operator to control the manipulation procedure. Furthermore, the underlying results can be used by a robot-controlled system to determine the corresponding positions of object and gripping tool. This is indispensable if an automatic manipulation is to be carried out. In Figure 5.23, the start and end positions of an approach by the gripper moving up towards the nanotubes are shown. Here, several different 3D views of the gripper and the tube at the start of the approach are seen in the left image area. It can be seen that the nanotube is lying at an angle (about 45°) and the gripper is below it. On the right side of Figure 5.23, more 3D views are shown. Here, the end position of the approach can be seen. This shows that the gripper jaws are positioned at the same height as the upper part of the nanotube.

After the gripper has been set up, the gripping process can begin. A voltage is applied to it in order to close the jaws. This way, the nanotube is gripped. Now, the nanopositioning unit can be moved and the position of the TEM network can be controlled.

5.5.3 Application for the Handling of Crystals

A further manipulation process involves the sorting of crystals. In this process, the crystals are moved with the help of a so-called STM tip. The difficulty here is that in the manipulation process, the tip has to lie against an outer wall of the crystal, otherwise the crystal cannot be moved. If the tip is too high, it misses the crystal. If it is too low and touches the base, it may become damaged and unusable. So the operator needs 3D information in this task, too. Similar to the process for manipulating nanotubes, with the help of a 3D imaging system, an anaglyph image and a corresponding 3D plot are available to the operator. This makes it possible for the operator to control the tip and manipulate the crystal. Figure 5.24 shows a stereo image pair and a 3D plot of such a manipulation process. The crystals have a size of about 20 µm and the STM tip shown has a diameter of about 2 µm. It can be seen here that the tip is positioned under the crystal. The crystal can therefore be moved and sorted by the tip.

5.6 Conclusions

5.6.1 Summary

In this chapter, various possibilities for obtaining 3D information from SEMs with respect to nanohandling are discussed.

For this, basic concepts such as the principle of stereoscopic image approaches in the SEM, mathematical basic and biological vision systems are presented. In addition, current systems used to obtain 3D information using SEMs are presented and discussed. These systems have major disadvantages in reference to nano-handling actions. Therefore, a universally applicable 3D imaging system that fulfills the requirements in nanohandling is discussed in more detail. In this system, stereo images are generated by beam tilting followed by processing of the image data. One major part, the 3D module which consists of a vergence and a stereo system, is depicted in detail.

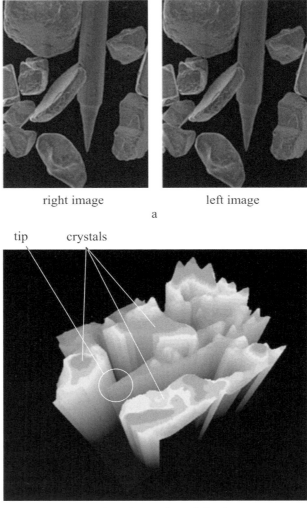

right image   left image

a

3D plot of crystals and the tip

b

Figure 5.24. a. input images and b. the 3D plot of the manipulation of crystals by an STM tip

The stereo system is based on a biologically motivated energy model with coherence detection. By means of the used stereo system, a sharp and high-density disparity map in sub-pixel accuracy can be calculated. Also, a dynamic filter is shown which gives the stereo system the capability to detect regions of low texture in the input images and thus to remove the noise of the disparity map in these regions. The vergence system with an enhanced solution for fixation point selection is presented. This enables the 3D module to supply accurate and robust results, even with disturbed input images (rotated, vertically shifted, or with different zoom scales).

The applicability of the 3D imaging system is shown by the handling task of nanoobjects such as nanotubes and crystals.

5.6.2 Outlook

In future work, the calculation time of the 3D imaging system will be further decreased. For this, an optimized implementation in hardware (*e.g.*, on a field programmable gate array, FPGA) for the 3D module will be developed. Therefore, it will be necessary to investigate some processing strategies for the hardware design. For example, a suitable partitioning of the vergence and stereo system has to be found.

The 3D imaging system will be able to calculate only a certain region of interest, so that the interesting parts (*e.g.*, the region between gripper and nanotube) can be observed in more detail.

5.7 References

[1]   Ullrich, S. 1999, 'Stereoskopische Tiefenbestimmung in Grauwertbildern mit unabhängigen Zuordnungsmodulen', Ph.D. thesis, Technische Universität Darmstadt.
[2]   Zitnick, C. L. & Kanade, T. 2000, 'A cooperative algorithm for stereo matching and occlusion detection', *IEEE Transactions on Pattern Analysis and Machine Intelligence*, vol. 22, no. 7, pp. 675–684.
[3]   Henkel, R. D. 2000, 'Synchronization, coherence-detection and three-dimensional vision', Technical report, Institut für Theoretische Neurophysik, Universität Bremen.
[4]   Ohzawa, I., DeAngelis, G. C. & Freeman, R. D. 1990, 'Stereoscopic depth discrimination in the visual cortex: neurons ideally suited as disparity detectors', *Science*, vol. 249, no. 4972, pp. 1037–1041.
[5]   Fleet, D. J., Jepson, A. D. & Jenkin, M. R. M. 1991, 'Phase-based disparity measurement', *Computer Vision, Graphics, and Image Processing; Image Understanding*, vol. 53, no. 2, pp. 198–210.
[6]   Mallot, H. A. 2000, *Sehen und die Verarbeitung visueller Informationen*, Vieweg Verlag.
[7]   Reimer, L. 1998, *Scanning Electron Microscopy*, Springer.
[8]   Piazzesi, G. 1973, 'Photogrammetry with the scanning electron microscope', *Journal of Physics E: Scientific Instruments*, vol. 6, pp. 392–396.
[9]   Hein, L. R. O., Silva, F. A., Nazar, A. M. M. & Ammann, L. J. 1999, 'Three-dimensional reconstruction of fracture surfaces: area matching algorithms for automatic parallax measurements', *Scanning: Internat. Journal of Scanning Electron Microscopy and Related Methods*, vol. 21, no. 4, pp. 253–263.

[10]   Abe, K., Kimura, K., Tsuruga, Y., Okada, S., Suzuki, H., Kochi, N., Koike, H., Hamaguchi, A. & Yamazaki, Y. 2004, 'Tilting and moving objective lens in CD-SEM (II)', *Proceedings of SPIE*, vol. 5375, pp. 1112–1117.

[11]   Thong, J. T. L. & Breton, B.C. 1992, 'A topography measurement instrument based on the scanning electron microscope', *Science Instrument*, vol. 63, no. 1, pp. 131–138.

[12]   Lee, K. W. & Thong, J. T. L. 1999, 'Improving the speed of scanning electron microscope deflection systems', *Measurement Science and Technology*, vol. 10, pp. 1070–1074.

[13]   Julesz, B. 1960, 'Binocular depth perception of computer-generated patterns', *Bell Systems Technical Journal*, vol. 39, pp. 1125–1162.

[14]   Hubel, D. H. & Wiesel, T. N. 1962, 'Receptive fields, binocular interaction and functional architecture in the cat's visual cortex', *Journal of Physiology*, vol. 160, pp. 106–154.

[15]   Pollen, D. A. & Taylor, J. H. 1971, 'How does the striate cortex begin the reconstruction of the visual world?', *Science*, vol. 173, pp. 74–77.

[16]   Fleet, D. J., Wagner, H. & Heeger, D. J. 1996, 'Neural encoding of binocular disparity: energy models, position shifts and phase shifts', *Vision Research*, vol. 36, no. 12, pp. 1839–1857.

[17]   Fatikow, S., Eichhorn, V., Wich, T., Hülsen, H., Hänßler, O. & Sievers, T. 2006, 'Development of an automatic nanorobot cell for handling of carbon nanotubes', *IARP – IEEE/RAS – EURON Joint Workshop on Micron and Nano Robotics,* Paris, France, October 23–24, online.

[18]   Kasaya, T., Miyazaki, H., Saito, S. & Sato, T. 1998, 'Micro object handling under SEM by vision-based automatic control', *Proceedings of SPIE*, vol. 3519, pp. 181–192.

[19]   Bürkle, A. & Schmoeckel, F. 2000, 'Quantitative measuring system of a flexible microrobot-based microassembly station', *4th Seminar on Quantitative Microscope-*Semmering, Austria, pp. 191–198.

[20]   Drzazga, W., Paluszynski, J. & Slowko, W. 2006, 'Three-dimensional characterization of microstructures in a SEM', *Measurement Science and Technology*, vol. 17, pp. 28–31.

[21]   Kuchler, G. & Brendel, R. 2003, 'Reconstruction of the surface of randomly textured silicon', *Progress in Photovoltaics: Research and Applications*, vol. 11, no. 2, pp. 89–95.

[22]   Lacey, A. J., Thacker, N. A., Crossley, S. & Yates, R. B. 1998, 'A multi-stage approach to the dense estimation of disparity from stereo SEM images', *Image and Vision Computing*, vol. 16, pp. 373–383.

[23]   Scherer, S., Werth, P., Pinz, A., Tatschl, A. & Kolednik, O. 1999, 'Automatic surface reconstruction using SEM images based on a new computer vision approach', *Institute of Physics Conference Series*, vol. 161, no. 3, pp. 107–110.

[24]   Jähnisch, M. & Fatikow, S. 2007, '3D vision feedback for nanohandling monitoring in a scanning electron microscope', *International Journal of Optomechatronics*, vol. 1, no. 1, pp. 4–26.

[25]   Sievers, T., Jähnisch, M., Schrader, C. & Fatikow, S. 2006, 'Vision feedback in an automatic nanohandling station inside an SEM', *6th Int. Optomechatronics Conference on Visual/Optical Based Assembly and Packaging,* SPIE's Optics East, Boston, MA, U. S. A., 1–4 October, pp. Vol. 6376, 63760B.

[26]   Jähnisch, M. & Schiffner, M. 2006, 'Stereoscopic depth-detection for handling and manipulation tasks in a scanning electron microscope', *IEEE Int. Conference on Robotics and Automation (ICRA),* Orlando, Florida, U. S. A., May 15–19, pp. 908–913.

[27] Adelson, E. H. & Bergen, J. R. 1985, 'Spatiotemporal energy models for the perception of motion', *Journal of the Optical Society of America A*, vol. 2, no. 2, pp. 284–299.

[28] Henkel, R. D. 1997, 'Fast stereovision by coherence detection', *Proceedings of CAIP'97 LCNS 1296*. Kiel, eds. G. Sommer, K. Daniilidis and J. Pauli, Springer, Heidelberg, pp. 297–303.

[29] Daugman, J. G. 1985, 'Uncertainty relation for resolution in space, spatial frequency, and orientation optimized by two-dimensional visual cortical filters', *Journal of the Optical Society of America A*, vol. 2, no. 7, pp. 1160–1169.

[30] Qian, N. 1994, 'Computing stereo disparity and motion with known binocular cell properties', *Neural Computation*, vol. 6, pp. 390–404.

[31] Henkel, R. D. 1999, 'Locking onto 3D-structure by a combined vergence- and fusionsystem', *2nd International Conference on 3D Digital Imaging and Modeling, 3DIM'99,* Ottawa, IEEE Computer Society Press, Los Alamitos, pp. 70–76.

[32] Mølhave, K. & Hansen, O. 2005, 'Electro-thermally actuated microgrippers with integrated force-feedback', *Journal of Micromechanics and Microengineering*, vol. 15, pp. 1265–1270.

[33] Teo, K. B. K. *et al.* 2003, 'Plasma enhanced chemical vapour deposition carbon nanotubes/nanofibres – how uniform do they grow?', *Nanotechnology*, vol. 14, pp. 204–211.

6

Force Feedback for Nanohandling

Stephan Fahlbusch

EMPA, Swiss Federal Laboratories for Materials Testing and Research
Laboratory for Mechanics of Materials and Nanostructures
Thun, Switzerland

6.1 Introduction

The development of microrobots for nanohandling, with ever-growing demands regarding flexibility and automation, raises the problem of appropriate process control. With dimensions of the parts to be handled of sometimes considerably less than 100 nm and with a typical positioning accuracy in the nanometer range, nanohandling applications have detached several orders of magnitude from the operators' macroscopic realm of experience. Powerful sensory feedback is required to overcome this scale gap [1] and to be able to handle and manufacture such small devices. The extraction and transfer of process information from the micro- and nanoworld into the macroworld are great challenges. However, they are important preconditions for closed-loop microrobotic control systems and thus a key to reliable nanohandling [2].

Obviously, visual process monitoring using different kinds of microscopes is essential (Chapters 2, 4, and 5). Apart from image processing, force feedback is the most important sensor application in nanohandling [3–5]. The main challenge is the control of the interaction between the part and the microgripper or end-effector [6]. A lot of micro- and nanoscale parts are very sensitive to gripping or contact forces due to their brittle materials, filigree structures, or small gripping areas. They can easily be damaged or even destroyed [3]. The integration of force sensors into grippers and end-effectors provides essential information on the nanohandling process for both the operator working with telemanipulation devices and the control system operating in automatic mode.

In the first part of this chapter, an overview of the fundamentals and principles of micro/nano force measurement will be given, with an emphasis on the special requirements in force feedback for nanohandling by microrobots. The state-of-the-art in force sensing using robot- and atomic force microscopy (AFM) based nanohandling systems is presented in the second part of this chapter.

6.2 Fundamentals of Micro/Nano Force Measurement

One basic principle for the measurement of a force is to compare it with a previously known force. This has been done for several thousands of years and, nevertheless, is still an important topic of current research activities [7]. Classical mechanics provides the corresponding theory and addresses the interaction between macroscopic objects. It is based upon the research and theory developed by Sir Isaac Newton (1643–1727).

Newton's Second Law. If an external force acts on a body, the body is accelerated in the direction of the force. Thus force can be defined as a vector. Isaac Newton discovered that the acceleration a is proportional to the force F and inversely proportional to the mass of the body m, a scalar:

$$\vec{a} = \frac{\vec{F}}{m}.$$

(6.1)

This equation is known as Newton's Second Law. It enables the definition of mechanical units. According to the SI system, mass (kg), length (m) and time (s) are the basic units. Acceleration and force are units derived from the basic units. The unit of force is the newton (N). One newton is defined as the force which accelerates a mass of 1 kg with 1 m/s^2:

$$1 \ \mathrm{N} = 1 \ \mathrm{kg} \times \frac{\mathrm{m}}{\mathrm{s}^2}.$$

(6.2)

In the following, the main principles of force measurement are presented and the requirements for force feedback in nanohandling are described.

6.2.1 Principles of Force Measurement

Altogether, there are six principles of force measurement, which can be classified as follows [8]:

1. Comparison of the unknown force with the force of gravity of a known mass.
2. Measurement of the acceleration of a known mass on which the unknown force acts.
3. Comparison of the force with an electromagnetically generated force.
4. Conversion of the force into pressure (of a liquid) and measurement of the pressure.
5. Direct measurement of the (absolute) deformation of an elastic body caused by the force.
6. Measurement of the (relative) strain in an elastic body caused by the force.

The last principle mentioned is the one mostly used for force sensors. Also, the fifth principle and – with some limitations – principles 3 and 4 are sometimes used for force measurement. The measurement of a force by measuring the acceleration of a known mass is not of practical importance for nanohandling.

In a physical sense, force is a vector which acts on a point. When measuring a force, its vector has to be transformed into a scalar. In practice, the force to be measured cannot act on a body directly as materials can only withstand limited mechanical stress. To be exact, there is no force measurement technique but rather a technique to measure mechanical stress fields which are caused by an external force (principle 6). In the following, this will be explained in more detail.

A force acting on a fixed body will deform this body and cause strain and mechanical stress. If the stress σ is equally distributed, it can be calculated dividing the force F by the cross-section of the body A:

$$\sigma = \frac{F}{A} \, .$$

(6.3)

The relative elongation or strain ε is defined as elongation Δl caused by the force F divided by the length of the body l in the direction of the force:

$$\varepsilon = \frac{\Delta l}{l} \, .$$

(6.4)

The mechanical stress and thus the required force is determined from strain measurement. Hooke's Law provides the basis for this.

Hooke's Law. In the case of a uniaxial stress condition, within the proportional area of the stress–strain curve, Hooke's Law applies for a uniaxially stressed beam:

$$\sigma = E \cdot \varepsilon \, .$$

(6.5)

Young's modulus E corresponds to the slope of the straight line of the stress–strain curve up to the proportional limit. It is obvious that force sensors may only operate within this elastic area to measure reliably.

If the strain of a body is measured for instance with strain gages, the force acting on it can be calculated as:

$$F = \varepsilon \cdot E \cdot A \, .$$

(6.6)

If the amplitude of a force is determined by directly measuring the deformation x of a force-sensitive element with known stiffness k (principle 5), as for example in case of the deflection of an AFM cantilever, the force F can be calculated according to:

$$F = \frac{x}{k} \, .$$

(6.7)

To put it simply, for the development of a force sensor a suitable physical principle for the measurement of either the mechanical stress σ (Equation 6.3) or the strain ε (Equation 6.6) or the deformation x (Equation 6.7) has to be chosen and technically realized. The parameters Young's modulus E, cross-section A, and stiffness k of the sensor are predefined by its material properties and geometry.

6.2.2 Types of Forces in Robotics

In robotics, force sensors are used to measure the interaction forces between gripper/end-effector, the part to be handled, and the environment. Typically, forces are divided into two classes: When gripping a part, the **gripping forces** which occur between gripper and part are of special interest. When manipulating a part, *e.g.*, during assembly, moving on a surface, or mechanical characterization, the **contact forces** have to be known. Sensors to measure the gripping forces are integrated into the gripper – ideally into the gripper jaws. Contact force sensors can be mounted between gripper/end-effector and manipulator (Figure 6.1).

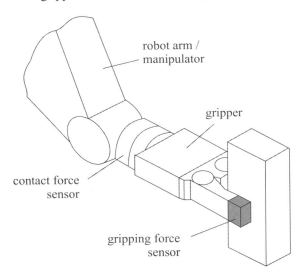

Figure 6.1. Sensors for gripping and contact force measurement integrated into the gripper and the manipulator, respectively

6.2.2.1 Gripping Forces
A gripper has to ensure a fixed position and orientation, *i.e.*, pose of the part with respect to the robot's last joint. In this manner, the gripper has to exert and withstand forces and torques. The main disturbing factors are forces of inertia due to the robot's acceleration as well as contact forces between the part and other objects within the working space. In micro- and nanohandling, forces of inertia play only a minor role, due to much smaller weights of the parts with respect to their gripping area, whereas adhesion forces prevail. However, it has to be ensured

that the microgripper grips the parts well enough to prevent slipping, while at the same time not damaging them.

The gripping force acts from the gripper on the gripped part. It has to counteract a total load resulting from several single forces and torques, including static holding forces as well as dynamic and process-related loads. The required magnitude of the gripping force depends on the geometries of gripper and part as well as – in the case of force-fit gripping – the friction coefficient between gripper jaw and part.

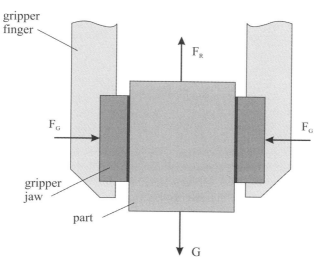

Figure 6.2. Forces acting on a gripped part

The forces occurring during force-fit gripping are shown in Figure 6.2. The gripping force F_G acts on the part as normal force and, thus, generates the friction force F_R. According to Coulomb's Law, the friction force acts against the direction of motion and, thus, against the weight G of the part. If the gripped part is not moving, the following equation applies:

$$\vec{G} = \vec{F}_R = n \cdot \mu \cdot \vec{F}_G , \tag{6.8}$$

with the static friction coefficient μ between gripper jaw and part and the number of gripper jaws n. Resolving this equation for the required gripping F_G force leads to:

$$\vec{F}_G = \frac{m \cdot \vec{g}}{n \cdot \mu} , \tag{6.9}$$

with the mass m of the part and the gravity g. When determining the required gripping force, it has to be considered that, on the one hand, the gripping force has to be high enough to compensate for the weight of the part and firmly grip it, and, on the other hand, the surface pressure during the gripping must not exceed the maximum tolerable value for the material of the part. This problem can be overcome by, for instance, integrating gripping force sensors and implementing a suitable gripping force control.

6.2.2.2 Contact Forces

Forces occurring between (i) the gripped part or the gripper and the environment or (ii) between the part and the end-effector are called **contact forces**. If two parts have to be assembled, or, for instance, the membrane of a biological cell has to be penetrated with a pipette, their relative position and orientation have to be known. Uncertainties regarding this information occur mainly due to measurement errors, robot inaccuracies, and disturbing external influences. Therefore, the assembly can fail or parts can even be damaged by assembly forces being too high.

By measuring contact forces, damage can be prevented, *e.g.*, during an unforeseen contact due to positioning inaccuracies, as well as gaining additional information about the relative position of the part and its join partner. Contact force measurement can not only complement gripping force measurement, but also provide additional information for the robot control and the operator, which may complement or even supersede image processing. A typical example of the potential benefit of contact force measurement is the so-called **peg-in-hole** benchmark.

6.2.3 Characteristics of the Micro- and Nanoworld

In principle, force measurement techniques known from the macroworld can be used in the micro- and nanoworld as well. However, there are a few characteristics and constraints to be taken into account, which can limit the usability of these techniques at small scales.

Nanohandling differs clearly from classical, *i.e.*, macroscopic, handling and assembly because of different scales; four different scaling areas can be distinguished. Table 6.1 gives an overview of these areas, together with suitable actuators and force sensors. The definition of the different ranges is not to be taken strictly but rather to give an indication. Regarding the object size and the accuracy required for positioning and handling, two different areas have been introduced because the accuracy often has to be better than the object size by a factor of 100 to 1000.

Handling and assembly in the macroscopic domain, usually carried out using industrial robots, is state-of-the-art today. This area is dominated by friction and inertia forces which lead to a predictable behavior of the parts. Numerous handling strategies for force- and form-fit gripping as well as techniques for a systematic gripper design have already been developed [9].

The situation changes during handling of parts with dimensions smaller than 1 mm and with a mass smaller than 1 mg. Due to an increasing ratio between surface area and volume, forces proportional to the surface area increase compared to forces proportional to the volume of the object (Figure 6.3). Friction and inertia forces dominating during macroscopic manipulation decrease compared to surface forces, and the parts tend to either stick to the end-effector or show a behavior difficult to predict, like jumping. Among the prominent surface or adhesion forces are surface tension, electrostatic forces, and van der Waals forces. They have been described in detail in the literature [10, 11].

Table 6.1. Scale domains in handling and assembly

Scale	Object size	Accuracy	Dominating forces	Actuators	Force sensors
Macro	> 1 mm	> 10 μm	gravity, friction force	micrometer screws, electric motors	state-of-the-art, commercially available
Meso	0.1–1 mm	< 10 μm	friction and surface forces	micrometer screws, electric motors, solid state actuators (*e.g.,* slip–stick)	partly state-of-the-art, forces in mN to μN range
Micro	1–100 μm	< 100 nm	surface forces	solid state actuators (*e.g.,* slip–stick)	difficult to develop, forces partly in μN range
Nano	< 100 nm	< 10 nm	surface and atomic forces	solid state actuators (*e.g.,* flexible hinges)	difficult to develop, forces partly in nN range, AFM technologies required

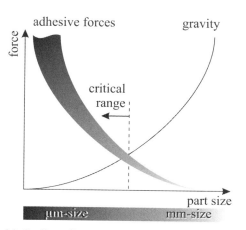

Figure 6.3. Scaling effects caused by surface and volume forces

Due to the dominating surface forces, the concepts and techniques for the handling and development of grippers known from the macroworld cannot be applied without modifications. One of the challenges in nanohandling consists in the modeling and reduction or, even better, in the utilization of surface forces. Different approaches have already been investigated [12, 13], but their practical use is still limited. A systematic method for the control of adhesion forces does not exist.

Possible ways to reduce adhesion forces include amongst others:

- adaptive design to minimize the contact area between gripper/end-effector and object;
- use of "rough" surfaces within the contact area;
- reduction of gripping forces by using sensors to measure gripping forces;
- use of hard materials in the contact area, especially for the gripper/ end-effector;
- reduction of surface tension in the vacuum, *e.g.*, in the scanning electron microscope (SEM), because of reduced humidity;
- minimization of electrostatic attractive forces by using electrically conducting and grounded grippers and objects.

It is the goal of the first four methods to reduce the actual surface area and thus limit the most important factor for surface forces. However, at the same time, these methods lead to an increased surface pressure between gripper/end-effector and object. The methods to be used are always a compromise between small adhesion forces and a load which is still tolerable for the part. Handling in a scanning electron microscope has the advantage of reduced surface tension due to a lower humidity in vacuum; at the same time, problems can occur because of charging effects and electrostatic forces.

Handling in the micro- and nanoworld has further characteristics and constraints which are, however, not directly related to the development of sensors for force feedback for nanohandling. These include, *e.g.*, high accuracy requirements for positioning, the ability to handle a large variety of geometries and parts materials, sensitivity against environmental influences, three-dimensional assembly, *etc.* These characteristics have already been discussed in detail in the literature [4, 14, 15].

6.2.4 Requirements on Force Feedback for Nanohandling

From the constraints in the micro- and nanoworld described above, specific requirements arise for force feedback during nanohandling. As existing commercial sensors cannot be applied, the use of novel microgrippers and end-effectors requires the development of special micro force sensors, which take the following aspects into account:

Limitation of surface pressure. Small gripping areas and small tolerable surface pressures require high measurement resolution and a low response threshold, *i.e.*, minimum detectable force. To limit the surface pressure during the first contact between gripper/end-effector and object, a force measurement system with high dynamics and real-time capability is required. Force measurement using image processing, as could be used in a scanning electron microscope, is often not precise enough and especially is not real-time capable.

Design of microgrippers. The development of micro force sensors is affected by a set of additional parameters. The accuracy which can be achieved by a specific gripping force depends on the positioning accuracy of the gripper jaws or end-effectors. To achieve maximum accuracy, the gripper's transmission system, which transfers the actuators' motion into a movement of the gripper jaws, has to be free from backlash. Another constraint results from the size of the objects to be handled. Micro force sensors with very small dimensions are required because of the dimensions of the grippers/end-effectors. The sensor itself has to be within the same range.

Integration of sensors. Ideally, micro force sensor and gripper should be connected in such a way as to enable a simple exchange of either the gripper jaws or the sensor. At the same time, a safe, force-fit, and perhaps even form-fit connection between gripper and sensor has to be ensured. The problem here is that the sensor should be placed close to the point of force transmission and a certain modularity has to be preserved – such as the exchangeability of end-effectors/jaws, a low number of connecting cables, *etc*. These factors have to be considered during the first development phase of the micro force sensor and microgripper. Otherwise, the integration of the sensor could become very difficult, and the risk of damage to the mechanically sensitive micro sensor could increase.

As already mentioned, micro force sensors can either be used as separate, mountable sensors or they can be directly integrated into the force-sensitive element of the gripper. The advantage of the first solution is a universal force sensor that can be used for different grippers/end-effectors and within different applications. This can often be found in case of industrial robots but it is not yet available for nanohandling. In the case of the second solution, an example would be strain gages that are glued onto the end-effectors, or force-sensitive material which is deposited on the microgripper. This setup solves the problem of a possibly reduced stiffness as in case of a separate sensor. However, the usability of these techniques largely depends on the design of the microgripper and is not as modular.

Use in clean rooms and scanning electron microscopes. Further constraints have to be taken into consideration when the micro force sensors are used within a clean room. During manufacturing of micro- and nanostructures, a production environment with constant temperature and humidity is required to reduce the number and the size of defects. Particles like dust can influence the functionality of these structures. Hence, robots, grippers, and sensors have to be clean room compatible and should produce almost no particles by abrasion or the like.

When using the micro force sensors inside a scanning electron microscope, particular conditions imposed by the vacuum and the electron beam have to be considered. These will be described in more detail in Section 6.2.5.

Technical specifications. Although nanohandling imposes several constraints on the development of micro force sensors, the technical specifications of the sensors cannot be disregarded. Measurement ranges between several 100 mN down to a few micronewtons are required. The resolution should be in the range of several tens of micronewtons down to several tens of nanonewtons – sometimes even below.

A study of the available literature, applications, and a survey of the partners of a European research project[1] resulted in the following specifications for force measurement in nanohandling:

- **Handling of micro-/nanomechanical parts**: An example handling task mentioned several times is to grip a silicon part with a force-fit gripper. An Si cube with an edge length of 1 mm (density: 2.329 g/cm^3, mass: 2.329 mg) has a weight G of 22.85 µN. With a stiction friction coefficient of 0.1, the minimal required gripping force F_G according to Equation 6.9 is 114 µN. A sensor with at least micronewton resolution is required to measure this force. To accomplish additional handling tasks, like assembly or joining, larger gripping forces are required, which are still in the range of millinewtons. The upper limit of the gripping force is defined by the tolerable surface pressure of the part's material. For example, a force of 1 mN causes a pressure of 0.4 MPa on an area of 50 × 50 µm^2 , which is still safe taking into account the yield point of silicon (120 MPa) or nickel (30 MPa). However, the gripping area often cannot be used entirely because of very small structures on the part, such as edges from sawing the silicon; structures with membranes or cantilevers, *etc.*, which can reduce the area significantly.
- **Micro-/nanomechanical characterization**: The forces to be measured, *e.g.*, during nanoindentation, range from 10 nN up to 500 mN and the required resolution is 1nN and better.
- **Handling of biological objects**: The forces required to penetrate or deflect the membrane of a biological cell are in the micronewton range. A force of 0.215 µN to 11 µN leads to a deflection of 5 µm of a lipid double-layer [16]. These forces have to be measured with an adequate resolution in the nanonewton range.

Calculation of (static) friction forces. The model usually applied to calculate static friction from normal force and friction coefficient is not necessarily valid in the micro- and nanoworld. In this domain, the friction coefficient is increasingly determined not only by the material of the interacting parts but also by their surface topography. However, the theory of friction at the macro- and nanoscale is not fully understood yet, and the macroscopic friction model has been applied above for a rough estimation of the gripping forces.

Conclusions. From the specifications mentioned previously it is obvious that a compromise has to be found between a micro force sensor's resolution and measurement range. The reproducibility of the sensor should be high enough to prevent damage during handling and assembly of micro/nano parts. However, for some applications, the measured values might not have to be calibrated exactly, and an estimation can be sufficient. It is not possible to develop a "universal"

---

[1] EU FP5 Growth Project ROBOSEM (Development of a Smart Nanorobot for Sensor-based Handling in a Scanning Electron Microscope), 2002–2005.

micro force sensor which fulfills the wide spectrum of requirements. Ideally, several classes of force sensors for different force ranges will be developed, each with an appropriate resolution, measurement range, and size.

6.2.5 Specific Requirements of Force Feedback for Microrobots

When using microrobots for nanohandling, specific requirements have to be considered for the development of micro force sensors. These result, firstly, from the size and the motion principle of the robots (robot-specific requirements) and secondly from their application area and the corresponding environmental conditions (application specific requirements).

Robot-specific requirements. The robots' small size of only a few cubic centimeters, with correspondingly small grippers and end-effectors, requires the development of very small force sensors with at the same time high resolution. The integration of commercial force sensors is not possible due to performance limitations regarding size, measurement range, and resolution. In addition to the size, the microrobots' mobility has to be taken into account. Some of the force measurement principles require a considerable amount of instrumentation. Force-sensitive structures have to be in the field of view of a microscope, or, in case of AFM-like systems, a detection unit for deflection measurement has to be moved relative to the robots' motions. These requirements are difficult to meet in the case of mobile microrobots. Additionally, the stiffness of the cables required for actuators and sensors leads to a motion behavior which is difficult to model. Any additional cable should preferably be avoided.

Application-specific requirements. A second complex of specific requirements for the integration of micro force sensors into microrobots results from their application area. Especially in the case of scanning-electron-microscopy-based nanohandling applications, two main aspects have to be considered, namely the influences of the vacuum and the electron beam.

Vacuum. A standard SEM requires a vacuum of 10^{-3}Pa or better. Accordingly, for components being developed to operate inside the SEM, special guidelines have to be considered to avoid or at least to minimize a disturbance of the vacuum. This includes the choice of proper materials, design, manufacturing, and cleaning of vacuum-compatible components. The parts should not contain pores or cracks and should have a low gas pressure at room temperature. In particular, the outgassing of materials like glue has to be kept low. Additionally, hollow spaces must be avoided. The variation in heat conduction can be problematical for sensors and actuators used inside the SEM. In the vacuum, heat cannot be dissipated by convection, but only by conduction. Components can therefore easily heat up in the SEM.

Electron beam. Components put into the SEM have to be compatible with the electron optics and the electron beam. Magnetic fields deflect the electron beam and lead to a distortion of the SEM image. The materials used should ideally be non-magnetic. Also, electrical fields caused by current-carrying wires or surfaces electrically charged by the electron beam can interfere with the SEM image. A suitable shielding or grounding has to be ensured so that the parts and their surfaces can discharge. If this is not done, local voltage concentrations can produce fields which would deflect the electron beam and thus lead to image distortions. The electron beam can also interfere with components like force sensors. For instance, electrostatic forces caused by the electron beam can lead to problems during nanohandling due to charging of the parts, causing them to be attracted or pushed off or behave otherwise unpredictably. To solve these problems, the parts have either to be made of electrically conducting materials, or the non-conducting surfaces have to be coated with a thin gold layer.

6.3 State-of-the-art

In the 1930s, Ernst Ruska invented the electron microscope and created the basis for the development of the scanning electron microscope [17]. With this new technology, objects could be displayed with nanometer resolution for the first time. But it took another 50 years to develop additional tools and instruments that enabled research and manipulation of parts and materials at the nanometer scale. The invention of the scanning tunneling microscope (STM) and subsequently the atomic force microscope have especially to be mentioned; see Section 6.3.4 for more details.

In parallel with STM-/AFM-based manipulation, a robot-based nanomanipulation approach emerged from combined research in robotics, precision mechanics, and microtechnology. The first research in this field was reported from Japan, where in 1990 an SEM-based nanomanipulation system was presented at the University of Tokyo [18].

It soon became necessary not only to transmit and scale human operations from the macroworld to the nanoworld by graphical user interfaces, telemanipulation devices, and dedicated actuators, but also to transmit information from the nano- to the macroworld. Ideally, information addressing all human senses would be acquired by sensors, transferred, scaled, and appropriately presented to the operator. However, so far, mainly vision and force sensors have been developed for this purpose. An overview of the state-of-the-art in force feedback for micro- and nanohandling will be given in the following sections.

6.3.1 Micro Force Sensors

Force sensing and force control has been an important topic in robotics research for more than 50 years [19, 20]. Various basic theories, force sensors, and control algorithms have been presented, implemented, and experimentally verified.

In contrast, there are only a few micro force sensors available that can be used for robot-based micro- or even nanohandling. Existing force sensors can be differentiated according to the measurement principles, *i.e.,* piezoresistive, piezoelectric, capacitive, or optical sensors. An overview of these sensors is given in Table 6.2.

6.3.1.1 Piezoresistive Micro Force Sensors
Several research groups have been working on the development of three- and six-axes force sensors for microhandling [21, 22], the manipulation of biological cells

[23], for robotics [24], and for micromechanical characterization [25]. All sensors have been manufactured using silicon micromachining and are based on integrated, piezoresistive elements, so-called **piezoresistors**. The sensitive element is either a crosswise structure or a so-called **boss membrane**. Details such as resolution, sensitivity or linearity are often not available from the literature. The lateral dimensions of the sensors range from 2.4 × 2.4 mm^2 to 5 × 5 mm^2 and their maximum measurement range is between 0.01 N and 10 N.

Table 6.2. Examples of micro force sensors

Sensor type	Measurement principle	Maximum force	Resolution	Note	Reference
contact force	piezoresistive	1.6 N		3 axes	[21]
contact force	piezoresistive		1 mN	6 axes	[22]
contact force	piezoresistive	5–10 N		3 axes	[23]
contact force	piezoresistive	5 N		F_{min}=0.1N	[24]
contact force	piezoresistive	10–100 mN	sub mN	3 axes	[25]
gripping force	piezoelectric	2 N			[26]
contact force	capacitive	490–900 µN	0.01–0.24 µN	2 axes	[27]
force in general	optical		10 µN		[28]
force in general	optical		µN range		[29]
contact force	laser	500 µN	19 nN	1 axis	[30]
contact & gripping force	optical		±3nN, ±3 mN	1 axis	[31]
contact force	optical		0.5 nN	1 axis	[32]

The three-axes force sensor described in [25] is used for mechanical investigations of micromechanical structures (Figure 6.4a). The basic element is a boss membrane made out of silicon onto which a stylus has been glued. If a force acts on the pin, the membrane gets deformed and mechanical stress occurs in the membrane. The areas of maximum stress are at the border of the membrane and at the intersection between membrane and boss (Figure 6.4b). Into these areas, 24 piezoresistors have been implanted and connected to three Wheatstone bridges. Mechanical stress causes a change of the resistances and thus a change of the three bridge voltages. If the resistors are connected in a suitable way, the bridge voltages are proportional to the three components of the force acting on the pin. The dimensions of this sensor are $5 \times 5 \times 0.36$ mm^3.

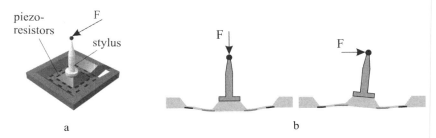

Figure 6.4. Piezoresistive three-axes micro force sensor. a. Schematic view of the sensor (from [25], © Springer-Verlag 2001, with kind permission of Springer Science and Business Media); b. working principle (from [33], courtesy of Shaker Verlag).

6.3.1.2 Piezoelectric Micro Force Sensors

A piezoelectric micro force sensor for microhandling was described for the first time in [26]. The dimensions of the sensor were $16 \times 2 \times 1$ mm^3. A prototype was built but was neither characterized nor integrated into a microgripper. Information about the performance of this sensor, such as resolution or accuracy, is not available.

6.3.1.3 Capacitive Micro Force Sensors

A two-axes capacitive force sensor for the handling and characterization of biological cells was presented in [27], Figure 6.5a.

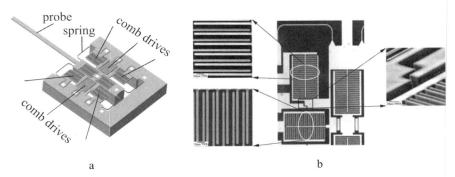

Figure 6.5. Capacitive micro force sensor for biological cell handling. a. Schematic representation of the sensor; b. detail of the comb drives (from [27], © 2002 IOP Publishing, with kind permission of IOP Publishing).

The outer fixed frame and the inner moveable plate are connected by four springs. A force acting on the tip causes a movement of the inner plate and thus a change of the distance between each pair of the comb-like capacitors. The force can be calculated from the total capacity. The stiffness of the sensor, and thus its measurement range, can be adjusted by changing the dimensions of the springs. To be able to measure forces in the x- and y-directions, several comb drives were arranged perpendicularly (Figure 6.5b). Two capacitors in the middle of the inner plate were designed as reference capacitors for the signal-processing circuit.

6.3.1.4 Optical Methods for Micro Force Measurement

Strain and force measurement can also be realized using optical methods. For instance, either passive micro strain gages which are mechanically amplified can be used [28] (Figure 6.6), or strain in the nanometer range is measured using a light-optical microscope and image-processing tools [29]. By using this method, it is possible to measure forces with a resolution in the micronewton range. Both methods have been used for the characterization of micromechanical materials.

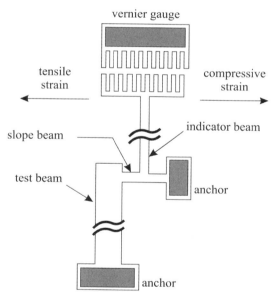

Figure 6.6. Optical force measurement using a passive, mechanically amplified micro strain gage

A force sensor for a teleoperated micromanipulation system is described in [30]. The sensor is made of a cylindrical glass tube (length: 10 cm, diameter: 500 µm) with a tip of 20 µm in diameter. The tube levitates between two plates made of diamagnetic material. Thus, mechanical friction can be avoided, and the resolution is improved. If a force acts on the glass tube, it is deflected and the deflection can be measured with a laser sensor. The resolution of 19 nN mentioned is only a theoretical value, which has been calculated from the stiffness of the sensor setup (19 nN/μm) and the resolution of the laser sensor (1 μm). The entire setup has dimensions of $17 \times 10 \times 7$ cm^3.

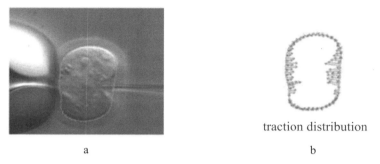

a   b

Figure 6.7. Vision-based force measurement. a. Tracking of the contour of a deformed biological cell; b. distribution of forces on the cell membrane (courtesy of the Institute of Robotics and Intelligent Systems, ETH Zurich, http://www.iris.ethz.ch).

A method of measuring the influence of a force acting on deformable, linear elastic objects is described in [31]. With a so-called template-matching algorithm, the degree of deformation of an object can be determined (Figure 6.7a). Using linear elastic theory, the force acting on the object can be calculated (Figure 6.7b). This method was validated using a silicon cantilever and a microgripper. With these test objects, a resolution of ±3 nN and ±3 mN, respectively, could be achieved.

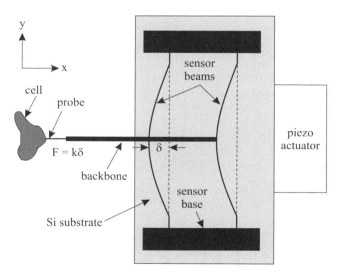

Figure 6.8. Micromechanical force sensor for the characterization of biological cells

In [32], a micromechanically manufactured force sensor for the characterization of living cells is presented. The sensor is made of a tip, which is fixed to frame via flexures (Figure 6.8). If the cells are indented with the tip, the corresponding force can be determined by measuring the deflection of the flexures using a light-optical microscope. The force can then be calculated as a product of the sensor's stiffness and the deflection. With a stiffness of a few nanometers per micrometer, the theoretical resolution of this method is 0.5 nN.

6.3.1.5 Commercial Micro Force Sensors

In spite of the importance of force feedback for nanohandling, only a few micro force sensors are commercially available. To some extent, this can be explained by the strong dependency between the development of microgrippers/end-effectors and micro force sensors. Due to integration issues, it is difficult – and to some extent impossible – to develop a force sensor without taking into account the design of the gripper/endeffector (Section 6.2.4). SensorOne Technologies Corporation, USA, is the only company offering a piezoresistive sensor based on silicon technology which can be used – after calibration – as a force sensor. The smallest commercially available multi-axes sensor is distributed by ATI Industrial Automation, USA. It is a six-axes force/torque sensor with a measurement range of up to 17 N/120 N·mm and a resolution of 0.78 mN/3.9 mN·mm. Due to its size, weight, and resolution, this sensor is unlikely to be useful for nanohandling. Its main application fields are robotics, micromanipulation [34], or dental research.

6.3.2 Microgrippers with Integrated Micro Force Sensors

Grippers and end-effectors are amongst the most important components of a handling and assembly system as they are in direct contact with the parts. The high number of publications dealing with the investigation of gripping principles demonstrates this importance: More than forty mechanical and micromechanical grippers have been developed worldwide within the last 15 years. Additionally, several suction and adhesive microgrippers, as well as contact-free handling methods, such as electrostatic grippers, have been reported. An overview of microgrippers with integrated force sensors is given in Table 6.3. Additional microgrippers, which have been integrated into robot-based micro- and nanohandling systems, are described in the Section 6.3.3.

The first investigations on force measurement for microhandling were carried out in Japan [35]. To measure the gripping force and the position of the end-effectors simultaneously, four strain gages were integrated into a microgripper driven by piezoelectric bimorph actuators. Also, the miniaturized grippers described in [36-38] were developed with integrated strain gages to measure the gripping force and, in case of the last two publications, with additional strain gages to measure the deflection of the gripper jaws.

A microgripper with an integrated force sensor based on a piezoresistive AFM cantilever was developed at the Fukuda Laboratory at Nagoya University, Japan [39]. In the center of the several-millimeters-long and 425 μm wide end-effector, the thickness was reduced to 50 μm. Within this area, four piezoresistive strain gages were implanted and connected into a full Wheatstone bridge. The achievable resolution of this force sensor is 0.1 μN.

Table 6.3. Overview of microgrippers with integrated force sensors

Sensor type	Measurement principle	Maximum force	Resolution	Note	Reference
gripping force	strain gages	15 mN			[35]
gripping force	strain gages	130 mN	±0.1 mN		[36]
gripping force	strain gages	several 10 mN			[37]
gripping force	strain gages	500 mN			[38]
gripping force	piezoresistive	2 mN	0.1 μN		[39]
gripping force	piezoelectric			qualitative measurement	[40]
gripping force	capacitive			qualitative measurement	Caesar[2], not published

Sensitive gripping of microobjects based on the piezoelectric effect was realized by [40]. The DC voltage required for opening and closing of the gripper via a piezoelectric actuator is superimposed with an AC voltage. This AC voltage has a smaller amplitude than the actuator signal and excites the end-effectors to oscillate at resonance frequency. If an object is gripped, the phase and amplitude of the sensor signal change and give an estimation of the relative change of force. One problem of this method is the visible oscillations, another one is the lack of an absolute force measurement.

A gripper with force feedback based on quartz tuning forks was developed at the Caesar Research Center, Germany. The gripper is driven by two piezoelectric bimorph actuators. A tuning fork is fixed on each of the actuators; as end-effectors, glass needles were glued at the end of each tuning fork. The drawback of force sensors based on tuning forks is the lack of quantitative measurements: the sensor cannot be calibrated but provides only tactile feedback. Another drawback is that the frequency change used for force feedback can also occur due to a temperature shift or due to deposits on the end-effectors, *e.g.*, from carbon contamination inside the SEM. Additionally, the vibrations of the tuning fork cause problems, since they are visible under a microscope and can hamper the handling of small parts.

6.3.3 Robot-based Nanohandling Systems with Force Feedback

Robot-based micro- and nanohandling is based on two different types of robots: (i) **industrial robots** (or: conventional robots) and (ii) **microrobots**, which can be

---

[2] Caesar Research Center, Germany, http://www.caesar.de/

either stationary or mobile. The latter have notable advantages regarding flexibility and modularity at the cost of difficult position sensing.

6.3.3.1 Industrial Microhandling Robots

Typical industrial robots used for microhandling are articulated robots, SCARA robots, and Cartesian robots. They have in common robust components like manipulators and grippers, a high stiffness, a high payload, and high positioning velocities. Their resolution and positioning accuracy is in the range of micrometers. Compared to microrobots, sensors, and CCD cameras can be integrated quite easily. The use of this type of robot is determined by industrial applications such as the automated assembly of hybrid micro-electro-mechanical systems (MEMS) or microoptical systems in small to medium quantities. Although these robots are not suitable for nanohandling, some implementations of force feedback have the potential for miniaturization and use in microrobotics-/nanohandling and are thus given in the following (Table 6.4).

Table 6.4. Overview of force feedback for industrial microhandling robots

Sensor type	Measurement principle	Maximum force	Resolution	Note	Reference
gripping force	fiber optical			qualitative measurement	[41]
gripping force	strain gages	538 mN		accuracy: ±14 mN	[42]
gripping force	piezoresistive	6 mN			[33]
contact & gripping force	fiber optical	40 mN/100 mN		linearity: 7.1% / 20%	[43]
force in general	piezoresistive	200 µN		AFM cantilever	[44]
portable gage	piezoresistive	10–3000 µN	a few µN		[45]
portable gage	piezoresistive	µN range		cantilever type	[46]

Fiber optical sensors for process control in microsystem assembly were developed by [41]. Two different solutions for integrated sensors were realized, and the usability of these sensors to measure gripping forces has been shown in principle (Figure 6.9). Using a grayscale picture, the presence of a gripped object and information about an increase or decrease of the gripping force could be gained. However, quantitative measurements were not possible.

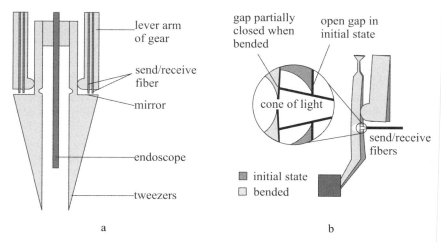

a   b

Figure 6.9. Two different concepts of a fiber-optic-based gripping force sensor measuring intensity changes by using a. a small mirror; b. a gap which changes its size during gripping

a   b

Figure 6.10. Gripping force sensor based on strain gages. a. Actuator integrated into the gripper; b. strain gages glued on the actuators' housings (from [42], courtesy of Shaker Verlag).

Hence, a gripping force sensor based on strain gages was developed for the same gripper by [42], Figure 6.10a. To enable a simple and fast exchange of the grippers' end-effectors without additional electrical connections, two strain gages have been glued on each of the two actuators' housings, Figure 6.10b. If an object is gripped, the strain caused by the deformation of the actuators' housing can be measured. Taking into account the change of the stiffness of this system, the gripping force can be calculated. This method has the drawback of low accuracy.

Several different microgrippers for the assembly of microparts were developed by [33] (Figure 6.11a). The grippers were manufactured using plasma-enhanced dry etching of silicon and deep UV lithography of SU-8. The joints of the grippers were designed as flexure hinges; they are actuated by shape memory alloys. A double cantilever beam structure for stress concentration was integrated into the gripper jaws, and piezoresistors were implanted in the areas with the highest mechanical stress (Figure 6.11b). The tensile and compression stresses during gripping cause a change of the resistance of the piezoresistors. A noticeable nonlinearity of the sensor's output presumably results from the wires to the piezoresistors. As the wires are made of a sputtered gold layer, they deflect together with the flexures and thus change their resistivity analogously to the piezoresistors due to mechanical stress.

a   b

Figure 6.11. Force feedback using piezoresistors. a. Microgripper with two integrated force sensors; b. double cantilever beam with stress concentrating structure (from [33], courtesy of Shaker Verlag).

The goal of the work described in [43] was the development of grippers with integrated sensors for gripping and contact force measurement. The gripper is actuated by piezoelectric bimorphs. A parallel movement of the gripper jaws is guaranteed by specially designed flexures. Force-sensitive elements and optical fibers have been integrated into the end-effectors. If a contact force acts on the gripped object, it is being transmitted through static friction forces (between object and end-effector) to the contact-force-sensitive elements (Figure 6.12a). Gripping forces are transmitted in a similar way to the gripping-force-sensitive elements (Figure 6.12b). Both elements transform the forces into deflections, causing the free end of the optical fiber to move relative to the fixed one. Thus, the intensity of the light interfacing with the receiving fiber is modulated by the force, and the force can be determined by measuring the change in intensity. Contact forces can only be measured correctly when the gripper's main axis is perpendicular to the working surface.

Micromachined cantilevers with piezoresistors have been used at the Physikalisch-Technische Bundesanstalt, Germany, for two different applications: (i) as a sensor in a coordinate measurement machine [44], and (ii) as a portable gage, *e.g.*, for the calibration of nanoindenters [45, 46]. For this purpose, both commercially available AFM cantilevers (Nanosensors GmbH, Germany; now: NanoWorld AG, Switzerland) and proprietary developments have been investigated and characterized. These investigations showed their principle usability as force sensors for the measurement of forces up to approximately 3 mN. The development of the portable gage showed that micro forces can be measured with high sensitivity and good linearity using micromachined cantilevers with integrated piezoresistors.

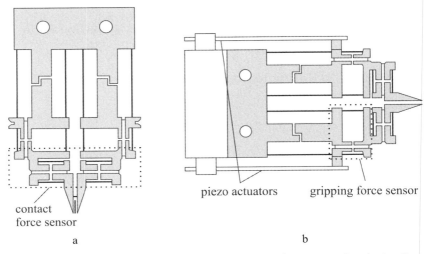

contact
force sensor

a

piezo actuators   gripping force sensor

b

Figure 6.12. Force sensors based on optical fibers. a. Contact force sensor; b. gripping force sensor.

6.3.3.2 Microrobots Outside the Scanning Electron Microscope

Microrobots are used for various applications from life science, microelectronics, materials research and, MEMS. Either commercially available, complete solutions are applied – *e.g.*, Kleindiek Nanotechnik, Klocke Nanotechnik, Zyvex, Sutter Instruments, Eppendorf, *etc.* – or custom-made solutions are developed, which are often based on commercial nanopositioning stages from, *e.g.*, Physik Instrumente, SmarAct, attocube, Nanomotion, *etc.* For these systems, a few microgrippers are available on the market. However, gripping or contact force measurement has not been integrated into any of them.

By contrast, there are several research publications describing the development of force sensors for micro- and nanohandling (Table 6.5). Three different types of force sensing have been implemented: (i) discrete strain gages, (ii) integrated piezoresistive strain gages, and (iii) piezoelectric force sensors based on PVDF films.

Discrete strain gages. For the measurement of gripping and contact forces in microrobotics, mainly discrete strain gages have been used so far. These are commercially available metal-foil or semiconductor strain gages which can directly be glued onto microgrippers and end-effectors.

Three publications describe the development of microhandling systems working under a light microscope [47–49]. Each of these systems uses piezoelectrically actuated microgrippers with integrated gripping force feedback based on semiconductor strain gages. The ideal positions of the strain gages, *i.e.*, the areas with the highest mechanical stress, have been determined using finite-element methods.

The measurement range is within millinewtons. Information about the achievable resolution is not available.

A different approach for microhandling was pursued at the University of California [53]. The "gripper" is made of two probes, each having one degree of freedom. These probes are mounted perpendicularly to each other. Three semiconductor strain gages have been glued onto them, two for measuring the position and one for measuring the force. With this setup, it is possible to grip parts smaller than 1 mm and rotate them around their vertical axis. At the same time, the gripping force can be measured with sub-millinewton resolution.

Table 6.5. Force feedback for microrobots operating outside the SEM

Sensor type	Measurement principle	Maximum force	Resolution	Note	Reference
gripping force	strain gage	22 mN			[47]
gripping force	strain gage	23 mN			[48]
gripping force	strain gage	69 mN			[49]
gripping force	piezoresistive	500 µN	0.2 µN		[50]
gripping force	piezoresistive	600 µN	2 nN		[51]
gripping force	piezoelectric	6 mN	several 10 µN		[52]
gripping force	strain gage	> 4 mN	< 0.4 mN		[53]
contact force	piezoelectric		µN range		[2]

Integrated strain gages. The smallest commercially available discrete strain gages have lateral dimensions around 1 mm. With that size, they are often already too large for integration into ever smaller grippers and end-effectors. Thus, piezoresistive strain gages are more and more directly integrated into areas with high mechanical stress by ion implantation. This work was pioneered by M. Tortonese, who developed the first piezoresistive AFM cantilever at IBM Almaden Research Center [54].

One of the first examples of the transfer of this technology to microrobotics is the tactile, silicon-based microgripper developed at ETH Zurich [50] (Figure 6.13a). The gripper is made of two fingers – a bimorph actuator and a piezoresistive force sensor – which both serve as end-effectors at the same time (Figure 6.13b). The length of the fingers is 1.5 mm, their width between 80 µm and 240 µm, and their thickness is 12 µm. Objects with a size up to 400 µm can be gripped. Two different types of this gripper were developed. One version had four piezoresistors connected into a full Wheatstone bridge and one smaller version had only two piezoresistors. An analog PI controller was implemented to allow for force-controlled gripping of microparts.

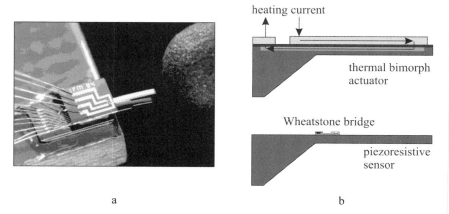

a   b

Figure 6.13. Tactile microgripper. a. Comparision to a match (from [50], courtesy of Georg Greitmann); b. cross.section of sensor and actuator finger

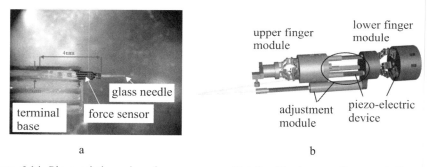

a   b

Figure 6.14. Piezoresistive micro force sensor. a. Detail with glass needle as endeffector; b. integrated into a two-fingered robot hand (from [51], © 2001 IEEE).

A similar force sensor was developed at the Mechanical Engineering Laboratory, Japan, in co-operation with Olympus [51]. By using semiconductor technologies and micromachining, two strain gages were integrated into a silicon cantilever (Figure 6.14a). Additionally, the cantilever was etched down to a thin membrane to amplify the mechanical stress. The strain gages have an opposite gage factor to increase the resolution of the sensor. Due to the lack of a suitable calibration, the resolution could only be calculated theoretically to be 2 nN. A thin glass needle was glued to the end of the force sensor as an end-effector. Two sensor/end-effector devices were integrated into a two-fingered micromanipulator (Figure 6.14b). By using two sensors and mounting them rotated by 90°, a force feedback in two axes could be realized.

Piezoelectric micro force sensors. The use of piezoelectric force sensors based on films of polyvinylidene fluoride (PVDF) is a new approach in microrobotics research. PVDF is a semi-crystalline polymer consisting of monomer chains of $(-CH_2-CF_2-)_n$. It has strong piezoelectric properties due to a large electronegativity of the fluorine atoms in comparison to the carbon atoms. PVDF creates electrical charges like other piezoelectric materials only when the film is subjected to mechanical, acoustic, or thermal stress.

At the Korea Institute for Science and Technology (KAIST), a force sensor based on PVDF has been integrated into a microgripper [52]. A commercially available film with nickel electrodes on both sides has been used (Measurement Specialties Inc., USA). The film was cut into 2×6 mm^2 slices and coated with a thin parylene layer for protection, as well as to enhance the mechanical stiffness. The resulting thickness of the sensor was 36 μm. The resolution was specified to be several tens of micronewtons (the load cell used for calibration had a resolution of 50 μN).

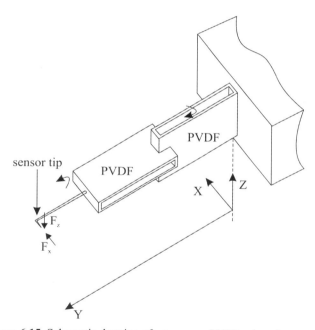

Figure 6.15. Schematic drawing of a two-axes PVDF micro force sensor

A commercial five-axes IC wafer probe was used as a micromanipulation system at Michigan State University, USA, to automate the assembly of micromirrors. The mirrors usually lie on the substrate's surface after production and have to be brought into an upright position where they are locked mechanically. A two-axes PVDF-based force sensor was developed to measure the forces during the lift process [2]. For each axis of the sensors, two PVDF films are connected in parallel, and the two axes are assembled in series and rotated by 90° (Figure 6.15). The entire sensor has a length of more than 4 cm and a height of more than 1 cm. It was calibrated by measuring its deflection under a light microscope; the stiffness required to calculate the force was determined from the Young's modulus and the geometry of the sensor.

6.3.3.3 Microrobots Inside the Scanning Electron Microscope

More and more applications in micro- and especially nanotechnology hit the light-optical microscope's resolution limit of about 400 nm. Smaller objects, such as carbon nanotubes or nanowires with dimensions of several hundreds of nanometers down to only a few nanometers, cannot be resolved individually with a light-optical microscope. The scanning electron microscope has been used for several decades now as a flexible microscope and analysis instrument. It combines a high resolution – down to about 1 nm for field-emission instruments – with a high depth of field. As a result, handling and characterization of parts is increasingly performed inside the SEM. Table 6.6 gives an overview of nanohandling systems with force feedback working in an SEM.

Table 6.6. Overview of microrobots inside the SEM with force feedback

Sensor type	Measurement principle	Maximum force	Resolution	Note	Reference
contact force	strain gage	0.5–3.5 N	130–180 µN	3 axes	[55]
contact force	strain gage	20 mN	14 µN	1 axis	[56]
contact force	strain gage	40 mN	309–391 µN	3 axes	[57]

The first research activities within this field were carried out at the University of Tokyo [18], where, for a duration of more than ten years, the so-called Nano Manufacturing World was developed. A microassembly system consisting of a miniaturized robot arm and a positioning stage with a total of twelve translatory and rotatory axes for coarse and fine positioning was installed in a stereo SEM. Forces occurring during assembly are measured with a three-axes strain gage contact force sensor based on parallel plates. The forces are transformed into acoustic signals; additionally, a haptic device provides force feedback for the operator [55].

A similar nanohandling system for automated handling of microobjects based on visual and force feedback is described in [56]. Additionally to the processing of the images from the SEM and a light-optical microscope, a force sensor has been installed inside the SEM. This sensor measures the normal force acting on the work plate. It is made of a cantilever (dimensions: $10 \times 8 \times 0.05$ mm^3) onto which four strain gages have been glued. The forces during assembly have values of about 1 mN and can be sufficiently resolved with this sensor.

At Tokai University another microhandling system has been installed inside an SEM in cooperation with Hitachi [57]. The system is made up of two identical micromanipulators each with three Cartesian axes for coarse and fine positioning

(Figure 6.16a). The sample is mounted on a stage with three Cartesian and one rotary axes. A three-axes contact force sensor was mounted on each of the micromanipulators (Figure 6.16b). The sensors are based on semiconductor and metal-foil strain gages, respectively. With these sensors, the contact forces between manipulator and environment or the manipulated part can be measured. The sensor information is transmitted to the operator via a haptic user interface.

 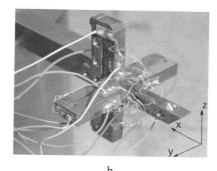

a   b

Figure 6.16. Microhandling system at Tokai University. a. Positioning stages integrated into an SEM; b. three-axes strain-gage-based contact force sensor (from [57], © 2003 IEEE).

Other microrobot-based handling systems inside the SEM have been developed, especially for the characterization of nanoobjects like carbon nanotubes [58-60]. Furthermore, different nanopositioning and nanohandling systems have been commercially available for several years. All these systems have in common piezoelectric actuators which can be controlled via teleoperation. Some have integrated position sensors; however, force feedback is not integrated into any of these devices.

6.3.3.4 Mobile Microrobots

For several years now, mobile microrobots have been an important topic of world-wide research activities. Their advantages compared to stationary microrobots described above are a greater flexibility, a larger working space and the possibility of using several cooperating robots to accomplish a task. Mobile microrobots have been developed, for instance, within European research projects (MiCRoN[3], I-Swarm[4]) as well as at the NanoRobotics Laboratory of the École Polytechnique de Montréal, Canada [61]. These projects focused on the development of miniaturized, autonomous robots with integrated electronics and wireless communication as well as the control of cooperating robots and robot swarms. The

---

[3] EU FP5 IST Project MiCRoN (Miniature co-operative robots advancing towards the nano-range), 2002–2005.
[4] EU FP5 IST FET-open Project I-Swarm (Intelligent Small World Autonomous Robots for Micro-manipulation), 2004–2008.

microrobots were not designed to work inside a scanning electron microscope. Micro force sensors have not been developed or integrated within these projects.

Mobile microrobots working inside an SEM have been developed within two European research projects (MINIMAN[5], ROBOSEM[6]) and at the Aoyama Laboratory in Tokyo [62]. Within MINIMAN, a prototype of a force-sensing system based on strain gages was realized and its repeatability, precision, and linearity quantified [63]. A simple method that links system parameters and system performance was presented and verified concerning cantilever-type end-effectors for microgrippers. The resolution obtained in an overall measurement range of 10 mN was 150 µN.

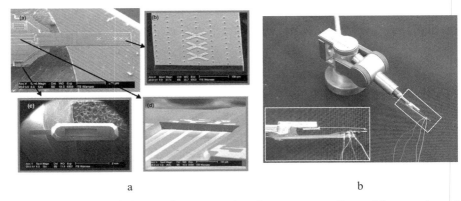

a   b

Figure 6.17. Piezoresistive cantilever-type micro force sensor. a. Top and bottom view of the sensor with piezoresistors, structured gripping area and cavity for mounting; b. sensor integrated into a microgripper, which is mounted on a micromanipulator (reprinted from [64], © 2004, with permission from Elsevier).

Within ROBOSEM, a set of sensors for the measurement of gripping forces in the range of 10^{-7} N to 10^{-1} N has been developed [64]. The force sensor consists of a silicon cantilever beam with force-sensing elements located at its support (Figure 6.17a). The piezoresistive circuit contains four piezoresistors in a full Wheatstone bridge setup with additional compensation of temperature and offset voltage. The cantilever beam was used as an end-effector of the gripping device. Sensors with different cantilever lengths ranging from 1 mm to 3 mm and thicknesses of up to 100µm were developed to cover a wide range of gripping forces. These sensors can easily be integrated into different types of microgrippers and end-effectors (Figure 6.17b). Tools equipped with these force sensors can be used for handling micro-objects under ambient and vacuum conditions.

---

[5] EU FP4 Esprit Project MINIMAN (Miniaturised Robot for Micro Manipulation), 1998–2001.
[6] EU FP5 Growth Project ROBOSEM (Development of a Smart Nanorobot for Sensor-based Handling in a Scanning Electron Microscope), 2001–2005.

6.3.4 AFM-based Nanohandling Systems

The manipulation of nanoscale objects and of matter at even the atomic level is a comparatively new research field. It has been opened up by the invention of the scanning tunneling microscope by Binning and Rohrer in the early eighties [65] and the subsequent invention of the atomic force microscope by Binning *et al.* in 1986 [66]. Even though both types of microscopes were designed for investigating surfaces of electrically conductive materials and insulators, respectively, it was soon discovered that they could also be used to modify these surfaces and manipulate objects as small as atoms (Chapter 1). Eigler and Schweizer were the first to manipulate atoms in an STM on a single-crystal nickel surface when they formed the IBM logo out of 35 xenon atoms [67].

Force measurement with nanonewton resolution and below is inherently integrated into AFMs. Although they were originally only intended for imaging, they are nowadays also used for nanohandling with force feedback. As described in the following sections, the design of standard AFMs limits their usability as nanohandling systems, but recent developments are likely to overcome these limitations and convert AFMs into flexible instruments for imaging and manipulation at the nanoscale.

6.3.4.1 Commercial and Custom-made AFMs for Nanohandling
Within the last few years, several nanohandling systems either based on commercial AFMs or on custom-made AFM instruments, have been developed [68–73]. At the Korea Institute of Science and Technology, an AFM has been built especially for nanohandling [74]. It is made up of a nanomanipulator with three degrees of freedom and piezoresistive AFM cantilevers.

a   b

Figure 6.18. Force-controlled pushing using an AFM cantilever. a. Side view of the experimental setup with AFM cantilever, workpiece, and reference object; b. top view of the configuration at the start (left) and end of the alignment process (right) (from [75], courtesy of SPIE).

Zesch *et al.* used a piezoresistive AFM cantilever for force-controlled pushing of sub-millimeter-sized silicon parts on a silicon substrate [75]. The substrate can be positioned using a precision table. The cantilever is mounted on a piezoelectric positioning stage (Figure 6.18a). With this setup, microobjects can be aligned laterally relative to each other (Figure 6.18b). By measuring the contact forces between object and AFM cantilever it is possible to draw conclusions as to whether the object is moved, if it got lost, or if it hit an obstacle. The measured forces are typically between of 0.8 μN and 3.5 μN during pushing and twice as high when static friction has to be overcome. The biggest problems of this method are instabilities during pushing, as well as non-predictable movements or jumping of the microobject.

The basic element of the nanohandling system developed at the University of Minnesota, USA, is an IC probing station into which a three-axes nanopositioning unit was integrated [76, 77]. The probing station was combined with an AFM head, which measures the deflection of an AFM cantilever using a laser beam. This system was mainly used to develop combined force/vision control and to investigate adhesion forces in nanohandling.

6.3.4.2 AFMs Combined with Haptic Devices and Virtual Reality

As identified in Chapter 1, the main drawback of standard AFMs is the lack of visual feedback of the manipulation process in real-time: each operation has to be designed offline based on a static AFM image and transferred to the AFM to carry out the task in open-loop. The success of the operation has subsequently to be verified by a new image scan. Obviously, the cycle of scan–design–manipulate–scan is very time consuming, because it usually takes several minutes to obtain an AFM image.

To overcome this problem, AFMs combined with haptic devices (e.g., NanoMan from Veeco, USA; NanoManipulator™ from 3rdTech, USA; NanoFeel™300 manipulator from NanoFeel, Switzerland; Omega Haptic Device from Force Dimension and Nanonis, Switzerland), and virtual reality (VR) interfaces to facilitate feedback during nanomanipulation, have been developed, e.g., [78, 79]. Haptic devices provide the operator with real-time force feedback but they cannot compensate for visual feedback. Virtual reality interfaces can display a static, virtual environment, and a dynamic position of the tip. However, they do not reflect any environmental changes in real-time, and the operator is still "blind", so to speak. Compared to standard AFMs, both techniques represent a significant improvement for the user, but they still do not eliminate the need for an SEM to visualize manipulation and processing in the nanoworld in real-time.

6.3.4.3 AFMs Integrated into Scanning Electron Microscopes

As mentioned in Chapter 1, AFMs integrated into SEMs combine the advantages of both tools into one instrument, e.g., allowing the combination of nanoscale chemistry, crystallography imaging via electron-matter interactions with information from tip-sample interactions like topography, or magnetic/electrostatic force imaging. Several of such hybrid AFM/SEM systems have been developed in the research [80–85]. Three commercial AFMs for integration into an SEM are or were available on the market. The Observer from Topometrix (now Veeco Instruments, USA) has been reported on twice in the literature [86, 87], but is not sold anymore. Nanonics Imaging Ltd., Israel, commercializes a modified version of their AFM for use inside an SEM. The third vendor is Kleindiek Nanotechnik

GmbH, Germany, who offer an AFM for SEMs based on one of their micromanipulators.

State-of-the-art is the manual positioning of an AFM tip with conventional atomic force microscopes integrated into large-chamber SEMs for topography imaging as well as crystallography, chemistry, and surface morphology analysis. Only Williams *et al.* [86] have reported on the use of the Topometrix Observer for the controlled placement of an individual carbon nanotube on a MEMS structure in an SEM. None of the above-mentioned systems was designed for manipulation or assembly.

6.4 Conclusions

The measurement of micro- and nanonewton forces is one of the central challenges in nanohandling. This is shown by the large number of publications and amount of research in that field. However, from the state-of-the-art it can be seen that no micro force sensor exists so far – neither in research nor on the market – that can fulfill the many requirements of microrobotics. Often, the sensors are just too large to be integrated into microgrippers or end-effectors. Mostly, their performance in terms of resolution or accuracy is poor. Force feedback solutions, which have been developed for industrial robotics, do not aim at nanohandling applications and are not suitable due to performance limits.

One of today's most promising solutions for force measurement in microrobotics is the use of piezoresistive sensors. They can be manufactured with the same semiconductor processes as microgrippers and thus be directly integrated into the grippers, requiring no additional assembly. Another possibility is to implant piezoresistors into cantilever-like end-effectors, which can then be used as micro force sensors and, at the same time, modular and easily exchangeable gripper jaws. Also, the results of force sensing based on capacitive sensors and image processing (Chapter 4) indicate the potential of these techniques for future developments in force feedback at the nanoscale.

AFMs can be used as nanohandling instruments with an already integrated, high-resolution force feedback. Combined with novel tools, such as dedicated micro-/nanogrippers, and integrated into scanning electron microscopes providing almost real-time vision feedback, AFMs promise to become a key instrument for the reliable handling and characterization of nanoscale objects.

6.5 References

[1]   Pfeifer, T., Driessen, S. & Dussler, G. 2004, 'Process observation for the assembly of hybrid micro systems', *Microsystem Technologies*, vol. 10, pp. 211–218.

[2]   Shen, Y., Xi, N., Song, B., Li, W. & Pomeroy, C. 2006, 'Networked human/robot cooperative environment for tele-assembly of MEMS devices', *Journal of Micromechatronics*, vol. 3, no. 3, pp. 239–266.

[3]   Benmayor, L. 2000, 'Dimensional analysis and similitude in microsystem design and assembly', Ph.D. thesis, École Polytechnique Fédérale de Lausanne, Lausanne.

[4]   Yang, G., Gaines, J. & Nelson, B. 2003, 'A supervisory wafer-Level 3D microassembly system for hybrid MEMS fabrication', *Journal of Intelligent and Robotic Systems*, vol. 37, no. 1, pp. 43–68.

[5]   Tichem, M., Lang, D. & Karpuschewski, B. 2004, 'A classification scheme for quantitative analysis of micro-grip principles', *Assembly Automation*, vol. 24, no. 1, pp. 88–93.

[6]   Fukuda, T., Arai, F. & Dong, L. 2001, 'Nano robotic world – from micro to nano', *Proceedings of the 2001 IEEE Conference on Robotics and Automation*, Seoul, Korea.

[7]   Williams, E. & Newell, D. 2003, 'The electronic kilogram', Technical report, National Institute of Standards and Technology (NIST). http://www.eeel.nist.gov/.

[8]   Doebelin, E. O. 1966, *Measurement Systems: Applications and Design*, McGraw-Hill, New York.

[9]   Nguyen, V.-D. 1988, 'Constructing force-closure grasps', *International Journal of Robotics Research*, vol. 7, no. 3, pp. 3–16.

[10] Fearing, R. S. 1995, 'Survey of sticking effects for micro-parts handling', *Proceedings of the IEEE Int. Conf. Robotics and Intelligent Systems (IROS '95)*, Pittsburgh, USA, pp. 212–217.

[11] Bhushan, B. (ed.) 2004, *Handbook of Nanotechnology*, Springer, chapter Part C: 'Nanotribology and nanomechanics', pp. 544–603.

[12] Arai, F., Andou, D. & Fukuda, T. 1996, 'Adhesion forces reduction for micro manipulation based on microphysics', *Proceedings of the International Workshop on Micro Electro Mechanical Systems*, San Diego, USA, pp. 354–359.

[13] Zhou, Y. & Nelson, B. 1998, 'Adhesion force modeling and measurement for micromanipulation', *Proceedings of SPIE 3519*, Microrobotics and Micromanipulation, pp. 169–180.

[14] Böhringer, K., Fearing, R. & Goldberg, K. 1999, *Handbook of Industrial Robotics*, 2nd edn, Wiley & Sons, chapter 'Microassembly', pp. 1045–1066.

[15] Fatikow, S. 2000, *Mikroroboter und Mikromontage*, Teubner.

[16] Sun, Y. & Nelson, B. 2001, 'Microrobotic cell injection', *Proceedings of the 2001 IEEE International Conference on Robotics and Automation*, Seoul, Korea, pp. 620–625.

[17] Knoll, M. & Ruska, E. 1932, 'Das Elektronenmikroskop', *Z. Physik*, vol. 78, pp. 318–339.

[18] Hatamura, Y. & Morishita, H. 1990, 'Direct coupling system between nanometer world and human world', *Proceedings of IEEE International Conference on Micro Electro Mechanical Systems*, Napa Valley, California, USA, pp. 203–208.

[19] Siciliano, B. & Villani, L. 1999, *Robot force control*, Kluwer Academic Publishers.

[20] Yoshikawa, T. 2000, 'Force control of robot manipulators', *Proceedings of the IEEE International Conference on Robotics and Automation*, San Francisco, USA, pp. 220–226.

[21] Horie, M., Funabashi, H. & Ikegami, K. 1995, 'A study on micro force sensors for microhandling systems', *Microsystem Technologies*, vol. 1, pp. 105–110.

[22] Jin, W. & Mote, C. 1998, 'A six-component silicon micro force sensor', *Sensors and Actuators A*, vol. 65, pp. 109–115.

[23] Arai, F., Sugiyama, T., Fukuda, T., Iwata, H. & Itoigawa, K. 1999, 'Micro tri-axial force sensor for 3D bio-micromanipulation', *Proceedings of the 1999 IEEE International Conference on Robotics and Automation*, Detroit, USA, pp. 2744–2749.

[24] Kim, J., Park, Y. & Kang, D. 2002, 'Design and fabrication of a three-component force sensor using micromachining technology', *VDI-Berichte*, vol. 1685, Düsseldorf, pp. 9–15.

[25] Bütefisch, S., Büttgenbach, S., Kleine-Besten, T. & Brand, U. 2001, 'Micromachined three-axial tactile force sensor for micromaterial characterisation', *Microsystem Technologies*, vol. 7, no. 4, pp. 171–174.

[26] Breguet, J.-M., Henein, S., Mericio, R. & Clavel, R. 1997, 'Monolithic piezoceramic flexible structures for micromanipulation', *Proceedings of the 9th International Precision Engineering Seminar*, vol. 2, Braunschweig, pp. 397–400.

[27] Sun, Y., Nelson, B., Potasek, D. & Enikov, E. 2002, 'A bulk microfabricated multi-axis capacitive cellular force sensor using transverse comb drives', *Journal of Micromechatronics and Microengineering*, vol. 12, no. 6, pp. 832–840.

[28] Lin, L., Pisano, A. & Howe, R. 1997, 'A micro strain gauge with mechanical amplifier', *Journal of Microelectromechanical Systems*, vol. 6, no. 4, pp. 313–321.

[29] Mazza, E., Danuser, G. & Dual, J. 1996, 'Light optical deformation measurement in microbars with nanometer resolution', *Microsystem Technologies*, vol. 2, pp. 83–91.

[30] Boukallel, M., Piat, E. & Abadie, J. 2003, 'Micromanipulation tasks using passive levitated force sensing manipulator', *Proceedings of the 2003 IEEE/RSJ Intl. Conference on Intelligent Robots and Systems*, Las Vegas, USA, pp. 529–534.

[31] Greminger, M. & Nelson, B. 2004, 'Vision-based force measurement', *IEEE Transactions on Pattern Analysis and Machine Intelligence*, vol. 26, no. 3, pp. 290–298.

[32] Yang, S. & Saif, T. 2005, 'Micromachined force sensors for the study of cell mechanics', *Review of Scientific Instruments*, vol. 76, pp. 44301–44309.

[33] Bütefisch, S. 2003, 'Entwicklung von Greifern für die automatisierte Montage hybrider Mikrosysteme', Ph.D. thesis, Technische Universität Braunschweig.

[34] Yang, G. & Nelson, B. 2003, 'Micromanipulation contact transition control by selective focusing and microforce control', *Proceedings of 2003 IEEE Int. Conf. of Robotics and Automation*, Taipei, Taiwan, pp. 200–206.

[35] Seki, H. 1992, 'Piezoelectric bimorph microgripper capable of force sensing and compliance control', *Transactions of the Japan/USA Symposium on Flexible Automation*, vol. 1, pp. 707–712.

[36] Henschke, F. 1994, 'Miniaturgreifer und montagegerechtes Konstruieren in der Mikromechanik', Fortschritt-Berichte vdi: Reihe 1, Konstruktionstechnik, Maschinenelemente ; 242, Technische Hochschule Darmstadt.

[37] Chonan, S., Jiang, Z. W. & Koseki, M. 1996, 'Soft-handling gripper driven by piezoceramic bimorph strips', *Smart Mater. Struct.*, vol. 5, pp. 407–414.

[38] Goldfarb, M. & Celanovic, N. 1999, 'A flexure-based gripper for small scale manipulation', *Robotica*, vol. 17, pp. 181–187.

[39] Arai, F., Andou, D., Nonoda, Y., Fukuda, T., Iwata, H. & Itoigawa, K. 1998, 'Integrated microendeffector for micromanipulation', *Transactions on Mechatronics*, vol. 3, no. 1, pp. 17–23.

[40] Qiao, F. 2003, 'Biologisch inspirierte mikrotechnische Werkzeuge für die Mikromontage und die minimal-invasive Chirurgie', Ph.D. thesis, Technische Universität Ilmenau, Ilmenau.

[41] Bröcher, B. 2000, 'Faseroptische Sensoren zur Prozeßüberwachung in der Mikrosystemtechnik', Berichte aus der Produktionstechnik, RWTH Aachen.

[42] Petersen, B. 2003, 'Flexible Handhabungstechnik für die automatisierte Mikromontage', Ph.D. thesis, RWTH Aachen, Aachen.

[43] Ma, K. 1999, 'Geif- und Fügekraftmessung mit faseroptischen Sensoren in der Mikromontage', Ph.D. thesis, Technische Universität Braunschweig, Essen.

[44] Hoffmann, W., Loheide, S., Kleine-Besten, T., Brand, U. & Schlachetzki, A. 2000, 'Methods of characterising micro mechanical beams and its calibration for the application in micro force measurement systems', *Proceedings of MicroTEC 2000*, Hannover, pp. 819–823.

[45] Doering, L., Brand, U. & Frühauf, J. 2002, 'Micro force transfer standards', *VDI-Berichte*, vol. 1685, Düsseldorf, pp. 83–88.

[46] Behrens, I., Doering, L. & Peiner, E. 2003, 'Piezoresistive cantilever as portable micro force calibration standard', *Journal of Micromechanics and Microengineering*, vol. 13, no. 4, pp. S171–S177.

[47] Carrozza, M., Eisinberg, A., Menciassi, A., Campolo, D., Micera, S. & Dario, P. 2000, 'Towards a force-controlled microgripper for assembling biomedical microdevices', *Journal of Micromechatronics and Microengineering*, vol. 10, no. 2, pp. 271–276.

[48] Kim, S., Kim, K., Shim, J., Kim, B., Kim, D.-H. & Chung, C. 2002, 'Position and force control of a sensorized microgripper', *Proceedings of the International Conference on Control, Automation and Systems*, Jeonbuk, Korea, pp. 319–322.

[49] Albut, A., Zhou, Q., del Corral, C. & Koivo, H. 2003, 'Development of a flexible force-controlled piezo-bimorph microgripping system', *Proceedings of the 2nd World Microtechnologies Congress (MICRO.tec)*, München, pp. 507–512.

[50] Greitmann, G. 1998, 'Micromechanical Tactile Gripper System for Microassembly', Ph.D. thesis, ETH Zurich.

[51] Tanikawa, T., Kawai, M., Koyachi, N., Arai, T., Ide, T., Kaneko, S., Ohta, R. & Hirose, T. 2001, 'Force control system for autonomous micro manipulation', *Proceedings of the 2001 IEEE International Conference on Robotics & Automation*, Soul, Korea, pp. 610–615.

[52] Kim, D.-H., Kim, B. & Kang, H. 2004, 'Development of a piezoelectric polymer-based sensorized microgripper for microassembly and micromanipulation', *Microsystem Technologies*, vol. 10, pp. 275–280.

[53] Thompson, J. & Fearing, R. 2001, 'Automating microassembly with ortho-tweezers and force sensing', *Proceedings of the 2001 IEEE/RSJ International Conference on Robots and Systems*, vol. 3, Maui, USA, pp. 1327–1334.

[54] Tortonese, M., Yamada, H., Barrett, R. C. & Quate, C. F. 1991, 'Atomic force microscopy using a piezoresistive cantilever', *Proceedings of the 6th International Conference on Solid-State Sensors and Actuators*, vol. 91, San Francisco, USA, pp. 448–451.

[55] Mitsuishi, M., Sugita, N., Nagao, T. & Hatamura, Y. 1996, 'A tele-micro machining system with operational environment transmission under a stereo-SEM', *Proceedings of the 1996 IEEE International Conference on Robotics and Automation*, Minneapolis, USA, pp. 2194–2201.

[56] Kasaya, T., Miyazaki, H., Saito, S. & Sato, T. 1999, 'Micro object handling under SEM by vision-based automatic control', *Proceedings of the 1999 IEEE International Conference on Robotics and Automation*, Detroit, USA, pp. 2189–2196.

[57] Yamamoto, Y., Konishi, R., Negishi, Y. & Kawakami, T. 2003, 'Prototyping Ubiquitous Micro-Manipulation System', *Proceedings of the 2003 IEEE/ASME International Conference on Advanced Intelligent Mechatronics*, Kobe, Japan, pp. 709–714.

[58] Yu, M., Dyer, M., Skidmore, G., Rohrs, H., Lu, X., Ausman, K., von Ehr, J. & Ruoff, R. 1999, 'Three-dimensional manipulation of carbon nanotubes under a scanning electron microscope', *Nanotechnology*, vol. 10, pp. 244–252.

[59] Akita, S., Nishijima, H., Nakayama, Y., Tokumasu, F. & Takeyasu, K. 1999, 'Carbon nanotube tips for a scanning probe microscope: their fabrication and properties', *Journal of Applied Physics*, vol. 32, pp. 1044–1048.

[60] Dong, L., Arai, F. & Fukuda, T. 2001, '3D Nanorobotic manipulation of multi-walled carbon nanotubes', *Proceedings of the 2001 IEEE International Conference on Robotics and Automation*, Seoul, Korea, pp. 632–637.

[61] Martel, S. & Hunter, I. 2002, 'Nanofactories based on a fleet of scientific instruments configured as miniature autonomous robots', *Journal of Micromechatronics*, vol. 2, no. 3-4, pp. 201–214.

[62] Aoyama, H. & Fuchiwaki, O. 2001, 'Flexible micro-processing by multible micro robots in SEM', *Proceedings of the 2001 IEEE International Conference on Robotics and Automation*, Seoul, Korea, pp. 3429–3434.

[63] Fahlbusch, S., Fatikow, S. & Santa, K. 1998, 'Force sensing in microrobotic systems: an overview', *Advances in Manufacturing: Decision, Control and Information Technology*, Springer, pp. 233–244.

[64] Domanski, K., Janus, P., Grabiec, P., Perez, R., Chaillet, N., Fahlbusch, S., Sill, A. & Fatikow, S. 2005, 'Design, fabrication and characterization of force sensors for nanorobot', *Microelectronic Engineering*, vol. 78-79, pp. 171–177.

[65] Binnig, G. & Rohrer, H. 1982, 'Scanning tunneling microscopy', *Helvetica Physica Acta*, vol. 55, pp. 726–735.

[66] Binning, G., Quate, C. & Gerber, C. 1986, 'Atomic Force Microscope', *Phys. Rev. Lett.*, vol. 56, pp. 930–933.

[67] Eigler, D. & Schweizer, E. 1990, 'Positioning single atoms with a scanning tunnelling microscope', *Nature*, vol. 344, pp. 524–526.

[68] Junno, T., Deppert, K., Montelius, L. & Samuelson, L. 1995, 'Controlled manipulation of nanoparticles with an atomic force microscope', *Applied Physics Letters*, vol. 66, no. 26, pp. 3627–3629.

[69] Schaefer, D., Reifenberger, R., Patil, A. & Andres, R. 1995, 'Fabrication of two-dimensional arrays of nanometer-size clusters with the scanning probe microscope', *Applied Physics Letters*, vol. 66, pp. 1012–1014.

[70] Requicha, A., Baur, C., Bugacov, A., Gazen, B., Koel, B., Madhukar, A., Ramachandran, T., Resch, R. & Will, P. 1998, 'Nanorobotic assembly of two-dimensional structures', *Proceedings of the IEEE International Conference on Robotics and Automation*, Leuven, Belgium, pp. 3368–3374.

[71] Hansen, L. T., Kuehle, A., Sørensenz, A., Bohr, J. & Lindelofy, P. 1998, 'A technique for positioning nanoparticles using an atomic force microscope', *Nanotechnology*, vol. 9, pp. 337–342.

[72] Li, G., Xi, N., Yu, M. & Fung, W. 2003, '3-D nanomanipulation using scanning probe microscopy', *Proceedings of the IEEE International Conferene on Robotics and Automation*, Taipei, Taiwan, pp. 3642–3647.

[73] Venture, G., Haliyo, D. S., Micaelli, A. & Regnier, S. 2006, 'Force-feedback coupling for micro-handling applications', *Micromechatronics*, vol. 3, no. 3-4, pp. 307–327.

[74] Kim, D.-H., Kim, K. & Hong, J. 2002, 'Implementation of a self-sensing MEMS cantilever for nanomanipulation', *Proceedings of the 4th Korean MEMS Conference*, Kyeongju, Korea, pp. 120–125.

[75] Zesch, W. & Fearing, R. 1998, *in* A. Sulzmann & B. Nelson (eds), *Microrobotics and Micromanipulation*, Bellingham, Washington, USA, vol. 3519, Boston, USA, pp. 148–156.

[76] Nelson, B., Zhou, Y. & Vikramadity, B. 1998, 'Sensor-based microassembly of Hybrid MEMS Devices', *IEEE Control Systems*, vol. 18, no. 6, pp. 35–45.

[77] Zhou, Y., Nelson, B. & Vikramaditya, B. 2000, 'Integrating optical force sensing with visual servoing for microassembly', *Journal of Intelligent and Robotic Systems*, vol. 28, no. 3, pp. 259–276.

[78] Sitti, M. & Hashimoto, H. 1998, 'Tele-nanorobotics using scanning probe microscope', *Proceedings of the IEEE/RSJ International Conference on Intelligent Robots and Systems*, Victoria, Canada, pp. 1739–1746.

[79] Guthold, M., Falvo, M., Matthews, W., Paulson, S., Washburn, S., Erie, D., Superfine, R., Brooks, F. & Taylor, R. 2000, 'Controlled manipulation of molecular samples

with the nanomanipulator', *IEEE/ASME Transactions on Mechatronics*, vol. 5, no. 2, pp. 189–198.

[80] Kikukawa, A., Hosaka, S., Honda, Y., & Koyanagi, H. 1993, 'Magnetic force microscope combined with a scanning electron microscope', *Vacuum Science and Technology*, vol. 11, no. 6, pp. 3092–3098.

[81] Stahl, U., Yuan, C., de Lozanne, A. & Tortonese, M. 1994, 'Scanning probe microscope using piezoresistive cantilevers and combined with a scanning electron microscope', *Applied Physics Letters*, vol. 65, no. 22, pp. 2878–2880.

[82] Ermakov, A. & Garfunkel, E. 1994, 'A novel SPM/STM/SEM system', *Review of Scientific Instruments*, vol. 65, no. 9, pp. 2853–2854.

[83] Troyon, M., Lei, H., Wang, Z. & Shang, G. 1997, 'A scanning force microscope combined with a scanning electron microscope for multidimensional data analysis', *Micros. Microanal. Microstruct.*, vol. 8, pp. 393–402.

[84] Fung, C., Elhajj, I., Li, W. & Ning, X. 2002, 'A 2-D PVDF force sensing system for micro-manipulation and micro-assembly', *Proceedings of the IEEE International Conference on Robotics and Automation*, Arlington, USA, pp. 1489–1494.

[85] Joachimsthaler, I., Heiderhoff, R. & Balk, L. 2003, 'A universal scanning-probe-microscope based hybrid system', *Measurement Science and Technology*, vol. 14, pp. 87–96.

[86] Williams, P., Papadakis, S., Falvo, M., Patel, A., Sinclair, M., Seeger, A., Helser, A., Taylor, R., Washburn, S. & Superfine, R. 2002, 'Controlled placement of an individual carbon nanotube onto a microelectromechanical structure', *Applied Physics Letters*, vol. 80, no. 14, pp. 2574–2576.

[87] Postek, M., Ho, H. & Weese, H. 1997, 'Dimensional metrology at the nanometer level: combined SEM and PPM', *Proceedings of SPIE 3050*, Metrology, Inspection, and Process Control for Microlithography XI, pp. 250–263.

Characterization and Handling of Carbon Nanotubes

Volkmar Eichhorn[*] and Christian Stolle[**]

[*] Division of Microrobotics and Control Engineering,
Department of Computing Science,
University of Oldenburg, Germany
[**] Division of Microsystems Technology and Nanohandling,
Oldenburg Institute of Information Technology (OFFIS), Germany

7.1 Introduction

The progressive reduction of integrated circuits is reaching the limit of structures that are realizable by lithographic methods. Intel is announcing 45 nm transistors for the year 2007. Using electron beam lithography, structures down to 10 nm are producible. But this will be the absolute lower limit for lithographic structuring, so that new materials have to be developed for a further miniaturization of structures.

Since their discovery in 1991 by [1] of NEC laboratory in Japan, **carbon nanotubes** (CNTs) have turned out to be the most promising material for computer science and nanotechnology. Due to their remarkable structure and properties, CNTs allow for new and unique applications. After 15 years of intensive research and development, the commercialization has already begun in some domains. Application areas with the highest patent applications are field emitters, nanoelectronics, polymer composites, and new electrodes for batteries and fuel cells.

Great progress has also been achieved in characterizing carbon nanotubes [2]. In addition to different methods such as atomic force microscopy (AFM), scanning tunneling microscopy (STM), transmission electron microscopy (TEM), and scanning electron microscopy (SEM), many spectroscopic and diffractional methods are used for a comprehensive characterization of CNTs. For the meantime, first characterization devices based on fluorescence and absorption spectra are commercially available. However, a systematic and reproducible analysis of the mechanical and electrical properties of individual CNTs is still missing in many cases.

This chapter focuses on SEM-based **nanomanipulation** and the **characterization** of individual carbon nanotubes by a microrobot-based nanohandling station. Mechanical properties such as the Young's modulus, as well as electrical properties such as the conductivity, are analyzed. Therefore, different end-effectors like AFM cantilevers and electrothermal nanogrippers are used and combined with real-time SEM image processing, resulting in a flexible and multifunctional nanohandling station. At the same time, automation of particular nanohandling steps requires the development of suitable control system architecture.

7.2 Basics of Carbon Nanotubes

7.2.1 Structure and Architecture

Carbon nanotubes consist of a two-dimensional sheet of carbon atoms coiled up to a cylinder. The carbon atoms are hexagonally arranged and σ-bonded. CNTs appear in one of two configurations (Figure 7.1).

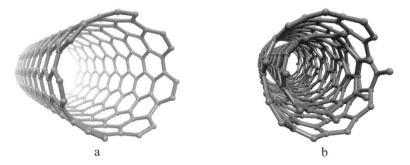

a   b

Figure 7.1. Schematic models for CNTs: a. SWCNTs; b. MWCNTs

Firstly, there are **single-wall carbon nanotubes** (SWCNTs) that are made up of one single carbon layer (Figure 7.1a). The typical diameter of SWCNTs is between 0.4 and 3 nm. SWCNT ropes can also be generated through a self-organizing process. Here, van der Waals forces keep the individual SWCNTs together, forming a triangular lattice with a lattice constant of 0.34 nm. Secondly, there are **multiwall carbon nanotubes** (MWCNTs) containing several coaxial carbon layers with a typical diameter of 2 up to 100 nm (Figure 7.1b). The length varies from several nanometers up to several millimeters for both configurations.

An SWCNT is uniquely characterized by its chiral vector (Figure 7.2):

$$\vec{C} = n\vec{a}_1 + m\vec{a}_2, \tag{7.1}$$

where $\vec{a}_1$ and $\vec{a}_2$ are the two unit vectors of the carbon sheet and (n, m) a set of two integers. SWCNTs are built by rolling up the carbon lattice such that the starting point A and the ending point B of the **chiral vector** $\vec{C}$ are superimposed. The chiral angle θ is defined with respect to the unit vector $\vec{a}_1$ and runs from $0°$ to $30°$. The lattice vector $\vec{T}$ is shown as well. The diameter D of an SWCNT is given by:

$$D = \frac{|\vec{C}|}{\pi} = \frac{a\sqrt{n^2 + m^2 + nm}}{\pi}, \qquad (7.2)$$

where a is the length of the unit vector and amounts to 0.249 nm.

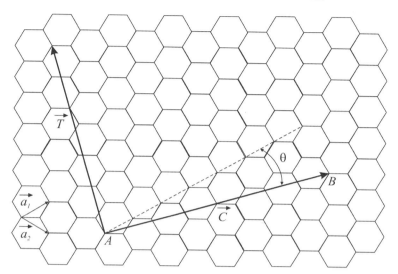

Figure 7.2. Definition of the chiral vector within a carbon lattice

SWCNTs with $(n, 0)$ are called **zigzag** nanotubes, while SWCNTs with (n, n) are **armchair** nanotubes. All other tubes with arbitrary pairs of integers (n, m) are referred to as **chiral** nanotubes. Figure 7.3 shows the three different types of SWCNTs.

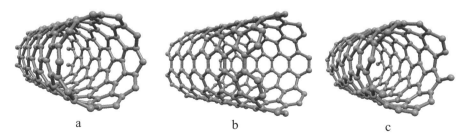

Figure 7.3. The three different types of SWCNTs: a. zigzag; b. armchair; c. chiral

7.2.2 Electronic Properties

The electronic properties of carbon nanotubes are a matter of particular interest due to their high application potential in nanoscale electronic devices. The ideal symmetric structure of carbon nanotubes is leading to unique electronic properties like quantum effects and ballistic electron transport.

SWCNTs can have **metallic** or **semiconducting** characteristics, depending on their chirality. Theoretical calculations have shown that SWCNTs are metallic conducting, if:

$$(n-m) = 3k, \tag{7.3}$$

where $k = 0, 1, 2, \dots$.

This means that one third of all possible SWCNTs are metallic and two thirds are semiconducting. For example, all armchair SWCNTs show metallic conductivity. The exact band gap of semiconducting SWCNTs depends on their diameter and can be calculated by:

$$E_{gap} = \frac{2a_{cc}\gamma_0}{D}, \tag{7.4}$$

where a_{cc} is the nearest neighbor distance and amounts to 0.142 nm and γ_0 is the nearest neighbor transfer energy with $\gamma_0 = 2.7$ eV [3].

The typical band gap for SWCNTs is below 1 eV. **SWCNTs** show **ballistic** electron transport without any electron scattering over the whole CNT length. So the ideal resistivity of an SWCNT is independent of the length. In reality, however, the ballistic transport of electrons and thus the conductivity of SWCNTs depends on the number of defects in the CNT. Defects can act as scattering centers and thus decrease the mean free path L_m of an electron, given by:

$$L_m = v_F \tau_m, \tag{7.5}$$

where v_F is the Fermi velocity and τ_m the characteristic relaxation time.

For MWCNTs, the determination of the conductivity is more complex, since it consists of multiple walls and thus depends on the chirality of every individual shell. An MWCNT can be considered as a bundle of parallel conductors, each having a different band structure. First experimental measurements have produced different results.

[4] identified that an MWCNT at low bias shows ohmic behavior and the electrical conductivity decreases analogously to the number of outer shells being removed from the MWCNT by current-induced oxidation. This effect of multilayer conductivity can be explained by shell-to-shell interaction. The coupling between the inner and outer shells is shown to be completely frozen out for temperatures below 90 K. Typical MWCNT resistivity ranged between 5 and 15 kΩ. [5] presented the fact that the conductivity of **MWCNTs** is independent of length, and that the electron transport seems to be **quasi-ballistic**, similar to SWCNTs.

Recapitulating, the conductivity of an MWCNT depends on three parameters: the contact resistance between CNT and electrodes; the resistivity of individual shells; the shell-to-shell resistivity.

7.2.3 Mechanical Properties

Due to the strong covalent σ-bonding between the neighboring carbon atoms forming a nanotube, the tensile strength and **Young's modulus** of CNTs exceed those of steel by magnitudes, which makes CNTs excellent for reinforced composite materials.

Theoretical and experimental results have attested a Young's modulus of about 1 TPa and a tensile strength of about 150 GPa [6-11]. On the one hand, the Young's modulus is reported to be independent of the tube diameter, tube chirality, and number of tube layers by using an empirical force-constant model [7]. But on the other hand, newer publications show effects of diameter, chirality, and number of layers by means of a molecular structural mechanics approach [9]. Thus, for a thorough understanding of the elastic properties of carbon nanotubes, **systematic experimental measurements** have to be carried out.

MWCNTs can be seen as nested SWCNTs. The interactions between adjacent layers are mainly due to van der Waals forces. Compared to the covalent bonds between the carbon atoms within the graphene sheet, the van der Waals interaction is rather weak. As a result, relative slippage between the layers in an MWCNT can occur, and MWCNTs can act as low-friction nanoscale linear bearings [12]. The structural flexibility was investigated by means of observations and atomistic simulations of the bending of SWCNTs and MWCNTs under mechanical stress [13]. The results show that single and multiple kinks can be observed, dependent on the bending curvature. Despite the occurrence of kinks, CNT bending is fully reversible and elastic up to very large bending angles of about 110°.

Regarding a cantilevered, free-standing MWCNT that can be assumed as a cylinder with given diameter (Figure 7.4), one can obtain a relation between the **deflection** δ of the MWCNT and its Young's modulus E. Using a standard beam deflection formula according to [14], the Young's modulus E is given by:

$$E = \frac{F(\delta)l^3}{3I\delta}, \tag{7.6}$$

where l is the length of the CNT, δ the deflection of the CNT, $F(\delta)$ the appendant force acting on the CNT, and I is the geometrical moment of inertia of a solid cylinder given by:

$$I = \pi d^4 / 64, \tag{7.7}$$

where d is the diameter of the CNT.

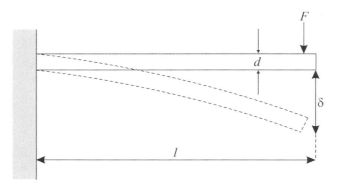

Figure 7.4. Deflection of a cantilevered CNT

7.2.4 Fabrication Techniques

Since the discovery of CNTs, when they were fabricated by arc discharge, a lot of additional techniques have been developed for the production of carbon nanotubes. The most common production techniques today are **arc discharge** [15, 16], **laser ablation** [17, 18], and **chemical vapor deposition** [19, 20].

7.2.4.1 Production by Arc Discharge
Arc discharge is a relatively simple and low-priced production technique for high-quality CNTs. For CNT production by arc discharge (Figure 7.5), two graphene electrodes are arranged in a vacuum chamber filled with helium. By impressing a voltage on the electrodes and decreasing the displacement between cathode and anode, an arc discharge occurs with a current flow of about 100 A. Due to this arc discharge, the carbon of the electrodes is evaporated and precipitates in the form of CNTs on the cathode. The yield obtained from arc discharge is up to 30%. Both SWCNTs and MWCNTs with only partial defects are created. By adding a metallic catalyst, the growths of SWCNTs can be enhanced. The main disadvantages of this method are the short length of nanotubes and their random size and alignment. The huge portion of byproducts require a following complex cleaning process.

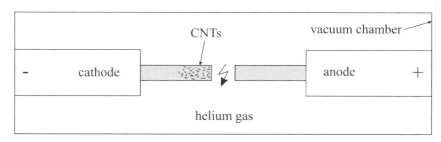

Figure 7.5. Production of CNTs by arc discharge

7.2.4.2 Production by Laser Ablation

Another method for producing CNTs called **laser ablation** is shown in Figure 7.6. For this method, a laser beam is directed at a graphite target, vaporizing parts of it. The vaporized carbon particles are carried by an argon gas flow within the vacuum chamber towards a copper collect vessel, where the carbon nanotubes grow. The yield of this method is up to 70% by weight. Mainly defect-free SWCNTs or ropes of SWCNTs are created. By varying the process temperature, the diameter of the CNTs can be affected. The disadvantages are the high costs due to the necessity of having a high-powered laser and, as in the case of arc discharge, the lack of control of the deposition area and the huge portion of byproducts.

Figure 7.7 shows an SEM micrograph of typical MWCNTs produced by arc discharge and laser ablation. The CNTs are available in the form of powder forming a lump of CNTs. The separation of individual CNTs is, therefore, hard to achieve.

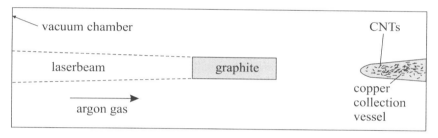

Figure 7.6. Production of CNTs by laser ablation

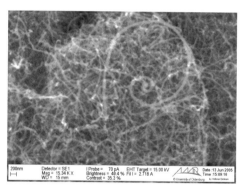

Figure 7.7. MWCNTs produced by arc discharge and laser ablation with diameters ranging from 60 nm up to 100 nm and length from 0.5 μm up to 600 μm

7.2.4.3 Production by Chemical Vapor Deposition (CVD)

Chemical vapor deposition (Figure 7.8) uses a chemical reaction, which transforms gaseous molecules, called precursors, into a solid material.

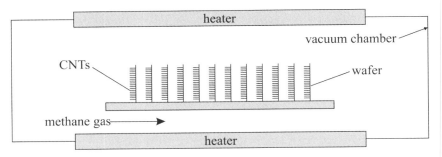

Figure 7.8. Production of CNTs by chemical vapor deposition

For producing CNTs, small nickel particles with a diameter of about 30 nm are brought onto a silicon substrate (wafer). A carbon-containing gas, such as methane, flows over the wafer surface. At temperatures of about 600°C, the gas disintegrates and the released carbon atoms form tubes using the nickel particles as condensation seeds. The yield can be up to 100% by weight. Length, alignment, and position of the CNTs can be affected by varying the process parameters. In this way, long and **perfectly aligned CNTs** can be achieved (Figure 7.9). Another advantage is the purity of the CNTs, as the CVD method produces pure CNTs without any byproducts. However, due to the low process temperature, the CNTs often exhibit defects.

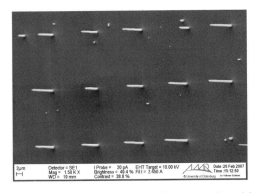

Figure 7.9. Array of MWCNTs produced by chemical vapor deposition. CNT diameter is about 500 nm, length 6 μm, and pitch 10 μm.

7.2.5 Applications

A multitude of possible applications for CNTs [21] arise from their remarkable physical properties, such as their high Young's modulus, high electrical conductivity, and high aspect ratio. For example the tensile strength of CNTs is in the order of 10^9 Pa, which corresponds to the multiple tensile strength of high-modulus steel. CNTs can carry a current density of about 10^9 A/cm^2, which is 1000 times as much as that of copper. In the following sections, the four most realistic applications, besides the development of **CNT-based sensors** [22], are presented.

7.2.5.1 Composites

The high Young's modulus of about 1 TPa and the high electrical conductivity make CNTs excellent for **reinforced** and **conducting composites**. For example, such conducting composites can be used for electrostatically applying paint onto car components or can act as antistatic shieldings. Another application is transparent electronic conductors in the display industry for the development of flexible displays on plastic substrates.

Due to their mechanical properties, CNTs have attracted much interest for light-weight structural composites and are likely to exceed the performance of carbon-fiber, reinforced materials. Today, such CNT-reinforced composites are already being applied in the serial production of tennis rackets, in order to improve their rigidity and damping features.

7.2.5.2 Field Emission

The high aspect ratio and the huge current density make CNTs ideal field emitters under an intense electric field. CNTs are, therefore, the perfect electron source to build new high-resolution and very flat **field-emissive displays** (FEDs). The principle is similar to CRT (cathode ray tube) displays. Electrons are accelerated towards a phosphor layer, creating a light spot (pixel). But in contrast to old CRT displays in a FED, every pixel has its own electron source – a multitude of vertically aligned CNTs. Figure 7.10 shows the schematic of a FED.

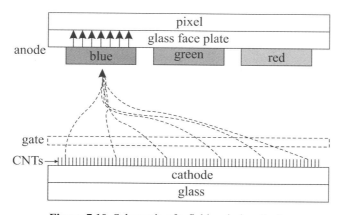

Figure 7.10. Schematic of a field emissive display

Between the cathode and gate-electrode a discharge voltage is applied, so that the CNTs start to emit electrons. By applying an acceleration voltage between the gate-electrode and anode, these electrons are directed towards the subpixel. Three subpixels form a pixel. The advantages of FEDs over LCDs are higher power efficiency and a wider operating temperature range.

Another application for field-emitting CNTs are electron guns for scanning electron microscopes (SEMs) and transmission electron microscopes (TEMs). The small energy spread of the electrons and the high current density will improve the lateral resolution and the brightness of the next generation of microscopes.

7.2.5.3 Electronics

Field-effect transistors (FETs) manage the current between source and drain through an electric field created by the gate. The electric field alters the shape of the conducting channel. In the case of **CNT-based field-effect transistors** (CNTFETs), the semiconducting SWCNT becomes conducting as soon as a certain voltage, and hence an electric field, is applied. Due to the small size of SWCNTs, it is possible to reduce the size of FETs and to build up CNTFETs that can switch 1000 times quicker than FET processors today [23].

The continuous miniaturization of silicon-based integrated circuits requires new **vertical interconnects** (VIAs) that can carry larger current densities. The mandatory vertical growth of CNTs can be realized by chemical vapor deposition. An additional application for CNTs is to act as horizontal interconnects. Also, the horizontal growth of CNTs can be realized by CVD [24]. A significant problem for applications in nanoelectronics is the band gap dependence on the CNT's chirality. Today's fabrication techniques generate a mixture of metallic and semiconducting CNTs. For industrialization, the growth of SWCNTs of a specific type, in a defined direction, and at a defined position, is required.

7.2.5.4 AFM Cantilever Tips

Atomic force microscopy has become an important technique in a wide range of research areas due to its ability to image and characterize nanoscale structures. The quality of information strongly depends on the size and shape of the AFM probes used. Commercially available AFM probes are usually made of micromachined silicon cantilevers with integrated pyramidal tips. Such tips have a typical radius of about 10 nm. Conventional AFM probes are, therefore, unable to penetrate high-aspect-ratio structures and to exactly profile surfaces with a complex topography.

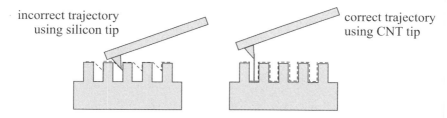

incorrect trajectory using silicon tip

correct trajectory using CNT tip

Figure 7.11. Comparison of standard AFM tips and CNT-based supertips

CNT-based supertips with superior characteristics (stability, resolution, lifetime, *etc.*) can overcome the limitations of micromachined silicon tips (Figure 7.11). Supertips, however, cannot only be used as the main sensing component ultimately responsible for the quality of AFM imaging. Their extremely small size and their high conductivity also make them suitable to function as ultra fine nanoelectrodes able to electrically contact nanoscale structures in nanoelectronics.

7.3 Characterization of CNTs

In order to realize the above-mentioned applications and to make use of carbon nanotubes in nanotechnology products, a complete and correct charaterization of carbon nanotubes must be achieved. The production of carbon nanotubes in a controlled way, in large amounts, and with well-defined properties, encounters problems that remain to be solved. First of all, it is essential to identify all the **physical properties** of CNTs. A multitude of techniques are available to fully characterize carbon nanotubes. An overview of the existing characterization techniques will be given in the following section.

7.3.1 Characterization Techniques and Tools

Many properties of nanotubes are directly influenced by the way in which the graphene sheets are wrapped. The most important properties of carbon nanotubes are:

- length and diameter of the CNT,
- distinction between metallic and semiconducting SWCNTs,
- appointment of the chirality of SWCNTs (zigzag, armchair, or chiral),
- measurement of sample purity and alignment of CNTs,
- features of MWCNTs (outer and inner radius, intershell spacing).

To determine all these properties, different characterization methods can be used [2]. As mentioned in Section 7.1, the most common tools and methods are:

- transmission electron microscopy,
- scanning tunneling microscopy,
- atomic force microscopy,
- scanning electron microscopy,
- spectroscopy (photoluminescence, X-ray, photoelectron, infrared, and Raman),
- neutron and X-ray diffraction.

7.3.1.1 Microscopic Characterization Methods
TEM is useful for the characterization of the structure of MWCNTs. For example, the outer and inner radius, the number of layers, and the intershell distance of an MWCNT can be obtained [25]. High-resolution transmission electron microscopy (HRTEM) offers an even higher resolution than TEM, so that the **atomic structure** of CNTs becomes visible. For example, the nucleation point of an SWCNT or the atomic structure of both the catalyst particles and the SWCNTs can be imaged. By applying inverse fast Fourier transformation to HRTEM images, the chirality of SWCNTs can be determined [26].

STM is based on the nanoscopic phenomena of the quantum tunneling effect. STM images can resolve both the atomic structure and the **3D morphology** of a CNT and mirror the convolution of the STM tip shape and the sample surface. Furthermore, the chiral angle of nanotubes can be determined by the analysis of STM images [27].

As described in Chapter 1, AFM can be used both as an imaging and manipulation tool [28, 29]. AFM-based measurement of the **Young's modulus** of SWCNT bundles can be achieved. SWCNT bundles are deposited across the pores of a polished alumina membrane. Afterwards, an AFM probe is used to bend the bundles at their centers and to measure how the deflection, which is inversely proportional to the Young's modulus, varies with the applied force [30].

As an advantage, images taken by an SEM can be recorded almost in real-time (Chapter 4). The **geometric data** of MWCNTs can be extracted from SEM images (Figure 7.12), and moreover, the relatively huge vacuum chamber allows the analysis of large samples and provides enough space for robotic nanomanipulation systems.

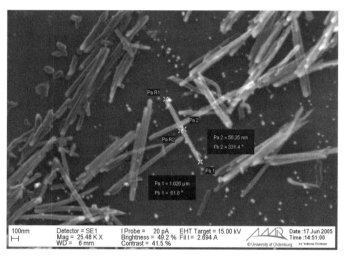

Figure 7.12. SEM micrograph of MWCNTs for measuring the geometric dimensions. MWCNT diameter is about 50 nm and length 1 μm.

7.3.1.2 Spectroscopic Characterization Methods

For analyzing the different CNT properties, various spectroscopic characterization methods are available. One of these methods is the so-called photoluminescence spectrocopy. For this, individual SWCNTs are needed. The nanotubes are excited by laser light, and a close infrared E_{11} emission of the SWCNTs can be measured. Using this method, the **conductive nature** (semiconducting or metallic), the geometry, and the diameter are accessible. In the case of semiconducting SWCNTs, the gap energy is directly related to the chirality and approximately proportional to the tube diameter [2].

X-ray photoelectron spectroscopy (XPS) can be used to investigate nitrogen doping effects on the chemical structure of MWCNTs. By controlling the NH_3/C_2H_2 flow ratio during the CVD process, different nitrogen concentrations are obtained. According to XPS, the increasing nitrogen concentration leads to an increase of the N-sp^3 C bonds [31].

Infrared spectroscopy can be applied to determine the symmetry of SWCNTs. There are 7-9 infrared active modes in SWCNTs, which depend on the symmetry: chiral, zigzag, and armchair. Also, impurities remaining from the synthesis of carbon nanotubes can be determined. For some applications of CNTs, molecules are attached to the surface of CNTs, in order to functionalize them. Infrared spectroscopy offers a way to characterize these functionalizing molecules [32].

Raman spectroscopy is one of the most powerful characterization techniques. It offers a fast and non-destructive analysis without sample preparation. An SWCNT sample is excited by laser light and a characteristic Raman spectrum is recorded. The so-called radial breathing mode (RBM) is directly dependent on the CNT's diameter. The **symmetry** (zigzag, armchair, or chiral) and the conductive nature (semiconducting or metallic) are reflected in the position, width, and relative intensity of the characteristic G-line [2].

7.3.1.3 Diffractional Characterization Methods

X-ray diffraction is not useful for the differentiation of microstructural details, but it can be used to determine the **sample purity**, the number of layers, the interlayer spacing, and the alignment of CNTs. The intensity and width of the characteristic peak in a typical X-ray diffraction spectrum is related to the number of layers and the alignment of the CNTs. The presence of catalyst particles (Co and Mo) remaining from the fabrication process can also be detected [2].

Neutron diffraction can be applied for the determination of **structural features** of CNTs, such as bond length and possible distortion of the hexagonal network. Even for the smallest diameters, a separation between armchair, zigzag, and chiral CNTs is possible. Moreover, the adsorption of He atoms on an SWCNT bundle can be analyzed [33].

7.3.2 Advantages of SEM-based Characterization of CNTs

In Section 7.3.1, an overview of the existing characterization tools and methods was given. Having in mind the title of this book, we will now refocus on the **microrobot-based** handling and characterization of CNTs within the SEM.

In order to observe nanoobjects and especially CNTs, the use of an SEM is necessary to overcome the limited angular resolution of optical microscopes. The developed SEM image processing (Chapter 4) allows for real-time SEM imaging and the recognition and tracking of nanoobjects using SEM images. Additionally, a 3D imaging system is available to determine depth information and thus the full 3D pose of nanoobjects (Chapter 5).

Besides the possibility of imaging objects with sizes down to some nanometers, the SEM offers a relatively spacious vacuum chamber compared to other electron microscopes such as TEM or HRTEM. SEM therefore is the ideal type of microscope for developing and installing microrobot-based handling and characterization systems (Chapters 1 and 2).

7.4 Characterization and Handling of CNTs in an SEM

Due to the remarkable physical properties of CNTs they are qualified for many applications, especially in nanoelectronics. However, the realization of these applictaions requires the handling and characterization of individual CNTs. SEM-based nanohandling robot stations can be used for the prototyping of nanoelectronic devices. The big advantage of visualizing the handling and characterization processes of CNTs has led to the development of various micro- and nanorobotic systems for the use within the SEM.

For example, [34] developed a **nanolaboratory**, a prototype nanomanufac-turing system based on a 16-DoF nanorobotic manipulation system for the assembly of nanodevices with MWCNTs. This system can use up to four AFM probes as end-effectors. The position resolution is of subnanometer order, and strokes are centimeter scale. *In situ* property characterization, destructive fabrication, shape modifications, and CNT junctions have already been presented. However, picking up CNTs with an AFM probe and by the use of dielectrophoresis or EBiD is reported to be difficult for strongly rooted CNTs and for transferring CNTs to another structure.

[35, 36] use a **hybrid AFM/SEM** system for the placement of an individual CNT onto a predetermined site on a microelectromechanical structure (MEMS). A single CNT is then retrieved with the AFM tip from a CNT cartridge using van der Waals forces between tip and CNT. Afterwards, the CNT is transferred to the desired site on the MEMS and then placed across a gap between a stationary structure (reticle) and a thermal actuator (pointer). The CNT is welded to the reticle and pointer by electron-beam-induced deposition of carbonaceous material. Considerations about electrical resistivity versus mechanical strain measurement on a CNT are then discussed.

Another system for three-dimensional manipulation and mechanical as well as electrical characterization has been realized by [37]. Here, a custom piezoelectric vacuum manipulator achieves positional resolutions comparable to scanning probe microscopes (SPM), with the ability to manipulate objects in one rotational and three linear DoF. This system is used for the manipulation of CNTs in order to explore their electromechanical properties under real-time SEM inspection. CNTs are stressed while measuring their conductivity, and CNTs are attached to AFM tips such that forces applied to the CNTs can be determined by the cantilevers' deflection. In addition, bending, kinking, breaking, and bundling of CNTs have been observed.

The nanomanipulator-assisted fabrication and characterization of CNTs inside an SEM has been reported by [38]. Here, two nanomanipulators are installed which can travel about 20 mm with a minimum increment of 1 nm. Electrochemically etched polycrystalline tungsten wires (STM tips) were used as end-effectors. This setup was used to construct a CNT transistor, to attach an MWCNT onto an AFM tip, to observe the elongation of a CNT, to modulate the shape of a CNT for tool fabrication (CNT hook), and to show *in situ* characterization of the electrical breakdown of MWCNTs. The latter was mentioned to provide a new method for characterizing the nature of conductivity of MWCNTs. It is worth mentioning that this method is a destructive one, so that the characterized CNTs cannot be used for further assembly.

Progress towards the **nanoengineering** of CNTs for nanotools has been made by [39]. The authors use AC electrophoresis for aligning MWCNTs at a knife-edge, forming a CNT cartridge. These MWCNTS are then transferred from the CNT cartridge onto a substrate in an SEM, in order to assemble nanotube AFM tips and nanotube tweezers that can operate in SPM. The fabricated nanotube AFM tips have been tested in AFM observation of biological samples and traces with deep crevices. The resulting nanotube tweezers are installed in an SPM to manipulate SiO_2 particles and CNTs. Moreover, the electron ablation of an MWCNT has been realized to adjust its length and to sharpen the MWCNT, having its inner layer with or without an end cap at the tip. The measurement of the sliding force for the inner layer with diameter 5 nm showed that the force amounts to approximately 4 nN, independent of the overlapping length.

The nanomanipulation of nanoobjects and CNTs in an SEM is also described in [40]. The authors present a sensor-based manipulation and processing system and demonstrate the gripping of micro-sized powder particles and the attachment of CNTs on tips for atomic force microscopes.

Great progress in the field of nanoengineering has been achieved by the Department of Micro and Nanotechnology (MIC) of the Technical University of Denmark. Besides the electrical and mechanical characterization of CNTs [41, 42], the development of **microfabricated grippers** [43] has advanced the pick-and-place nanomanipulation of nanowires and nanotubes [44, 45]. Moreover, the integration of CNTs into microsystems with electron beam lithography and *in situ* catalytically activated growth has been realized [46].

Several other methods for the characterization and handling of CNTs have been demonstrated. For example, the manipulation inside a TEM [47, 48] is advantageous because of the availability of diverse characterization tools down to atomic resolution (Section 7.3.2.1). However, TEM offers a very small vacuum chamber that is not easily accessible for robotic systems.

Another possibility is to use AFM or STM to manipulate CNTs [28, 49], nanoobjects [29, 50], or even single atoms [51, 52]. The big disadvantage of such SPM-based techniques is that the handling task can be carried out in only two DoF and must be performed "blindly", since during the manipulation, the SPM tip is not available for imaging the scene, so that SPM images can only be acquired before and/or after the manipulation.

Contact-free manipulation techniques have also been developed. Firstly, multidimensional manipulation of CNTs with optical tweezers has been reported [53], secondly, dielectrophoresis has been applied to the assembly of MWCNTs [54], and thirdly self-assembly has been used for the assembly of SWCNTs into hierarchical structures with controlled nanotube orientation [55]. However, for these contact-free methods, the CNTs must be dispersed in a liquid solution.

Recapitulating, SEM-based nanohandling robot systems are the most promising method for a controlled and reliable three-dimensional manipulation and characterrization of CNTs and thus for assembling them into new prototypical nanostructures and -devices.

7.5 AMNS for CNT Handling

Following the generic concept of the automated microrobot-based nanohandling station (AMNS) introduced in Chapter 1, a robot station was developed that can be integrated into an SEM and can use different end-effectors for the non-destructive characterization and reliable three-dimensional handling of CNTs. Either the tip of a piezoresistive AFM probe for the mechanical characterization of CNTs or an electrothermal nanogripper [43] for CNT handling is applied.

7.5.1 Experimental Setup

A mechanical drawing of the experimental setup of the **nanohandling robot station** is shown in Figure 7.13. The setup is mounted on a base plate and contains two different manipulators. There is a three-axes micromanipulator MM3A [56] that is equipped with an end-effector holder and is used for the coarse positioning between end-effector and CNT sample.

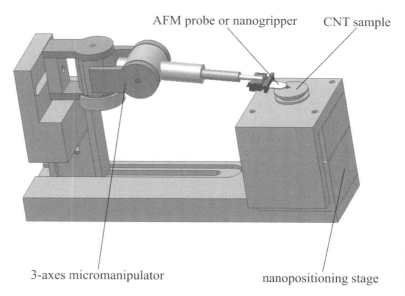

AFM probe or nanogripper   CNT sample

3-axes micromanipulator   nanopositioning stage

Figure 7.13. Mechanical drawing of the nanohandling robot station

The micromanipulator offers a theoretically possible resolution of 5 nm for the two rotational axes and 0.25 nm for the linear axis. A nanopositioning piezo stage with three degrees of freedom carries the CNT sample. The resolution of the nanopositioning stage [57] is limited by the 16-bit D/A converter of the control module and amounts to 1.55 nm in closed-loop mode.

For **nanohandling inside the SEM**, the whole setup is installed onto the positioning stage of a LEO 1450 SEM [58] (Figure 7.14).

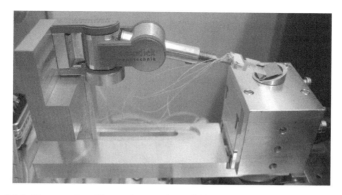

Figure 7.14. The nanohandling robot station within the vacuum chamber of the SEM

Figure 7.15 shows the overall station's **control system** that is currently used for the teleoperated nanomanipulation of CNTs in the SEM. By operating the SEM stage, the whole setup is adjusted within the vacuum chamber, getting it into the SEM focus, so that the manipulation scene can be observed by using the SEM imaging software. After adjusting the SEM image, the three-axes micromanipulator is used for the coarse positioning of the end-effector close to the CNT. For the fine positioning and thus the real manipulation of CNTs, the nanopositioning stage is controlled in order to move the stage in 1.55 nm steps. The stage controller delivers an encoder signal, giving the exact value of the stage's displacement, so that the stage can be operated in **closed-loop mode**. The nanopositioning stage is operated by control software written in LabVIEW and the connected control input.

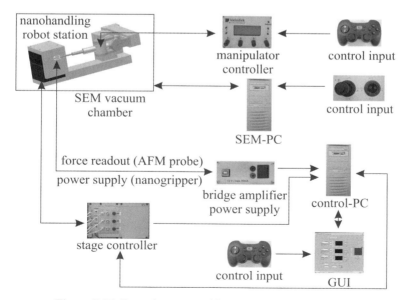

Figure 7.15. Control system architecture for telemanipulation

7.5.2 Gripping and Handling of CNTs

The nanomanipulation, meaning the **controlled gripping** and handling of carbon nanotubes, is of great importance for nanotechnology. Therefore, new gripping tools have to be designed that are suitable for the handling of nanoobjects and CNTs [44, 59]. The electrothermal nanogrippers used for the CNT handling below have been provided by MIC, Denmark. The **electrothermal gripper** can be actuated in various modes by applying voltages to the different gripper beams [43]. The width of the gripper arms and the gripper opening, is 2 μm.

As mentioned above, the capability to observe the nanohandling process in real-time with high magnification is achieved by operating the nanohandling robot station within an SEM. A frequent nanomanipulation task is the gripping of a carbon nanotube and placing it onto a TEM grid for further analysis and characterization in TEM, or placing it onto an electrode structure for electrical characterization. Currently, the gripping process itself is done by teleoperation. Automatic positioning of gripper and CNT sample will be realized using a special control system described in Section 7.6. Such automation of nanohandling processes will be the next step towards the integration of new, promising nanomaterials into conventionally micromachined parts and thus the prototyping of nanoelectronic components. Figure 7.16 shows the process flow of gripping an MWCNT grown on silicon substrate by chemical vapor deposition.

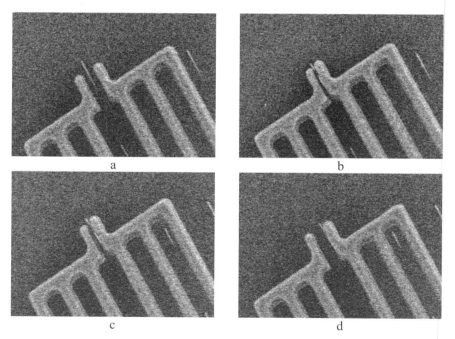

Figure 7.16. Gripping sequence of an MWCNT: a. coarse and fine positioning between gripper and CNT has been realized; b. gripper actuation; c. removing the CNT from the substrate; d. reopening the gripper (in collaboration with MIC)

The arrays of MWCNTs have been provided by the Engineering Department, University of Cambridge, United Kingdom [20]. The diameter of the handled MWCNT is about 200 nm and the length is about 6 μm. In Figure 7.16a, the coarse and fine positioning of the nanogripper close to the CNT has been realized in order to grip the CNT. By applying a voltage to the nanogripper, the gripper is actuated in order to grab the CNT (Figure 7.16b). By moving the CNT with the nano-positioning stage and keeping the position of the nanogripper fixed, the CNT is removed from the substrate and can be positioned in three dimensions (Figure 7.16c). After reopening the nanogripper, the CNT sticks to the right gripper jaw due to **adhesion forces** (Figure 7.16d). The controlled put-down of the CNT on a well-defined location, like a TEM grid or a electrode structure, will be the next step of exploration. Electron-beam-induced deposition (Chapter 10) will be used to fix the CNT on the target structure in order to overcome these adhesion forces.

7.5.3 Mechanical Characterization of CNTs

The nanohandling robot station can also be used to perform **deflection measurements** of MWCNTs, in order to calculate their Young's modulus E. Figure 7.17 shows the principle of the bending experiments using a tipless piezoresistive AFM probe.

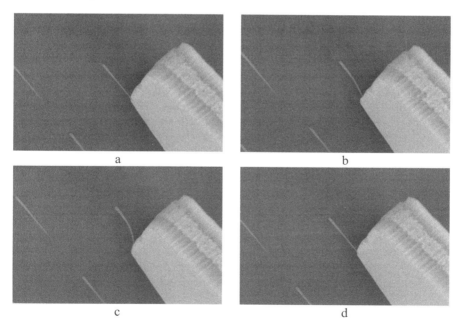

a   b

c   d

Figure 7.17. Sequence of piezoresistive AFM probe-based deflection of an MWCNT: a. coarse and fine positioning between AFM probe and CNT has been realized; b. beginning of CNT bending; c. maximum CNT bending; d. coming full circle

In Figure 7.17a, the coarse and fine positioning of an MWCNT and an AFM probe has been realized. The MWCNT used for the deflection experiments has a length of 8.8 μm and a diameter of 370 nm at the bottom and 160 nm at the top. By moving the nanopositioning stage and keeping the AFM probe fixed, the MWCNT counteracts the AFM probe and bends (Figure 7.17b-c). The deflection of the CNT is measured with the help of the position sensor of the nanopositioning stage in closed-loop mode, and the bending force is measured by the piezoresistive readout of the AFM probe deflection and on the basis of the following AFM probe calibration.

The first step of **AFM probe calibration** is to measure the geometric parameters of the AFM probe beam at different points using the SEM. The average of these values can be used to calculate the probe's stiffness. The geometric parameters (length l, width w, and thickness t) of the piezoresistive AFM probe for these measurements are:

$$l = 507.5\,\mu m, \; w = 157\,\mu m, \; t = 7\,\mu m. \tag{7.8}$$

The stiffness S of the AFM probe can be calculated by:

$$S = \frac{Ewt^3}{4l^3}, \tag{7.9}$$

where E is the Young's modulus of silicon and the parameters w, t, and l are the geometric values of the AFM probe. Thus, the calculated stiffness amounts to $S = 17.08$ N/m.

The sensor signal of the piezoresistive Wheatstone bridge integrated into the AFM probe is a differential voltage, which has to be calibrated to the corresponding force. Calibration can be done easily by deflecting the AFM probe against a hard reference material such as a hard silicon specimen. By recording the output voltage and the deflection of the AFM probe, the calibration factor K that connects voltage and force can be calculated. Figure 7.18 shows the measured voltage-deflection curve during AFM probe calibration. A linear fit gives a voltage-deflection factor V of 3.76 mV/μm. With this factor V and the stiffness S, the calibration factor K of the AFM probe can be calculated by:

$$K = \frac{S}{V}. \tag{7.10}$$

The calculated stiffness of $S = 17.08$ N/m and the voltage-deflection factor of $V = 3.76$ mV/μm lead to a calibration factor of $K = 4.537$ μN/mV. By multiplying the (offset corrected) voltage signal obtained from the piezoresistive AFM probe by this factor K, the effective force F can be calculated. For the final CNT bending experiments, the deflection of the MWCNT is performed by moving the nanopositioning stage in an **automatic ramp function**.

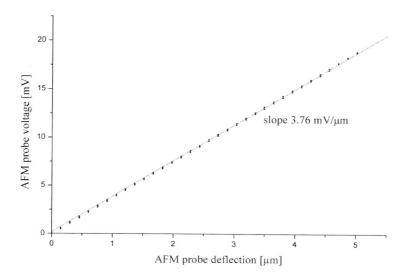

Figure 7.18. Calibration curve for the piezoresistive AFM probe

The associated force-deflection curves are therefore generated automatically. Figure 7.19 shows the measured force depending on the CNT deflection for the realized automatic CNT bending. The maximum CNT deflection is approximately 1.5 µm, with a resulting force of 2.5 µN. The correspondence between loading and unloading emphasizes the **elastic behavior** of MWCNTs.

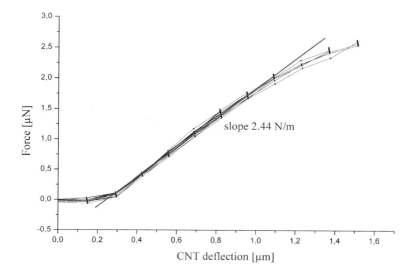

Figure 7.19. Force-deflection curve for MWCNT bending

The slope of the linear region is evaluated as $m = 2.44$ N/m. According to Equation 7.6 and Equation 7.7, the Young's modulus of the MWCNT can be calculated by:

$$E = \frac{64ml^3}{3\pi d^4},$$
(7.11)

where l is the length of the MWCNT, given by the position of the nanotube having contact with the AFM probe, E is the Young's modulus, and d the diameter of the MWCNT.

The multiwall carbon nanotube has different diameters at the bottom and at the top of $d_b = 370$ nm and $d_t = 160$ nm. The **Young's modulus** has been calculated for both values, and the mean value leads to $E = 0.95$ TPa, with $l = 4.17$ μm and $m = 2.44$ N/m. This presented result is in good conformity with theoretical calculations of the Young's modulus of MWCNTs predicting values of about 1.05 TPa [8, 9].

7.6 Towards Automated Nanohandling of CNTs

The previous sections have introduced techniques and tools to handle and characterize CNTs. However, performing these experiments on a large number of CNTs is a very time consuming and tedious process. Even though manual handling might be sufficient to achieve good research results on physical material constants, it is not good enough for industrial purposes or material tests that require a large number of samples. For the industrial usage of CNTs, reliable and repeatable handling of nanostructures at high throughput is required, which is accomplished for macro objects by automated manufacturing and quality control. However, the single automation cells (*i.e.*, setup and control) in nano- and microfactories need to be flexible enough to be reused for several different automation tasks.

7.6.1 Levels of Automation

Different degrees of automation can be considered. The lowest level is **tele-operation**, where all movements are performed by a human operator. It is the responsibility of the operator to avoid hitting objects, to perform the operation in the right order to accomplish the assembly goal and to literally move the actors from one to another position.

The next level of automation is **semi-automation**. Low-level closed-loop control is available for some, or even all operations. A human operator can trigger a movement. For example an actor should move into a predefined parking position on a single operator command. The execution of the task is handled by the responsible closed-loop controller automatically. A human operator is responsible for selecting the correct input sensors and sensor models. In addition, the operator needs to check whether a closed-loop goal has been achieved or not and resolve possible errors.

The highest level of automation is **full automation**. At this level, the full assembly sequence can be run without human intervention, and all assembly steps can be separately performed as closed-loop operations. A full automatized assembly sequence requires high reliability of all single subtasks and knowledge of the assembly situation before and after the execution of every single task, in order to be able to handle errors. Collision avoidance is achieved by the abstract planning of end positions for every step. These end positions get split into intermediate steps, starting from the current position such that the robot cannot collide with other objects. The intermediate steps introduced by the high-level controller need to avoid collisions, and they are executed one by one by the corresponding low-level controller (LoLeC).

7.6.2 Restrictions on Automated Handling Inside an SEM

Automation in microrobotics encounters many of the problems that have been thoroughly studied for decades in the domains of industrial robotics and autonomous service robotics. Some of the problems are, *e.g.*, collision avoidance in path planning, error handling due to uncertainty of operations [60], or timing constraints which need to be met for successful automation.

However, there are some **environmental challenges** while operating in the vacuum chamber of an SEM. It takes several minutes to generate a high vacuum, which is a serious time constraint. Therefore, all automation cell parts, or even a complete manufacturing line, need to be inside the SEM before the operation starts, including tools and objects to be handled. The workpieces need some kind of a depot, and so do the end products. Tools, sensors, and objects have to be vacuum-compliant. Depending on the kind of manipulation, the materials need to be selected such that no contamination of the preprocessed workpieces (*e.g.*, wafers) and the clean room can occur.

Physics in microrobotics works the same way as in macrorobotics. There are, however, some differences. In contrast to large-scale objects, it is harder to release an object than to grip it. The reason for this "**sticky finger**" effect [61] is that adhesive forces are stronger at the scale of the gripper jaws and samples than gravity.[1] The released objects most often do not settle directly below the gripper. They jump around unpredictably, attracted by local surface forces. Therefore, pick-and-place operations need to be carefully planned. There is as yet no perfect solution for this problem. However, one possible solution is to reduce the contact area between gripper and object by reshaping the gripper or by stripping off the object at an edge. Another possibility that gives more control to the surface end position of the object is to use EBiD. One end of the gripped CNT is fixed to the specimen holder by the deposited material of the EBiD process, just before releasing it from the gripper (Chapter 10).

---

[1] In fact, gravity is the least significant force below a scale of 20 µm compared to van der Waals, electrostatic, and capillary forces.

The sizes of both the tools and objects also introduces some automation challenges to sensors. The AMNS setup presented in Section 2.6 introduced SEM images as the most important source of control information due to their high scanning rate and resolution. For automation, SEM vision fulfills several important tasks. One vision task is real-time position tracking of end-effectors using the SEM. In closed-loop control, the image acquisition and processing time is one of the major constraints. If the images are generated too slowly, the robot may overshoot its position or, even worse, never reach it. High scanning rates can only be achieved with the tradeoff of noisy image data in an SEM (Chapter 4).

Another important vision task is to obtain **three-dimensional information** on the automation environment. Without the additional depth information, one can only grip an object using a trial-and-error approach: *"Move the gripper and close it; test if a CNT has been gripped, if not reopen gripper and retry"*. Collision avoidance also needs reliable depth information. How to add the third dimension to SEM images is described in more detail in Chapter 5. As a last task, quality control through SEM vision is important. Many actors lack sufficient feedback. SEM images are one way of gaining knowledge about the outcome of an operation. Here, vision acts as a classifier for questions like *"Was the last placement operation successful?"* and *"Is the object at the right position?"*.

7.6.3 Control System Architecture

The AMNS setup for handling TEM lamellae (Section 2.6) is the first implementation of AMNS control architecture for automated nanohandling. However, it has some major limitations that make modifications necessary for the presented CNT characterization tasks. The old architecture cannot handle linear actors and consists of one low-level controller only, which does not scale up to setups with a larger number of robots working in parallel. In addition, integrated robot sensors cannot be used, since the low-level controller is not designed as a sensor program. Therefore, a new system architecture has been developed for the CNT handling.

The new system consists of **high-level control**, sensor, vision, and several **low-level control** servers (Figure 7.20) following a "black box design" [62]. Every server offers an individual service, which is defined by a public interface. Therefore, every component in this modular control system can be easily replaced or updated independently of the other components.

Visual feedback is provided by the vision server. It is responsible for collecting images from different sources and extracts position and orientation data (poses) of tracked objects in real-time. These poses are transmitted to the sensor server. All sensor data (*e.g.*, pose, touchdown, force, or temperature) are collected by the sensor server, which is supplied by different sensor service programs (*e.g.*, vision server). These data are provided to low- or high-level control. It therefore represents an abstraction layer for sensors and builds a common interface for further processing. The different sensors may acquire their data at different update rates.

In this architecture, there is one low-level control server for every single actor. Every individual low-level control server retrieves the data needed for closed-loop control from the sensor server. For this purpose, the sensor server provides two data retrieval models, a data push and a data pull model. The data push model sends new data to the requesting low-level controller if it is available, limited by a maximum rate threshold. Low-level controllers using the pull model periodically ask the sensor server for new data. There is a uniform interface for commands from high-level control. The advantage of this lean low-level control server approach is that single servers can be easily distributed among several PCs. Their good maintainability is another important feature. Furthermore, it is easy to include **different robot platforms** (*e.g.*, stationary and mobile), because only the internal structure of the low-level controller has to be changed. On the high-level control level, still only poses have to be submitted through the command interface. Actors with internal sensors are included in this architecture in a way that the low-level control server provides the internal sensor readings to the sensor server and receives high-level commands.

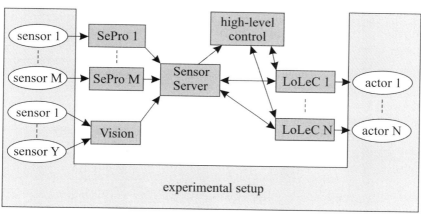

Figure 7.20. Software architecture connection chart. Rectangles are servers, and ellipses are hardware components. The automation module is part of the high-level control server.

The high-level control server (Figure 7.21) processes an automation sequence or receives teleoperation commands from a user via graphical user interface (GUI). Input information is translated into low-level command tokens called tasks and steering signals for the selection of certain sensors. High-level control is responsible for **automation, path planning, error handling**, and the **parallel execution** of tasks.

In order not to adapt the high-level control server to every change in an automation sequence, a script language has been developed, which is described in more detail in Section 7.6.5. These scripts are interpreted by the high-level control server and then mapped to low-level control tasks and steering signals. Low-level control tasks can either be closed-loop for positioning or open-loop (*e.g.*, gripping objects) tasks. The latter have to be monitored to check the operation results.

Every task has a defined set of pre- and postconditions. Preconditions need to be met before a task can be executed. The postconditions do hold after successful task completion. High-level control decides, based on required resources (*e.g.*, sensors and actors) and pre- and postconditions, whether two consecutive tasks can run in parallel. These postconditions should not be contradictory. If resource conflicts arise, a barrier approach is taken, so that all parallel tasks are terminated before automation of the resource-critical task starts (Figure 7.22). For SEM automation, these **resource-critical tasks** are often tasks where the SEM is used as a tracking sensor for two objects at different heights, or where the SEM acts as sensor and actor at the same time (*e.g.*, EBiD and a concurrent positioning task).

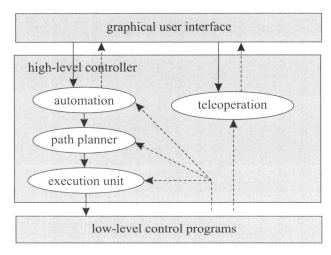

Figure 7.21. High-level controller consisting of automation, path planning, execution, and teleoperation unit

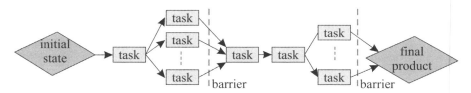

Figure 7.22. Schematic view on an arbitrary automation sequence. Parallel tasks are terminated before resource critical tasks get started.

Path planning can be done either during the automation sequence design or online. Online path planning is suitable for simple automation sequences, but it requires more knowledge about the work scene for collision avoidance. For online path planning, only goal positions would be included in the automation sequence, and the high-level controller has to compute intermediate positions and low-level control tasks. At the current development stage, only offline path planning is supported, which is sufficient in most cases, but delegates more responsibility to the designer of the automation sequence.

One of the major problems in micro- and nanorobotics is the high error rate of single automation tasks. Due to **environmental and scale effects**, operations that are simple in macro automation (*e.g.*, positioning) have a high error rate. There might be endless positioning attempts because the low-level control error threshold is too low, due to changes in humidity or temperature.

As in macro automation, the **error of a sequence** ε_{seq} is the multiplied error rate of every single automation task ε_t :

$$\varepsilon_{seq} = 1 - \prod_{t \in tasks} (1 - \varepsilon_t).$$

(7.12)

For example for a sequence consisting of seven tasks, each having an error rate of 10%, the error rate of the automation sequence is higher than 50%.

The first step in dealing with these high error rates is to achieve reliable error detection, which is the task of the feedback unit in the high-level controller. This unit receives all error conditions raised by components of the system (*e.g.*, a LoLeC) and presents them to the right abstraction level. One possibility for dealing with such error situations would be to jump back to an earlier task of the automation sequence. The preconditions of the task need to meet the current system state.

The proposed system architecture uses the Common Object Request Broker Architecture (CORBA) as the communication framework, which is object-oriented-middleware defined by the Object Management Group (OMG) and which is platform- and language-independent. There are several implementations of the CORBA standard, including some for real-time applications. CORBA's Interface Definition Language (IDL) is used as the common language for all communication interfaces between different network components. Therefore, the servers and clients in the system architecture merely need to implement the required IDL interfaces to be able to communicate with each other via simple method calls. A major advantage of IDLs is that they can be translated into different programming languages, which enables **heterogeneous software design**. The communication overhead in a closed-loop cycle has been evaluated as low enough (< 5 µs) in a local fully switched Ethernet network. The limiting time factor remaining is, therefore, the image acquisition time.

The interfaces of the client-server architecture are designed to provide asynchronous communication. All control commands return unique process IDs, so that delayed control feedback can be matched with the corresponding command. This part of the control feedback is crucial for **successful, reliable automation**. The system also has a common time base enabling low-level control servers to decide whether sensor data is outdated or not. The synchronization is periodically triggered by a master clock.

Every network component is designed in such a way that it can be run on different PCs, which makes the system fairly flexible. The distribution of low-level control servers is especially useful for control cycles that run in parallel, as the distributed controllers do not compete for the same PC hardware resources. The control architecture may contain several sensor servers for different data to overcome possible data acquisition bottlenecks. It is possible to organize the sensor data traffic, shaping a minimum update interval, so that the inbound traffic of any single sensor server can be controlled.

The problem of keeping several low-level control programs maintained is tackled by common low-level server templates, which are available for different actor types. This also enables rapid design and the integration of new actors. Similar templates will also be provided to the sensor data acquisition server.

7.6.4 First Implementation Steps

The general system architecture presented in the last subsection can easily be tailored to meet the requirements of the **nanohandling robot station** presented earlier in this chapter (Section 7.5.1). The single system components are illustrated in Figure 7.23. For this setup, three low-level control programs need to run in parallel, one for each actor group. Our low-level controller for the nanopositioning actor [57] implements the high-level control interface as well as the provider interface of the sensor server. This is due to the fact that the actor has an integrated sensor. This sensor data is made accessible to high-level control through the sensor server. The closed-loop cycle, of course, is performed directly inside the LoLeC.

After defining the system components, the modeling of a suitable automation sequence for a given handling task can be performed in three steps. Firstly, the hardware setup has to be defined according to the requirement analysis. Then, robot-based process primitives and their pre- and postconditions have to be defined (*e.g.*, move robot to target position if it is in range of the vision sensor). Finally, an automation sequence has to be found, which meets all pre- and post-conditions of the process primitives, avoids collisions, and eventually accomplishes the **automation task**. Additional constraints, as, *e.g.*, executing time, can be taken into account as well.

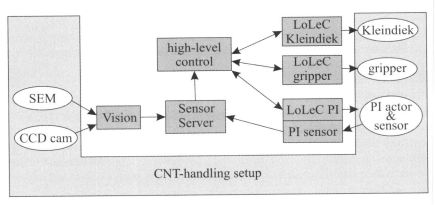

Figure 7.23. Tailored control architecture for the experimental setup in Figure 7.13. LoLeC PI acts as low-level controller for high-level control and provides actor-integrated sensor information.

A flexible **script language** has been developed, following the script-based approach of Thompson and Fearing [63]. The different commands of the language are the process primitives themselves. They are implemented as subclasses of a common task base class. This avoids reimplementation of common concepts such as error handling or message protocols. Composition and iteration are provided as general language constructs. Based on this script language, arbitrary automation sequences on the predefined operators can be defined (Figure 7.24).

```
lift(S_TOUCHDOWN_DIST);
move(EC,E_WORKING_POS);
grip(TRUE);
lift(S_SECURITY_DIST);
move(S_DROP_POS);
```

Figure 7.24. Automation sequence that lifts the specimen holder, moves the robot into the focus of the electron beam, and grips an object

A sequence file gets interpreted and executed by the high-level controller. In this way, the automation sequences can take advantage of the underlying closed-loop control. The concept enables rapid development of different automation sequences for the same set of process primitives, while the high-level control program remains unchanged.

The language design has been chosen with regard to future application of planning algorithms such as Metric-FF [64] to find the **optimal automation sequence** for a given automation task. Error handling can be performed online as a local refinement of the offline plan. For error handling, reliable error detection is crucial. The outcome of single process primitives has, therefore, to be monitored to provide feedback to the high-level controller, reflecting the current system state.

7.7 Conclusions

Since the discovery of CNTs about 15 years ago, intensive research has been successfully carried out on their basics (Section 7.2) and characterization techniques (Section 7.3). Nevertheless, another couple of years will be required to make use of CNTs in nanotechnology products. The **integration of individual CNTs** into micro- and nanomechanical systems will be one of the key issues. To achieve this objective, SEM-based microrobotic handling systems have proven to be a promising technique for the manipulation and characterization of individual CNTs.

An overview of the existing systems and the state-of-the-art has been given in Section 7.4. Most of the existing systems merely allow for destructive handling and characterization of CNTs. Thus, the characterized CNTs cannot be used for further assembly of prototypical CNT-based devices.

The nanohandling robot station presented in Section 7.5 permits **three-dimensional handling** and **non-destructive characterization** of CNTs. As for pick-and-place manipulation of CNTs, the main problem of parasitic adhesion forces acting between CNT and gripper jaw has to be solved. Therefore, special handling strategies will be developed, and/or EBiD will be used to overcome the adhesion forces and to allow a defined put-down. Non-destructive mechanical characterization was presented in Section 7.5.3. Future research activities will also focus on the realization of a non-destructive electrical characterization of CNTs. This would lead to a non-destructive mechanical and electrical *in situ* characterization of CNTs, so that the characterized CNTs, with their well-known physical properties, can be used for further **assembly of prototypical CNT-based devices**.

The progress towards automated nanohandling was presented in Section 7.6. A control system architecture was introduced, and first implementation steps were shown, including a high-level automation language. The system is capable of being adapted to the arbitrary hardware setups of AMNS. A major future task will be to identify and build a set of automated handling tasks that can be easily combined for arbitrary research and manufacturing tasks.

From the authors' point of view, the concept of automated microrobot-based nanohandling stations is the most promising concept to overcome the limitations suggested by fundamental research on CNTs and to allow for the industrial utilization of CNTs. Thus, the above mentioned *in situ* **characterization**, the **three-dimensional handling**, and the **automation** of particular manipulation and characterization processes have to be realized. Of course, bulk production cannot be reached by automated microrobot-based nanohandling stations, but this concept will enable the systematical prototyping of CNT-based nano-electro-mechanical systems.

7.8 References

[1]   Iijima, S. 1991, 'Helical microtubules of graphitic carbon', *Nature*, vol. 354, pp. 56–58.

[2]   Belin, T. & Epron, F. 2005, 'Characterization methods of carbon nanotubes: a review', *Material Science & Engineering B*, vol. 119, pp. 105–118.

[3]   Wildöer, J. W. P., Venema, L. C., Rinzler, A. G., Smalley, R. E. & Dekker, C. 1998, 'Electronic structure of atomically resolved carbon nanotubes', *Nature*, vol. 391, pp. 59–62.

[4]   Collins, P. G. & Avouris, Ph. 2002, 'Multishell conduction in multiwalled carbon nanotubes', *Applied Physics A*, vol. 74, pp. 329–332.

[5]   Frank, S., Poncharal, P., Wang, Z. L. & de Heer, W. A. 1998, 'Carbon nanotube quantum resistors', *Science*, vol. 280, pp. 1744–1746.

[6]   Treacy, M. M. J., Ebbesen, T. W. & Gibson, J. M. 1996, 'Exceptionally high Young's modulus observed for individual carbon nanotubes', *Nature*, vol. 381, pp. 678–680.

[7]   Lu, J. P. 1997, 'Elastic properties of carbon nanotubes and nanoropes', *Physical Review Letters*, vol. 79, pp. 1297–1300.

[8]   Tu, Z.-C. & Ou-Yang, Z.-C. 2002, 'Single-walled and multiwalled carbon nanotubes viewed as elastic tubes with the effective Young's moduli dependent on layer number', *Physical Review B*, vol. 65, no. 23, 233407.

[9]   Li, C. & Chou, T.-W. 2003, 'Elastic moduli of multi-walled carbon nanotubes and the effect of van der Waals forces', *Composites Science and Technology*, vol. 63, pp. 1517–1524.

[10]   Lukic, B., Seo, J. W., Bacsa, R. R., Delpeux, S., Beguin, F., Bister, G., Fonseca, A., Nagy, J. B., Kis, A., Jeney, S., Kulik, A. J. & Forro, L. 2005, 'Catalytically grown carbon nanotubes of small diameter have a high Young's modulus', *Nano Letters*, vol. 5, no. 10, pp. 2074–2077.

[11]   Nakajima, M., Arai, F. & Fukuda, T. 2006, 'In situ Measurement of Young's modulus of carbon nanotubes inside a TEM through a hybrid nanorobotic manipulation system', *IEEE Transactions on Nanotechnology*, vol. 5, no. 3, pp. 243–248.

[12] Cumings, J. & Zettl, A. 2000, 'Low-friction nanoscale linear bearing realized from multiwall carbon nanotubes', *Science*, vol. 289, pp. 602–604.

[13] Iijima, S., Brabec, C., Maiti, A. & Bernholc, J. 1996, 'Structural flexibility of carbon nanotubes', *Journal of Chemical Physics*, vol. 104, no. 5, pp. 2089–2092.

[14] Czichos, H. & Hennecke, M. 2004, *Hütte - Das Ingenieurwissen*, Springer.

[15] Iijima, S. & Ichihashi, T. 1993, 'Single-shell carbon nanotubes of 1-nm diameter', *Nature*, vol. 363, pp. 603–605.

[16] Ebbesen, T. W. & Ajayan, P. M. 1992, 'Large-scale synthesis of carbon nanotubes', *Nature*, vol. 358, pp. 220–222.

[17] Guo, T., Nikolaev, P., Thess, A., Colbert, D. T. & Smalley, R. E. 1995, 'Catalytic growth of single-walled nanotubes by laser vaporization', *Chemical Physics Letters*, vol. 243, pp. 49–54.

[18] Thess, A., Lee, R., Nikolaev, P., Dai, H., Petit, P., Robert, J., Xu, C., Lee, Y. H., Kim, S. G., Rinzler, A. G., Colbert, D. T., Scuseria, G. E., Tománek, D., Fischer, J. E. & Smalley, R. E. 1996, 'Crystalline ropes of metallic carbon nanotubes', *Science*, vol. 273, pp. 483–487.

[19] Fan, S., Chapline, M. G., Franklin, N. R., Tombler, T. W., Cassell, A. M. & Dai, H. 1999, 'Self-oriented regular arrays of carbon nanotubes and their field emission properties', *Science*, vol. 283, pp. 512–514.

[20] Teo, K. B. K., Chhowalla, M., Amaratunga, G. A. J., Milne, W. I., Hasko, D. G., Pirio, G., Legagneux, P., Wyczisk, F. & Pribat, D. 2001, 'Uniform patterned growth of carbon nanotubes without surface carbon', *Applied Physics Letters*, vol. 79, no. 10, pp. 1534–1536.

[21] Baughman, R. H., Zakhidov, A. A. & de Heer, W. A. 2002, 'Carbon nanotubes - the route toward applications', *Science*, vol. 297, pp. 787–792.

[22] Mahar, B., Laslau, C., Yip, R. & Sun, Y. 2007, 'Development of carbon nanotube-based sensors - a review', *IEEE Sensors Journal*, vol. 7, no. 2, pp. 266–284.

[23] Collins, P. G. & Avouris, P. 2000, 'Nanotubes for electronics', *Scientific American*, vol. 283, pp. 62–69.

[24] Hsu, C. M., Lin, C. H., Chang, H. L. & Kuo, C. T. 2002, 'Growth of the large area horizontally-aligned carbon nanotubes by ECR-CVD', *Thin Solid Films*, vol. 420-421, pp. 225–229.

[25] Kiang, C.-H., Endo, M., Ajayan, P. M., Dresselhaus, G. & Dresselhaus, M. S. 1998, 'Size effects in carbon nanotubes', *Physical Review Letters*, vol. 81, no. 9, pp. 1869–1872.

[26] Zhu, H., Suenaga, K., Hashimoto, A., Urita, K., Hata, K. & Iijima, S. 2005, 'Atomic-resolution imaging of the nucleation points of single-walled carbon nanotubes', *Small*, vol. 1, no. 12, pp. 1180–1183.

[27] Park, M. H., Jang, J. W., Lee, C. E. & Lee, C. J. 2005, 'Interwall support in double-walled carbon nanotubes studied by scanning tunneling microscopy', *Applied Physics Letters*, vol. 86, 023110.

[28] Rubio-Sierra, F. J., Heckl, W. M. & Stark, R. W. 2005, 'Nanomanipulation by atomic force microscopy', *Advanced Engineering Materials*, vol. 7, no. 4, pp. 193–196.

[29] Sitti, M., Aruk, B., Shintani, H. & Hashimoto, H. 2003, 'Scaled teleoperation system for nano-scale interaction and manipulation', *Advanced Robotics*, vol. 17, pp. 275–291.

[30] Salvetat, J. P., Briggs, G. A. D., Bonard, J.-M., Bacsa, R. R., Kulik, A. J., Stöckli, T., Burnham, N. A. & Forró, L. 1999, 'Elastic and shear moduli of single-walled carbon nanotube ropes', *Physical Review Letters*, vol. 82, no. 5, pp. 944–947.

[31] Jang, J. W., Lee, C. E., Lyu, S. C., Lee, T. J. & Lee, C. J. 2004, 'Structural study of nitrogen-doping effects in bamboo-shaped multiwalled carbon nanotubes', *Applied Physics Letters*, vol. 84, no. 15, pp. 2877–2879.

[32] Chopra, N., Majumder, M. & Hinds, B. J. 2005, 'Bifunctional carbon nanotubes by sidewall protection', *Advanced Functional Materials*, vol. 15, no. 5, pp. 858–864.

[33] Pearce, J. V., Adams, M. A., Vilches, O. E., Johnson, M. R. & Glyde, H. R. 2005, 'One-dimensional and two-dimensional quantum systems on carbon nanotube bundles', *Physical Review Letters*, vol. 95, 185302.

[34] Fukuda, T., Arai, F. & Dong, L. 2003, 'Assembly of nanodevices with carbon nanotubes through nanorobotic manipulations', *Proceedings of the IEEE*, Vol. 91, pp. 1803–1818.

[35] Guthold, M., Falvo, M. R., Matthews, W. G., Paulson, S., Washburn, S., Erie, D. A., Superfine, R., Brooks Jr., F. P. & Taylor II., R. M. 2000, 'Controlled manipulation of molecular samples with the nanomanipulator', *IEEE/ASME Transactions on Mechatronics*, vol. 5, no. 2, pp. 189–198.

[36] Williams, P. A., Papadakis, S. J., Falvo, M. R., Patel, A. M., Sinclair, M., Seeger, A., Helser, A., Taylor II., R. M., Washburn, S. & Superfine, R. 2002, 'Controlled placement of an individual carbon nanotube onto a microelectromechanical structure', *Applied Physics Letters*, vol. 80, no. 14, pp. 2574–2576.

[37] Yu, M., Dyer, M. J., Skidmore, G. D., Rohrs, H. W., Lu, X., Ausman, K. D., Von Ehr, J. R. & Ruoff, R. S. 1999, 'Three-dimensional manipulation of carbon nanotubes under a scanning electron microscope', *Nanotechnology*, vol. 10, pp. 244–252.

[38] Lim, S. C., Kim, K. S., Lee, I. B., Jeong, S. Y., Cho, S., Yoo, J.-E. & Lee, Y. E. 2005, 'Nanomanipulator-assisted fabrication and characterization of carbon nanotubes inside scanning electron microscope', *Micron*, vol. 36, pp. 471–476.

[39] Nakayama, Y. & Akita, S. 2003, 'Nanoengineering of carbon nanotubes for nanotools', *New Journal of Physics*, vol. 5, pp. 128.1–128.23.

[40] Fahlbusch, S., Mazerolle, S., Breguet, J. M., Steinecker, A., Agnus, J., Pérez, R. & Michler, J. 2005, 'Nanomanipulation in a scanning electron microscope', *Journal of Materials Processing Technology*, vol. 167, no. 2-3, pp. 371–382.

[41] Dohn, S., Molhave, K. & Boggild, P. 2005, 'Direct measurement of resistance of multiwalled carbon nanotubes using micro four-point probes', *Sensor Letters*, vol. 3, no. 4, pp. 300–303.

[42] Dohn, S., Kjelstrup-Hansen, J., Madsen, D. N., Molhave, K. & Boggild, P. 2005, 'Multi-walled carbon nanotubes integrated in microcantilevers for application of tensile strain', *Ultramicroscopy*, vol. 105, no. 1-4, pp. 209–214.

[43] Molhave, K. & Hansen, O. 2005, 'Electro-thermally actuated microgrippers with integrated force-feedback', *Journal of Micromechanics and Microengineering*, vol. 15, pp. 1265–1270.

[44] Molhave, K., Hansen, T. M., Madsen, D. N. & Boggild, P. 2004, 'Towards pick-and-place assembly of nanostructures', *Journal of Nanoscience and Nanotechnology*, vol. 4, no. 3, pp. 279–282.

[45] Molhave, K., Wich, T., Kortschack, A. & Boggild, P. 2006, 'Pick-and-place nanomanipulation using microfabricated grippers', *Nanotechnology*, vol. 17, pp. 2434–2441.

[46] Gjerde, K., Mora, M. F., Kjelstrup-Hansen, J., Schurmann, T., Gammelgaard, L., Aono, M., Teo, K. B. K., Milne, W. I. & Boggild, P. 2006, 'Integrating nanotubes into microsystems with electron beam lithography and in situ catalytically activated growth', *Physica Status Solidi (a)*, vol. 203, no. 6, pp. 1094–1099.

[47] Dong, L. X., Nelson, B. J., Tao, X. Y., Zhang, L., Zhang, X. B., Frutiger, D. R. & Subramanian, A. 2006, 'Nanorobotic manipulation of carbon nanotubes inside a transmission electron microscope', *MECHATRONICS 2006 - 4th IFAC-Symposium on Mechatronic Systems*, pp. 114–119.

[48] Wang, Z. L., Poncharal, P. & de Heer, W. A. 2000, 'Nanomeasurements of individual carbon nanotubes by in situ TEM', *Pure and Applied Chemistry*, vol. 72, no. 1-2, pp. 209–219.

[49] Albrecht, P. M. & Lyding, J. W. 2007, 'Lateral manipulation of single-walled carbon nanotubes on H-passivated Si(100) surfaces with an ultrahigh-vacuum scanning tunneling microscope', *Small*, vol. 3, no. 1, pp. 146–152.

[50] Requicha, A. A. G. 2003, 'Nanorobots, NEMS, and nanoassembly', *Proceedings of the IEEE*, Vol. 91, pp. 1922–1933.

[51] Eigler, D. M. & Schweizer, E. K. 1990, 'Positioning single atoms with a scanning tunnelling microscope', *Nature*, vol. 344, pp. 524–526.

[52] Martin, M., Roschier, L., Hakonen, P., Parts, U., Paalanen, M., Schleicher, B. & Kauppinen, E. I. 1998, 'Manipulation of Ag nanoparticles utilizing noncontact atomic force microscopy', *Applied Physics Letters*, vol. 73, no. 11, pp. 1505–1507.

[53] Zhang, J., Kim, H. I., Oh, C. H., Sun, X. & Lee, H. 2006, 'Multidimensional manipulation of carbon nanotubes bundles with optical tweezers', *Applied Physics Letters*, vol. 88, 053123.

[54] Chung, J., Lee, K.-H., Lee, L. & Ruoff, R. S. 2004, 'Toward large-scale integration of carbon nanotubes', *Langmuir*, vol. 20, pp. 3011–3017.

[55] Shimoda, H., Oh, S. J., Geng, H. Z., Walker, R. J., Zhang, X. B., McNeil, L. E. & Zhou, O. 2002, 'Self-assembly of carbon nanotubes', *Advanced Materials*, vol. 14, no. 12, pp. 899–901.

[56] Kleindiek Nanotechnik GmbH, Available at: http://www.nanotechnik.com.

[57] Physik Instrumente (PI) GmbH & Co. KG, Available at: http://www.physikinstrumente.com.

[58] Carl Zeiss NTS GmbH, Available at: http://www.smt.zeiss.com/leo.

[59] Wang, X., Vincent, L., Yu, M., Huang, Y. & Liu, C. 2003, 'Architecture of a three-probe MEMS nanomanipulator with nanoscale end-effectors', *IEEE/ASME International Conference on Advanced Intelligent Mechatronics*, pp. 891–896.

[60] Bennewitz, M. & Burgard, W. 2000, 'A probabilistic method for planning collision-free trajectories of multiple mobile robots', *Workshop Service Robotics - Applications and Safety Issues in an Emerging Market at the 14th European Conference on Artificial Intelligence (ECAI)*.

[61] Fearing, R. S. 1995, 'Survey of sticking effects for microparts handling', *International Conference on Intelligent Robots and Systems*, pp. 2212–2217.

[62] Fatikow, S., Wich, T., Hülsen, H., Sievers, T. & Jähnisch, M. 2006, 'Microrobot system for automatic nanohandling inside a scanning electron microscope', *International Conference on Robotics and Automation (ICRA'06)*, Orlando, FL, USA, pp. 1402–1407.

[63] Thompson, J. & Fearing, R. 2001, 'Automating microassembly with ortho-tweezers and force sensing', *IEEE/RSJ International Conference on Intelligent Robots and Systems (IROS)*, pp. 1327–1334.

[64] Hoffmann, J. 2003, 'The Metric-FF planning system: translating "ignoring delete lists" to numeric state variables', *JAIR*, vol. 20, pp. 291–341.

8

Characterization and Handling of Biological Cells

Saskia Hagemann

Division of Microsystems Technology and Nanohandling,
Oldenburg Institute of Information Technology (OFFIS), Germany

8.1 Introduction

This chapter will exclusively focus on the **manipulation** and **characterization** of biological cells by an atomic force microscope (AFM). Although other methods such as optical tweezers [1], dielectrophoresis [2], *etc.* exist, not only would detailed commenting on every branch of biohandling go beyond the scope of this chapter, but also most of these techniques are well established, while AFM–based characterization and manipulation is a strongly developing area. A brief comparison of AFM, dielectrophoresis and optical tweezer as manipulation and characterization methods for biological objects is given in Table 8.1.

Since its development in the mid–eighties by Binnig, Quate, and Gerber [3], the potential of AFM for characterizing biological objects has quickly been realized, and it has become a valuable tool in biological studies ever since. An AFM combines several abilities that make it a very attractive instrument for **high**– (sub–nanometer) **resolution** applicatons, comparable with a scanning electron microscope (SEM). The **ambient conditions**, under which the samples can be investigated and which allow for appropriate physiological conditions, are similar to working beneath an optical microscope. There is also the option to measure **different properties**, such as elasticity, conductivity, or friction, beyond the topological information. Because of the nature of the AFM, not only visualization, but also **manipulation** by its tip is an option. However, the ability to manipulate has limitations. Special care has to be taken that the scan–mode–dependent effect does not become a significant factor.

Because of these abilities, the AFM makes the observation or manipulation of processes on a **molecular to tissue level** of living specimens in a physiological environment possible. The first measurements included the mapping of the elasticity of bone [4] and the imaging of the activation process of human platelets [5]. Along with the development of different measurement modes, for instance the dynamic mode, the possibilities of biological characterization also evolved. In this way, the gating processes for intercellular communication could be imaged, and the electrical reaction of stimulated single hair cells in the inner ear could be recorded. The binding forces of receptor ligand pairs or the adhesion of cells, as required when leucocytes attach themselves to infected cells, can be measured. Most of these experiments would not be possible without the aid of an AFM and help to push the frontiers of research, *e.g.*, exploring cytological functions or laying a foundation for pharmaceutical accomplishments.

The following section will give an overview of AFM basics, with special emphasis on the relevant parameters for the characterization of biological material. A brief introduction to several biological matters of interest for AFM–based research is given as well, to provide a background for the technical terms used in the last section. Several applications of AFM–based handling and characterization of biological material will be described, with an emphasis on current developments that trace the frontiers of current research.

Table 8.1. Comparison of AFM, dielectrophoresis and optical tweezers as manipulation and characterization methods for biological objects

	AFM	**Dielectrophoresis**	**Optical tweezers**
Measurable and exerted forces	pN – μN	pN	pN
Requirements for the object	Immobilized	Dielectric (must have movable charges)	Dielectric, for objects large in comparision to the wavelength of the light transparency (act as lenses)
Environment	Vacuum, liquid, air	Vacuum, liquid, air	Vacuum, liquid, air, but surrounding medium must have smaller index of refraction than the object
Interaction between method and object	Force between object and tip, contact or non–contact, surface can be damaged	Electrical field, non–contact	Electromagnetic waves, non–contact, damaging due to heating possible
Possibility to manipulate	Movement, cutting, pushing	Movement, rotation, sorting	Movement, sorting, stretching, heating
Measurable characteristics	Elasticity, friction, conductivity, force measurements, topology, *etc.*	Polarizability	Viscoelastic or elastic properties, force measurements
Imaging ability	yes	no	no

8.2 AFM Basics

As several well–established books about AFM exist [6, 7], this section does not intend to explain an AFM in full detail. Nevertheless, an overview of its functionality, with special emphasis on handling of objects in a liquid environment, is given.

The functioning principle itself is quite simple: the heart of an AFM is a cantilever with a **very sharp tip** (usual tip radius of around 10 nm), which scans the area of interest, reminiscent of a record player as concerning its functionality. Due to interacting forces between tip and sample, the cantilever will eventually **bend**, which is detected. From the forces working on the cantilever, a topographical image of the sample can be generated. For scanning the sample, either the cantilever or the sample has to be moved with nanometer accuracy, which is usually carried out by piezostacks. The area of interest is limited. Normally, custom scanners can cover an area from 90×90 μm to 100×100 μm [8–10]. The scanning of an image usually takes some minutes, depending on scan rate, scan size, and mode. Similar to the SEM, scan speed and image quality are inversely proportional. Therefore, other methods to speed up the imaging process have been the subject of research, including AFM cantilevers with several tips at a fixed distance, as well as current developments on "video rate" AFMs, which will be covered in Section 8.2.6.

As already mentioned, a main characteristic of the AFM is that it can work under very diverse ambient conditions, from different liquids to vacuum. It is worth noting that each of these conditions requires different parameters of the measurement modes. For the handling of biological objects, the medium of choice will mostly be liquid, so that the cells can proliferate in a physiological environment. Another characteristic of the AFM concerns sample preparation. Since the tip "feels" the surface, no complex preparation is needed, for example, the need for drying and coating with a conductive layer when using an SEM. However, the samples have to be immobilized in respect to the tip.

8.2.1 Cantilever Position Measurement

For the detection of the cantilever bending, different methods can be deployed. In general, stationary systems – as are commercial AFMs – usually use **optical detection methods**, while mobile systems (such as autonomous robots) use detection methods based on **self–sensing options**. Only the most common method, that of laser beam deflection, and the robot–relevant method of piezoelectric or piezoresistive cantilevers, will be presented here. Other methods, such as interferometry, electron tunneling, or capacitance–based options (although the latter is widely used in purely indentation–oriented measurement setups or as displacement sensors for piezoelectric scanners) are omitted, as they are laborious, suffer from noise problems, or are unstable and thus will not be found in modern systems.

8.2.1.1 Optical: Laser Beam Deflection
The most common position detection method, used for almost every commercial system, is based on optical means. A laser beam is reflected by the cantilever. Every bending of the cantilever changes the position of the laser beam reflection, which is detected by a two– or four–segment **photodiode**. It is difficult to include this method in small and non–static systems, such as a mobile platform carrying a cantilever. Another drawback is the two–dimensional representation of a three–dimensional sample, which is not crucial when measuring forces in no more than two dimensions, for instance force–distance curves or mere imaging. It has its shortcomings, however, when trying to measure three–dimensional forces. Lateral forces can be detected, but whether a cantilever is bent because of forward/backward or down/uplifting forces cannot be specified by this detection method.

8.2.1.2 Self–sensing: Piezoelectric and Piezoresistive
The deflection of a cantilever can also be measured piezoelectrically/resistively by incorporating a **piezoelectric/resistive material** on the cantilever, which reacts with a specific potential difference or different resistance to a specific bending of the cantilever. A clever design of the resistors allows a cantilever to register the forces working on it in different directions, enabling even three–dimensional force measurements. Electrical noise can make this method more complicated to use, but a resourceful setup can make it a good alternative, especially for mobile platforms. Another problem can be the sensitivity, as these cantilevers have to be very thin or relatively very long, in order to be soft enough for contacting biological samples. While very thin cantilevers are difficult to fabricate, long cantilevers are noise–sensitive.

8.2.2 AFM Modes

Different modes can be used for image acquisition or the manipulation of different samples, all having their advantages and shortcomings in imaging or manipulating samples. To make use of the unique ability of the AFM to be able to work with living specimens at nanoscale resolution, the specimens are usually kept in a liquid suspension to generate a native environment. This atmospheric condition influences the **acting forces** between specimen, substrate, and tip, making the measuring in liquid different from measurements taken in ambient air or even in a vacuum.

In liquid, the acting force regime is characterized by small capillary forces and weak van der Waals forces. Additionally, the Q factor is lower in an aqueous medium, due to its higher density, than in air, thus dampening oscillations more. In the following sections, three common methods of measurement are described, with special emphasis on their use in an aqueous regime with soft specimens.

8.2.2.1 Contact Mode
During AFM development, the first imaging mode that was used was the **contact mode**. As the name implies, the cantilever tip is "in contact" with the surface of the sample, recording topography by the cantilever bending due to repulsive forces.

When scanning, the tip is dragged along the surface, which leads to obvious problems with softer materials, because of the lateral force applied. This mode can also be used for manipulation, *e.g.,* cutting chromosomes. Two different methods can be used for measuring in contact mode.

In the **constant height** method, the height of the cantilever in respect of the sample is kept constant. The bending of the cantilever due to variations in topology of the material is used for measuring the height. This method is obviously not very sensitive to the sample and is not suitable for very uneven samples, as the tip might lose contact, thus stopping the imaging or, conversely, crashing into the sample, which would lead to the destruction of either sample or cantilever.

Alternatively, the **constant force** method can be used, where the force acting on the cantilever (and thus the bending) is kept constant. To ensure this, the height of the cantilever is modified while moving over the specimen. The variation of height, which is detectable by the voltage applied to actuate the piezo–scanner, which either moves the probe or the scanner, therefore gives the necessary topological information. This method is more sensitive to the sample and, furthermore, is much more capable of adapting to uneven surfaces.

The different ambient conditions do not influence this measurement method at large scale, but inherent lateral forces make it problematic to use with soft materials, as the specimen can easily be moved or destroyed. Another problem is caused by the strong capillary forces in ambient air, which drag on the sample additionally to the lateral forces, when the tip moves in respect to it. When measuring in liquid, the capillary forces, at least, degrade substantially. Measuring soft samples is still possible in constant force mode. Cantilever force constants, however, should lie around 0.1 N/m, and preferably be even smaller. These very soft cantilevers exhibit another problem for as they are very sensitive to noise, the measurement setup has to be very well isolated from vibrations.

8.2.2.2 Dynamic Mode

In **dynamic mode**, the probe is oscillated near its resonance frequency above the sample. The relevant interacting forces are long–distance van der Waals forces, which influence and change the frequency of the cantilever. This mode can also be divided into two sub–modes, non–contact and intermittent.

For non–contact measurements, the cantilever oscillates above the sample, with a driving frequency lying slightly above the resonance frequency of the cantilever. This is chosen because the frequency of the cantilever decreases when brought into the force regime of the sample. This also leads to a decreasing amplitude, as in non–contact mode any contact of tip and sample is undesirable. Alternatively, the oscillation can be driven by a magnetic field. For this purpose, the cantilever is covered with a magnetic material. This approach is called **MAC mode**. The cantilevers for this mode can have a smaller spring constant as the amplitude vibrations in MAC mode are smaller.

The **intermittent mode** can be seen as a gentle combination of the two previously described methods of contact and non–contact mode, where the cantilever tip periodically comes in contact with the probe. Here, the cantilever is driven with a frequency slightly below its resonance frequency, thus leading to an increasing amplitude when approaching the sample. In this way, the tip periodically comes into contact with the sample. This method is often used for biohandling as is treats the sample much more sensitively, almost entirely avoiding disturbing lateral forces. As with the true non–contact method, the capillary forces, which can disturb the measurements, are not present when the sample is immersed in a liquid. When measuring in liquid, the otherwise disturbing capillary forces cease to exist, but additionally, the van der Waals forces this measurement method is based on are lowered (approx. 10 times lower than in air), along with the Q factor, resulting in a less sensitive system. The resonance frequency drops as well, to about a third of its value in ambient air. As a further drawback, the applied forces are not directly known, because the feedback is gained from the amplitude, not from the normal force. Both dynamic methods have the advantage of small (intermittent) or no (non–contact) lateral forces, thus being very gentle on the specimen.

8.2.2.3 Lateral Force Mode

This mode is a variation of the contact mode, where the cantilever is moved perpendicularly to its aligment for a line scan. **Lateral torsion** of the cantilever is detected. The cantilever can thus measure the friction of the sample surface, as well as its topology. Because of the obviously strong lateral forces, this method can only be recommended for relatively hard samples that are very well fixed to the substrate.

8.2.2.4 Jumping Mode / Force Volume Mode and Force–Distance Curves

The **jumping, or force volume, mode** is a series of force–distance curves evenly distributed over the desired specimen, coupled with feedback for approaching the sample in between each curve taken. The cantilever is brought into contact with the surface, with active feedback keeping it at the set point value. Upon reaching the set point, the cantilever is retracted with turned–off feedback, thus measuring the force–distance curve. At the maximum tip–sample distance, the tip is moved to the next measurement point, avoiding lateral forces on the sample. The force acting on the specimen is controlled by the set point. The characteristic shape of force–distance curves in distinct atmospheric conditions differs owing to the specific acting force regimes. In ambient air, the capillary forces are rather strong (Figure 8.1a). Having approached the surface, the cantilever jumps into contact, because of the attractive forces acting between tip and sample. Upon retracting, the cantilever "clings" to the sample surface until the retractional forces overcome the adhesive forces. This point is called the **snap–back point**. These forces are strong enough to disturb or even to destroy sensitive specimens. For specimens covered with one or more additional layers (*e.g.*, liquid layers are common), the part of the retraction curve where the cantilever still clings to the surface can be in the style of a step function. This is owing to the different snap–back points at which the cantilever breaks free from the current layer or specimen. While the capillary forces tend to be a problem in ambient air, they cease to exist in an aqueous solution, making the jumping mode a very sensitive method there. The curve in Figure 8.1b shows a typical force distance–curve in liquid, where the capillary forces on the cantilever are nearly non–existent. Thus, jumping mode allows the exertion of a controlled force on the specimen all the time, which is an improvement on the unknown forces applied in intermittent mode.

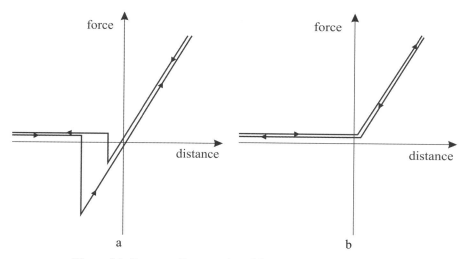

Figure 8.1. Force *vs.* distance, plotted for a: ambient air and b: liquid

Because of the additional effort of the force–distance curves, this method is rather slow in comparison to both the contact and dynamic mode. However, it is easier to implement than the dynamic mode, since no dynamic options have to be integrated in hard– or software. As a bonus, adhesion and elasticity information can be gained at the same time as topological information. The cantilever specifications are similar to contact mode, needing soft cantilevers to avoid damage to the specimen.

8.2.3 Measurements of Different Characteristics

Depending on cantilever, tip, and mode, different characteristics of a sample can be measured. The capability of the AFM not only to conduct topographical measurement, but also to measure different characteristics, ranging from elasticity to conductivity to chemical composition, is established this way. As topological information, including height profiles and general spatial measurements, is rather self–explanatory, it will not be dealt with explicitly.

8.2.3.1 Mechanical Characterization
By means of **force–distance curves**, several, mostly mechanical, characteristics of a sample can be measured. This includes, for example, the **Young's modulus**, stiffness, and adhesion.

When measuring the Young's modulus of a cell membrane or a biofilm, it has to be taken into account that the material itself will give way to the acting force. The force detected on the cantilever is, therefore, combined with the indentation depth:

$$F = k \cdot d(z) = k \cdot (z - \delta) \tag{8.1}$$

where F is loading force, k cantilever spring constant, $d(z)$ cantilever deflection, z piezo position, δ indentation depth.

The curves are generally not as steep as indentation curves on a hard substrate such as silicon (Figure 8.2). For calculating the Young's modulus of the substrate, a model developed by Hertz and enhanced by Sneddon [11] is used:

$$F_{cone} = \frac{2}{\pi} \frac{E}{\left(1-v^2\right)} \tan\left(\alpha\right)\delta^2 \,, \qquad (8.2)$$

where F is loading force, E Young's modulus, v Poisson ratio, α half opening angle of conical tip, δ indentation depth:

$$E = \frac{\pi \cdot k \cdot d(z) \cdot \left(1-v^2\right)}{2 \cdot \tan\left(\alpha\right) \cdot \left(z - d(z)\right)^2} \,, \qquad (8.3)$$

With a measured force–distance curve, this equation leaves only the Young's modulus E as a variable. Thus, Equation 8.3 has to be fitted to the linear part of the measured curve by the variation of E.

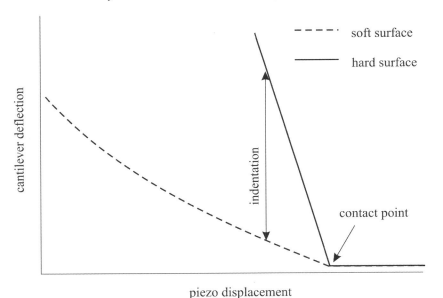

Figure 8.2. Cantilever deflection *vs.* piezo displacement, plotted for hard and soft surfaces

For the force–distance curves, sometimes the use of spherical tips is preferable to conical ones, due to a better defined contact area, more sensitive handling (damage by poking), and a smaller indentation depth. For spherical tips, the loading force is described by:

$$F_{paraboloid} = \frac{4}{3}\frac{E}{1-v^2} R^{1/2}\delta^{3/2},\tag{8.4}$$

where F is loading force, E Young's modulus, v Poisson ratio, R sphere radius, δ indentation depth.

The thickness of the material is another important parameter, as with very thin materials, the substrate below can influence the measurements, due to the indentation depth [12].

8.2.3.2 Magnetic Force Measurements

Measurements of **magnetic forces** are conducted in non–contact mode by a tip coated with a ferromagnetic material. As the magnetic forces have a larger range than the forces exerted by the surface, the two forces acting on the cantilever can be distinguished. The measurement is conducted in two steps. First, a topological image is taken. In a follow–up scan, the cantilever is raised to a certain height (below 100 nm) where only the longer–ranging magnetic forces influence the frequency shift of the cantilever.

8.2.3.3 Conductivity Measurements

In this mode, topological and conductive information can simultaneously be gained. **Conductivity measurements** can be conducted in contact or non–contact mode. In contact mode, an AFM tip with a conductive coating is used. A voltage bias is applied between cantilever and substrate and from the resulting current flow, a spatially resolved conductivity image can be measured. This is of interest, for instance, in visualizing ionic processes across a cell membrane.

For measuring the electrostatic characteristics of a sample, a voltage is applied between tip and sample, but the cantilever is not brought into contact while scanning the sample. The electrostatic forces acting between tip and sample influence its deflection, and thus charged domains can be detected. Another option for non–contact measurements is **Kelvin probe force microscopy**, in which a charged cantilever is scanned across a surface with applied potential. The potential is composed of an AC and a DC component. Potential differences of the DC part between cantilever and sample surface make the cantilever oscillate. By means of a feedback mechanism, the potential of the cantilever is then fitted to minimize the difference and thus the oscillation of the cantilever. As a result, the work function[1] of the sample surface can be calculated from the measurements.

---

[1] The work function describes the minimum energy that has to be applied to remove an electron from inside of a solid to a point directly outside of the solid's surface.

8.2.3.4 Molecular Recognition Force Measurements

By means of **functionalized cantilevers**, the chemical structure or **intermolecular binding forces** can be analyzed. The tip of the cantilever is coated with ligands or antibodies (Figure 8.3), which have a specific corresponding partner molecule in the sample. Usually, the sample is brought in contact with the tip (or *vice versa*, depending on the AFM setup) and, after reaching contact, is slowly retracted. The breaking of the intermolecular bonds can easily be measured by the snap–back point of the resulting force–distance curve.

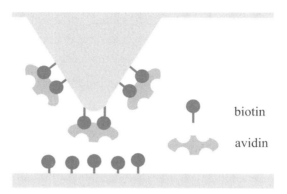

Figure 8.3. Measuring ligand–receptor binding forces with a functionalized cantilever

An improvement on this method has been developed [13], where an antibody was bound to the tip by a flexible part, allowing much better single binding of the respective molecules (Figure 8.4). The cantilever can be driven in dynamic mode, so that in addition to the binding force experiments and the recognition measurements, where the distribution of certain molecules in the sample can be visualized, topological measurements can be done at the same time.

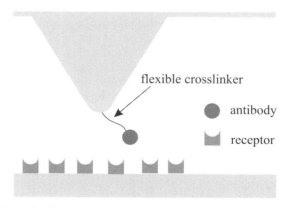

Figure 8.4. Measuring ligand–receptor binding forces with a functionalized cantilever

8.2.4 Sample Preparation

Although **sample preparation** is not as complex as for the SEM, which requires its samples to be dried and conductively coated, the sample has to be brought onto a substrate or otherwise **immobilized**. Different immobilization methods exist, from holding with a pipette to special, functionalized substrates.

A common method is to place the sample onto a substrate of freshly cleaved mica, which has a charged, hydrophilic surface, to which proteins and other biomolecules bind easily. Sometimes, the mica is treated with a special solution such as APTES,[2] which changes the surface charge. Additional coating of the mica with gelatin has proved to be effective when imaging bacteria parts [14]. Another practical method is a gold substrate coated with protein–reactive monolayers, making use of intermolecular binding forces for immobilization.

More restrictive physical methods enclose the sample in a gel–like agar–agar or a porous medium, to avoid displacement. For instantaneous measurements, these methods are advantageous, supporting the natural height of the samples. However, they become problematic when the samples begin to proliferate, as cell cultures usually do.

8.2.5 Cantilevers

Cantilevers are characterized by their material, tip shape and mechanical parameters such as **spring constant, resonance frequency, and quality (Q) factor**. For biological applications, cantilevers of silicon or silicon nitride are usually chosen. This material is chemically inert and can be doped to dissipate static charges. Also, the material allows a high Q factor for high sensitivity. When working in liquid, the pH value and electrolytes influence the interacting forces between tip and sample. For instance, under physiological conditions, silicon is charged negatively.

Different treatments exist to allow a specific surface charge or hydration properties, namely plasma treatment or silanization. Silanization is also used for modifying the surface chemistry of substrates (for instance DNA binding to mica treated with APTES, Section 8.2.4). Tips can be functionalized with chemical coating (often as a preparative measure for biological coating) or biological coating, which can cover different interactions between receptor and ligand, molecules and cells, or cells and cells. Another option of cantilever modification is a reflective coating of the cantilever, which can be beneficial, since the biological buffer solution can dim the reflected laser beam.

The **tip shape** is of crucial importance to the imaging quality. However, although a smaller tip radius yields more precise images, for imaging biological samples sometimes a duller (thus broader) tip is preferable, as it lessens the risk of damaging the sample. For elasticity measurements, besides the usual conical ones, spherical tips are also used, as they allow a smaller indentation depth and a better–defined contact area. Nevertheless, these tips usually measure an average elastic

---

[2] Aminopropyltriethoxysilane, an amino–functional chemical compound.

modulus, in comparison to the higher lateral resolution by conical tips, which allow local measurements at different regions of the cell.

8.2.6 Video Rate AFMs

Common AFMs are limited to a rather slow scan rate in the minute range, allowing the measurement only of static or slowly developing processes. Several institutes are independently developing a **high–speed AFM**, trying to overcome the mechanical limitations to the scan speed. By optimization of the mechanical performance of stages and cantilever, scan rates of 10 s per image can be reached.

However, just as for the mechanical limitations, resonant frequencies stemming from the utilized piezoelectric actuators could not be completely overcome and a new approach has, therefore, been tested. Instead of avoiding resonant frequencies by the use of lower driving frequencies, a microresonant scanner was used, which utilizes these frequencies. The cantilever was enhanced by passive means to maintain tip–sample interaction. In combination with dedicated high–speed electronics, a system has been built, which is able to image an area of 3×3 µm with a resolution of 256×256 pixels at 15 frames per second.

As the forces acting on the sample are much smaller, it is possible to depict even soft samples rapidly. The manufacturer claims that their system can be easily combined with existing commercial systems, upgrading them to video–rate capability, while preserving the standard AFM methods. [15].

8.2.7 Advantages and Disadvantages of AFM for Biohandling

In comparison to other microscopes, the AFM has several advantages when handling and imaging biological samples. It actually cherry–picks from the abilities of other microscopes, making it a very powerful characterization device. From the previous sections, the advantages of AFM–based biohandling can be summarized as follows:

- A major advantage of the AFM is the **variety of media** in which it can be used. Biological probes can be examined under physiological conditions, and offer the capability to work with living specimens.
- The possibility to use the AFM tip for **nanomanipulation** can be very useful. For example, higher–lying protein rings can be moved away, or the separation of chromosomes can be combined with the extraction of DNA. A cell membrane can be indented, or an ion channel can be activated by applying stress.
- The possible resolution lies in the **sub–nanometer range**.
- **Three–dimensional** topological information is offered.
- No special structure–changing **preparation** is needed. Samples do not have to be dried or electrically conductive, as for instance in SEM measurements, where the probes have to be dehydrated and coated with a conductive material, if they do not possess these characteristics naturally.
- **Different characteristics** of a cell can be measured, such as elasticity, conductivity, adhesion, *etc.*

However, owing to the functioning principle of the AFM, there also are some disadvantages, limiting the possible measurements, which are summed up as follows:

- **No real three–dimensional force measurements** can be conducted due to the construction of the AFM. While lateral forces (x–direction for convenience) can be measured, it is not possible to detect deviations in either the z– or y–directions separately.
- Another drawback is the **scanning speed**, which is usually much too slow for real–time imaging. The newly available video–rate AFMs might help to solve this problem, but for now, their scanning area is very small (approx. 3 μm).
- The possibilities of nanomanipulation can also be a drawback: soft materials can be **compressed**, rendering height information valueless, be **damaged**, or be **dragged** from their initial position.

8.3 Biological Background

A brief section on the **biological background** is inserted here to give an idea of the dimensions and forces in regard to biological samples. Some keywords are amplified to give better understanding of the application sections. The topics covered include the mechanical, chemical, and electrical characteristics of cells as well as a short description of bacteria, ion channels, and intermolecular binding forces, all of which can be found in this section. For more detailed information than in the highlights presented here, biological textbooks such as [16] are recommended.

8.3.1 Characteristics of Cells

Biological cells can be classified into two kingdoms: prokaryotes (bacteria) and eukaryotes (animal, fungi, plant, and other cells). They differ in structure and DNA storage. While eukaryotes have a nucleus and organelles, and can be both multicellular and single cellular, prokaryotes lack a nucleus or special organelles inside. Viruses do not belong to these two kingdoms, as they are not able to reproduce themselves, but need a cell host to do so. They are mainly capsules containing DNA information, which can dock to living cells and infect and alter them.

8.3.1.1 Mechanical Characteristics
The **mechanical characteristics** of eukaryotic (and some prokaryotic) cells are mainly defined by their **cytoskeleton**, which consists of actin–filaments, microtubules, and intermediate filaments. The cytoskeleton is responsible for cell motility, cytokinesis, [3] and the organization of the inner structure of cells. Moreover, it

---

[3] Cytokinesis: Cell division.

has great influence on the elasticity of the cell membrane. **Cell motility** is of special importance, *e.g,.* in wound healing, phagocytosis,[4] or the building of neuronal nets. Thus, cell mechanics and measuring cell elasticity have always been of interest and were examined by different means (cell poking, optical tweezers, *etc.)* even before the development of the AFM. The AFM has the advantage of a much higher lateral resolution in comparison to these methods. Certain parts of the cytoskeleton structure are of special importance in cell elasticity. For instance, elasticity changes due to **actin–filament** and microtubule degradation has been observed [17], where cells, after having been exposed to actin–degrading chemicals, showed a considerably increased elasticity. The cytoskeleton is enveloped by and attached to the cell membrane, which today is seen as a lipid bilayer with freely moving lipids and embedded proteins, as proposed in the fluid mosaic model by Singer and Nicolson, which is still used today, albeit with some modifications [18].

8.3.1.2 Electrical Characteristics

Neurons and muscle cells mainly exhibit certain **electrical characteristics**, which are of importance in, *e.g.,* neurological transmissions. The membranes of these cells are excitable and have a potential, which is regulated by the ion transfer in and out of the cell via certain ion channels. This so–called **action potential** is a rapid, self–progressing state of excitation of the membrane. It is caused by a depolarization of the membrane, which is followed by an opening of voltage–activated Na^+ ion channels. Opening these channels lets Na^+ ions pass into the cell to change the potential of the membrane surrounding the ion channel from -70 mV to $+50$mV. Besides voltage–activated ion channels, other types exist. A brief but more general description of ion channels is given in Section 8.3.3.

The fact that **neurological activity** is determined by electrical currents can be utilized in the artificial activation of these processes. Recent experiments attempted to stimulate cells by using single–wall nanotubes as an interface [19]. Carbon nanotubes and their electrical and mechanical characteristics are introduced in Chapter 7. Potential shifts can also be found for photoresponsive plant material, where they are the result of the light–induced production of indoleacetic acid, a growth hormone. This effect is used by the plant to grow towards the light. As the material is rather sensitive, touching the surface can result in erroneous signals because of stress, in wounding the cell, or other damage. For sensitive measurements, therefore, the non–contact method of Kelvin probe force microscopy is applied [20]. Another field of interest is the use of biological material merged with electronics for miniaturizing electrical circuits. For instance, in the case of DNA, it is not its electrical properties in a biological context that are of interest, but its being a building block for those circuits and for biosensors. Here, the conductivity measurements are usually conducted in contact mode.

---

[4] Enveloping of large mostly phatogenic particles by the cell membrane, followed by internalization into the cell.

8.3.1.3 Chemical Characteristics

The **chemical characteristics** of a cell are found within several areas, as in the building of the membrane (peripheral proteins), with special **ion channels**, which are ligand–activated and in **intermolecular binding forces**. Intermolecular binding forces, as experienced in antibody–antigen bindings, can help to examine *e.g.* the composition of cells. Because of the specificity of the binding pairs, they can be identified easily, and thus are used *e.g.* as marker proteins in fluorescence microscopy and other measuring methods. The membrane potential can additionally be determined by chemical differences (for instance Ca^+ concentration).

8.3.2 *Escherichia Coli* Bacterium

Escherichia coli is one of the most common objects in cell biology. This bacterium lives in the lower intestines of warm–blooded animals and is responsible for the digestion of food. It is very well analysed and its complete genome has now been completely decoded. While usually *E. coli* is a necessary and friendly bacterium, certain strains of the *E. coli* family are hazardous to humans and are able to infect previously harmless strains, making them pathological.

A common problem nowadays with bacteria is the **resistance to certain antimicrobial drugs**, for instance, antibiotics, as a resistant bacterium can spread its alterations by transferring its corrupted DNA to other bacteria. This is done by cell–to–cell contact; the process is called **conjugation**. For building the cell contact, *E. coli* has small "hairs" (actually plasma tubes) on its cell membrane, technically called fimbriaes, or pili for very short hairs. An examination of these hairs can give information about conjugation, and help to avoid resistance transfection.

While the bacterium itself measures 1–2 m in width and 2–6 µm in length (thus being well visible under an optical microscope), the fimbriaes are several µm long

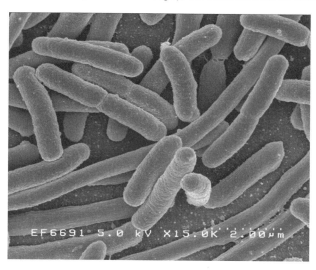

Figure 8.5. An SEM image of E. coli (Rocky Mountain Laboratories, NIAID, NIH)

but only of 3–10 nm in diameter, which excludes them from investigation with an optical microscope. Figure 8.5 shows an SEM image of the specimen.

8.3.3 Ion Channels

The cell membrane is considered as a lipid bilayer (about 5 nm thick) with **embedded protein molecules**. Some of these protein molecules function as gates in and out of the cell; the size of these proteins lies in the nm range (Figure 8.6). They are responsible for the absorption and dispensation of ions, resulting in their name **ion channel**. In this way, they **change the membrane potential**, which is used for signal transduction.

Ion channels are activated by different means. Voltage activation by the membrane potential, as for Na^+–channels, which are important for signal processing in neurons and muscle cells, has already been described in Section 8.3.1.2. Other channels are ligand–activated, for instance, the nicotinic acetylcholine receptors, which respond to the binding of nicotine. The stereocilia in the inner ear, which enable us to hear, contain ion channels that respond to mechanical stimuli (thus termed "mechanosensitive" or stress–activated) as produced by sound waves. And the photoreceptors in the eyes of vertebrates are sensitive to light; for the more sensitive rod cells[5] even a single photon can result in a detectable electrical response.

Ion channels are a favorite target in drug research, as they are involved in diverse processes which lead to rapid changes in a cell, such as in cardiac muscles, for t–cell activation, and insulin release, alongside some directly ion–channel–based diseases such as the shaker gene, the brugada syndrome, and some sorts of epilepsy. Some venoms, for example, work by "blocking" ion channels, thus

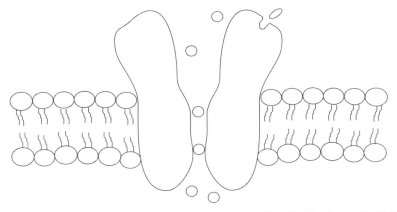

Figure 8.6. Scheme of a ligand–activated ion channel, embedded in the lipid bilayer

---

[5] Rod cells are brightness sensitive, enabling black and white to be seen even in dim light, while the cone cells, mainly working in brighter light, are responsible for the perception of colors.

prohibiting signal transduction, which results in paralysis. Stress–activated ion channels are of special interest in AFM handling, because due to the flexibility of using an AFM as a nanomanipulator in direct contact to the sample, these channels can be activated by the cantilever tip (Section 8.4.3.1).

8.3.4 Intermolecular Binding Forces

For different processes, the **key–lock principle** (Figure 8.7) is found in micro-molecular biology. As mentioned, some ion channels are activated by ligands. They possess a specific receptor, to which the ligand can bind, thus opening the ion channel.

Other important examples of intermolecular bindings are processes in the immune system of the body. Antigens and antibodies build similar bindings, thus deactivating viruses or bacterical toxins, or activating carrier and t–cells to take care of invading microorganisms or parasites. Studies of these binding forces can aid in the development of medicine, as it helps to understand the activation processes of immune cells. It may also be possible to block harmful cells or viruses from docking to healthy ones. The binding forces lie in a range of some pN, requiring very sensitive and noise–free measurement methods.

Binding forces between proteins are also responsible for **adhesion** between cells. These adhesion forces in cell–to–cell contact play a major role in biological processes, for instance in the immune system, when white blood cells attach to infected cells (Figure 8.8). The trapping centers needed for attachment build up or decompose in seconds. This is an effect which is also of interest in self–assembly tasks, in biological manipulation, and in characterization setups [21].

Figure 8.7. Key–Lock principle, antigen and antibody binding

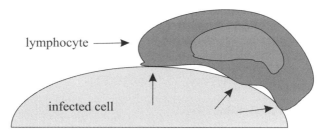

Figure 8.8. A lymphocyte attaches to an infected cell, owing to adhesive domains

8.4 AFM in Biology – State–of–the–art

This section gives an overview of the state–of–the–art in AFM–based biological research. Its versatility, as well as an insight into the areas which benefit from the newly available measurement methods, is presented, ranging from purely imaging tasks, over the characterization of different physical, chemical, or electrical properties, to the use of the AFM as a manipulator as well as the cooperation with other end–effectors.

8.4.1 Imaging

A major field of AFM application is **visualizing biological objects**. Due to the possibility of measuring with nanoscale resolution in aqueous solution, this option is highly attractive for research and can give insight into molecular processes. The possibility of working in a buffer solution with non–static samples offers the possibility to visualize processes as well as taking mere snapshots.

For example, the effect of poison on cell membranes has been examined by the CeNs group at the Ludwig–Maximilian University [22]. The poison contains an enzyme, which lets the **cell membranes degrade**. These enzymes dock to defects in the membrane, where they begin to react with the sample. The resulting rifts that gape in the membrane have a width of merely 10 nm, and holes of around 100 nm. Both can be imaged by the AFM. The knowledge of how enzymes work can be of benefit in drug development, for example, molecules similar to the imaged ones are found in the gastric acid.

Other investigations at the Max Planck Institute for molecular cell biology and genetics have visualized the reaction of the proteins responsible for **intercellular communications** [23]. These gap junctions exchange molecules and signals. At the connexine, 26 gap junction surface pores are formed, which close under the influence of Ca^{2+}.

Rare–earth metal ions are used as an additive in fertilizers for agriculture. To analyze the influence of the ions on living specimens, the reaction of *E. coli* bacteria exposed to La^{3+} ions has been measured by means of AFM and SEM imaging [24]. The imaging was conducted in MAC mode, where only small forces act on the specimen. From the **surface analysis** of the bacteria, it could be seen that the surface roughness and the permeability had increased after a treatment with

La^{3+} ions. These are structural alterations, on which the influence of La^{3+} in organisms is based.

8.4.2 Physical, Electrical, and Chemical Properties

8.4.2.1 Elasticity and Stiffness Measurements

The elasticity of a cell membrane is an indicator of several changes – for example, older cardiac cells tend to get stiffer, while cancer–infected cells generally expose a softer membrane. Other findings are the change of the whole cytoskeleton, which influences the cell motility – whichis significant, *e.g.*, when healing a wound in the skin. The specific conditions underlying these measurements have already been discussed.

For instance, **cell volume control** is influenced by the cytoskeleton. Experiments tackling this area were conducted by [25], where the influence of osmotic stress on the stiffness of kidney cells has been investigated. Physical changes such as the detachment of the actin cortex from the membrane due to swelling of the cell could be indicated by the measurements of the Young's modulus. Different measurements with two types of probes have been conducted, one with a spherical tip for an average measurement on a cell, or conical tips for local measurements in the cell's center or the periphery, and imaging. The mechanical changes of the cells due to osmotic stress were proven to be reversible. It has been found that the swelling of the cell due to osmotic reaction gets ahead of the cytoskeleton–based physical adaptation of the cell, thus rupturing the actin cortex or detaching the cytoskeleton from the membrane.

In the inner ear, mechanosensitive hair cells are responsible for converting sounds into electrical signals. The **elastic properties** of these hair cells are influenced by filament crosslinking between neighboring hairs and their cytoskeleton. The AFM allows unique experiments, where a single stereocilium[6] can be displaced. To ensure that the mechanical stimulus targeted at one hair is not stimulating neighboring hairs, the coupling of these hairs was investigated. The first experiment included the displacement of individual stereocilia with an AFM tip, in order to measure the stiffness of the stereocilia in relation to their position. The other experiment tested the **force transmission** to neighboring stereocilia. Up to two stereocilia were stimulated by an end–effector, while the whole hair bundle was scanned with an AFM probe. It was found that the influence of adjacent stereocilia was rather small, and the stimulation of a single stereocilium is, therefore, possible. This is a good basis for measuring the reaction of single transduction channels (which are mechanosensitive ion channels, Section 8.3.3), for which the exact knowledge of mechanical characteristics of the stereocilia is of primary importance [26].

---

[6] Stereocilia are mechanosensing organelles of hair cells, working for instance as an acoustic sensor in the cochlea of the human inner ear.

8.4.2.2 Intermolecular Binding Forces

Recognition force measurements for chemical analysis have, for example, been conducted by the University of Linz, where new methods for **antigen–antibody binding force measurements** were developed [27]. For these measurements, the tip of the cantilever was modified with a halved antibody, bound to it by a flexible interface. This modification ensured that binding with only one antibody at a time occurred. The cantilever can be used in dynamic mode, allowing it to take both topographical and recognition pictures at the same time. These methods were used in investigations concerning the localization of the well–studied single avidin– biotin interactions, to develop a stable setup for topographical and recognition imaging.

Another field of interest for the analysis of intermolecular forces is the study of biosensors. There are a lot of different **biosensors**, all of which are based on the principle of an analyte binding to a biological interface (receptor). These binding forces are of interest in the development of new sensors. For instance, there are receptors which change their structure when a ligand (analyte) binds to them. Using AFM measurements, these rigidity changes can be analyzed, and the data can be used for the optimization of existing biosensors or the development of new ones. The are some systems which as a sensor use nucleic acid, which is more sensitive and specific in its function. However, the isolation of nucleic acid has proven to be very complex and these systems are still, therefore, very challenging [28].

8.4.2.3 Adhesion Forces

For the measurement of cell adhesion, the AFM was utilized to advance experiments into the molecular range, where previous methods of adhesion measurements had failed to distinguish whether the bond was composed of many weak or a few strong bonds. At the Ludwig–Maximilian University, a setup for measuring **cell adhesion at the molecular level** has been built [29]. To provide a smooth surface for cell attachment, either the tip of a cantilever was removed, or a sphere– tip cantilever was used. Afterwards, the cantilevers were functionalized with adhesive molecules to let cells attach to them. For one experiment, cells were cultivated on the spherical tip, growing a cellular monolayer. The adhesion of these cells was tested with several different surfaces. Various parameters, such as contact time, were considered as well. For each surface, several force–distance curves were taken. Here, an increase of the adhesion load could be found, until the maximum adhesion force was reached. This is visible in the retraction part of the force– distance curve, indicated by the snap–back point (Section 8.2.2.4), followed by a series of detachment processes visualized by force steps. The setup was applicable to a multitude of different cells' adhesions.

More application–oriented measurements were conducted at Purdue University, where a material for small vascular grafts was researched to improve cell adhesion to this material [30]. The widespread atherosclerotic vascular disease is often treated with the implantation of synthetic vascular grafts, if the patient does not have sufficient vascular material of his own. In the course of five years, these synthetic grafts suffer from a strong decrease in potency of up to 95% in large– diameter, and 30% in small–diameter, vascular grafts. New materials are therefore researched, to help improve vascular cell functions. The fibronection binding to differently patterned polylactic–co–glycolic acid surfaces was measured. Different nm–sized features where found to have a different effect on the binding of cells to the substrate. By means of topological AFM imaging, the cell absorption on the substrate could be measured. The findings are not only of interest in vascular treatment, as the researched binding protein fibronectin is of major importance in cell adhesion for many different types of cells.

8.4.2.4 Cell Pressure

AFM–based **cell volume measurements** have been accomplished by [31]. These measurements aided the experiments concerning the influence of the hormone aldosterone, which controls the water and electrolyte balance in the kidney, on cells of the cardiovascular system. Volume information was gained via the topological images, which contained height profiles of the cells. It could be shown that the volume of the cells actually increases after cultivation in a medium containing aldosterone. This finding can help in treating cardiovascular dysfunctions.

8.4.2.5 Virus Shell Stability

Force measurements on virus shells were conducted by [32], followed by a specific structure analysis. The capsids of viruses are not only involved in protecting the viral DNA, but also in selectively packing and injecting it to hosts. For measuring the structure and topological information, a series of jumping mode images were taken that allow the application of a **controlled force** to the capsid. The contact mode proved to be problematic – because of the lateral forces, the capsids were rolled over the substrate. After the "top" of a capsid was identified by topological imaging, a force–distance curve was conducted at this point. An inhomogeneous shell structure was found, as the calculated Young's moduli where closely arranged arround two different values. Information about **resistance** and **stability** was gained by repeatedly pushing the capsids. It was found that the capsids recovered their original height after a few ms. These experiments provided new information about shell structure, which is – because of its clever design – of interest in design techniques in nanotechnology, and proved that the capsids give not only chemical but also mechanical protection.

8.4.2.6 Electrical Properties of DNA

In the miniaturization of electronics, there are attempts to find suitable components for molecular devices. DNA, being well known and suitably sized, seems to be a good candidate for serving as a **wire in nanoelectronics**. As the electrical characteristics of such small–scale objects are expected to considerably differ from macroscopic objects, the exact measurement of these properties is of great importance. **DNA conductivity** has been a frequently discussed topic in recent years. Sample preparation and measuring techniques seem to have an overwhelming influence, since heavily differing transport properties touching the whole spectrum from conductor, semiconductor, supra conductor, to insulator, have been reported. Recent research confirms the conductivity of dsDNA under special conditions, while a self–assembled monolayer of ssDNA has been discovered to serve as an insulator [33]. Here, different types of DNA were brought onto a gold substrate by coupling them with an thiol group. Some DNA strands were connected by means of another thiol group to a 10 nm gold nanoparticle. This particle was contacted by the gold–coated AFM tip, and current–voltage curves were recorded. Other experiments include measurements without the gold nanoparticle. It has yet to be seen whether variations in these measurement techniques will come to a repeatable and usable result.

8.4.3 Cooperation and Manipulation with an AFM

8.4.3.1 Stimulation and Recording of Mechanosenstive Ion Channels
The capability of the AFM tip to modify the sample at a nanometer level can be used to **stimulate mechanosensitive ion channels**. The exact knowledge of the mechanical properties of the channel–bearing structures are important. For instance, for displacing and characterizing single hair cells, the validity of AFM–based measurements has been clarified in [26]. The electrical response of one or more channels can be measured by an additional patch–clamp pipette [34]. Within the cochlea, individual stereocilia of hair cells were stimulated by displacement, analogously to the stiffness measurements described earlier. The electrical response, a transduction current, was simultaneously recorded via a patch–clamp pipette in a whole cell configuration. Even responses of single channels could be recorded, due to the ability to displace only a single stereocilium without significant influence on neigboring hair cells [35]. This method was not only shown to be useful in mechanical coupling, but also for the electrical response of the channels. These experiments can give a better understanding of the **gating processes in the inner ear**.

8.4.3.2 Cutting and Extraction Processes on Chromosomes
For the task of **dissecting chromosomes**, a sophisticated nanomanipulator was built, consisting of an AFM installed in parallel to a laser dissector [36]. The nano-manipulator not only allows interactive control via joysticks, but also haptic feedback, enabling the user to directly feel the applied forces. The abilities of both instruments to cut chromosomes were examined. Here, the cantilever tip was placed using optical control near to a spread of chromosomes. The chromosomes were then imaged, using the AFM, to find a suitable adjacent pair for dissection. The dissections were conducted by the user's input via joysticks. It was discovered that the cuts accomplished by the cantilever were more precise when the cut width was smaller. While the minimal cut width of the laser dissector was found to be 380 nm, the minimal cut width of a successful mechanical dissection was 280 nm. The force applied on the cantilever has to be chosen carefully, as insufficient force will result in a not fully dissected chromosome. Given an appropiate force, very precise cuts can be performed, *e.g.,* the dissection of only one chromatide ("arm") of a chromosome. An additional benefit is the fact that after cutting the chromosome, DNA material adheres to the cantilever tip. The extracted DNA has been used in biochemical analysis, where the validity of gaining DNA samples using this method was approved.

8.4.4 Additional Cantilever

For combining two independent AFM measurements, it would be very laborious to incorporate two custom AFMs. The use, preferably, of a **self–sensing cantilever**, mounted perhaps on a mobile platform, seems to be a viable alternative to an additional bulky custom AFM. In the framework of the MICRON project [37], such a preliminary platform has been developed, and first results proved to be promising. A robot bearing a self–sensing cantilever was able to image a compact disk surface. While the piezoelectric cantilever still has some problems reaching the accuracy of optical position measurements, some of these problems stem from electrical noise, which can be decreased by a clever design of circuits. Although the main idea of this project was a swarm of small, independent, cooperating robots carrying different tools (force sensor, pipette), further developments of AFM probes cooperating with a tool–carrying robot can be imagined. The advantage of such an independent second AFM cantilever could be the ability to use one of the cantilevers as a stress generator and the other as a sensor, in order to, *e.g.,* experiment on stress transmission of cells.

8.5 AMNS for Cell Handling

The AMNS concept introduced in Chapter 1 can be utilized for handling and characterizing biological cells. Analogously to the **AMNS** presented in Section 7.5, a setup for **mechanical cell characterization**, initially for measuring cell elasticity, has been developed. As an optical sensor, an inverted optical microscope is used instead of an SEM, as the samples used are of a size still visible with an optical microscope, and measurements with living cells are proposed. Nevertheless, operating this station, *e.g.,* in an ESEM (environmental SEM) is definitely possible. The force measurements are conducted with a piezoresistive cantilever. Similar to the experiments for CNT characterization, coarse positioning of the cantilever is conducted by the nanomanipulator, while fine positioning and the force–distance curves are carried out using the nanopositioning stages.

8.5.1 Experimental Setup

First measurements were taken using the setup depicted in Figure 8.9. The AMNS differed from other stations, as changes had to be made to adapt the station to the use with an **inverted optical microscope** and the handling of **biological samples**, which were available in containers different to, *e.g.,* the CNTs.

For this, the three degree–of–freedom piezo–scanning stage [38] was equipped with an attached holder for a glass slide or Petri dish containing the biological sample to be analyzed. The holder had to have a recess for the optical path and was positioned sideways to the piezo–scanning stage for the same reason. A modified base plate was mounted on the specimen stage of an inverted optical microscope [39], which was used as an optical sensor. The integrated capacitive sensors of the piezo stage allowed it to be operated in closed–loop mode.

Figure 8.9. Setup of the nanohandling station for characterizing biological cells

The three–axes nanomanipulator [40] was attached to the base of the scanning stage. It offers a theoretical resolution of 5 nm in the x– and y–directions and 0.5 nm in the z–direction. As an end–effector for the nanomanipulator, piezoresistive cantilevers [41] were used. These AFM probes have a pyramid tip with a radius below 10 nm and a height above 17 μm.

8.5.2 Control System

Figure 8.10 shows the control scheme used for the elasticity measurements conducted with the AMNS. The characterization of biological cells is performed by bringing the **piezoresistive AFM probe** in contact with the sample. The resulting bending of the cantilever due to interacting forces leads to a varying resistance, which is converted to a detectable voltage by a four–wire wheatstone bridge. The voltage signal is amplified and digitized by a custom–made bridge amplifier and transferred to the control PC via the universal serial bus. Different input possibilities are given. The user has the opportunity to interact with the system to **manually** perform the first experiments, beginning with the nanomanipulator positioning, which can be operated manually, by a joypad input device, or by the graphical user interface (GUI). Meanwhile, the image acquired from the optical microscope is giving visual feedback to the user. In addition to the height information gained from the focus scan of the microscope, the sensor data received from the piezoresistive cantilever are used as a safety mechanism, avoiding possible tip crashes. As a next step, the piezo–scanning stage holding the sample can be manually positioned to a desired location. Again, the force data gained from the piezoresistive cantilever are used to avoid a collision between the AFM tip and the sample surface.

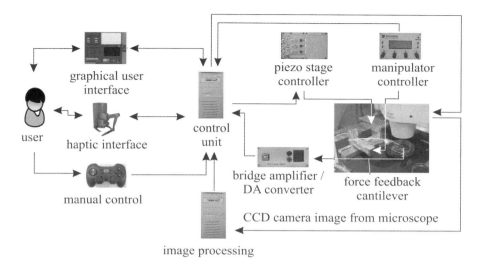

Figure 8.10. Control scheme of the AMNS for cell handling

The stage controller delivers an analog signal encoding the stage's exact displacement, so the stage can be operated in closed–loop mode. A haptic interface can be used as an auxiliary input method to move the scanning stage for fine positioning of the sample in respect to the cantilever. The forces acting on the AFM cantilever are rendered on the haptic device to give the user additional feedback of his manual operation.

To accomplish the highly automated measurements, a complex control system is needed. The system has to cover object recognition and tracking tasks, analyzing images gained by the inverted optical microscope. The use of an optical microscope as a visual sensor to recognize the three–dimensional position of end–effectors and biological objects has already been realized by a previously developed cell handling station [42]. This facilitates the application of path planning algorithms to avoid collisions during the performance of high level tasks. A client/server architecture has been implemented, allowing maximum possible flexibility.

8.5.3 Calculation of the Young's Modulus

This AMNS is designed for the mechanical characterization of biological cells, focusing on measuring the **Young's modulus** of the specimen. Before the measurements can be conducted, however, the cantilever must be calibrated. This calibration takes place on a hard substrate, similar to Section 7.5.3 and [43], and thus is not influenced by the soft specimen. For measurements concerning the specimen, owing to the softness of the material, special care has to be taken regarding indentation. Indentation results in force–distance curves considerably different from those taken on a hard substrate (Figure 8.2), and as mentioned there, the measured deflection of the cantilever is superposed with the indentation into the material. The formulae and the theoretical background describing the calculation of the Young's modulus for soft specimens was given in Section 8.2.3.1.

8.5.4 Experimental Results

First experiments were conducted in ambient air [44]. Dried tumor cells were chosen as specimens, which are adherent to the bottom of a Petri dish, forming a layer of cells and cell debris with a miscellaneous Young's modulus. The cells themselves have a diameter between 20 and 30 μm, but are densely embedded in the cell debris.

At different areas of the sample, force–distance curves were taken. The measurements indicate changing **adhesion** and **elasticity**, which is to be expected from the inhomogeneous layer. Mean values of a succession of ten force–distance curves were taken for evaluation, as depicted in Figure 8.11. Most areas show the characteristic behavior of **softer material**, the curves being slightly non–linear and having a smaller slope than theoretical curves for a hard substrate (*cf.* Figure 8.2). Additionally, large adhesion forces can be found, as, on the retraction of the cantilever, the cantilever sticks to the sample for some time until it finally breaks free. The force–distance curves show the characteristic **steps** in the retraction part of the curve for contaminated or coated samples in ambient air, indicating different layers successively detaching from the tip, which seems reasonable for the cell–debris–coated surface (Section 8.2.2.4). The measured Young's moduli lie between 577 kPa and 1.03 MPa, thus being close to gelatin, and seem reasonable for the gel–like biological material. Young's moduli of living cells are considerably smaller (50 kPa), while, *e.g.,* bone is much harder (30 GPa).

With the realization of a cell elasticity measurement station, a reasonable foundation for a versatile cell characterization station has been built. Further experiments are necessary to establish whether the required resolution and spring constants for measuring softer and more delicate specimens can be achieved by the piezoresistive cantilevers.

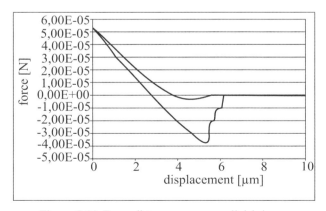

Figure 8.11. Force distance curve on cell debris

8.6 Conclusions

8.6.1 Summary

The versatility and effectiveness of AFM–based biocharacterization has been shown in the previous sections. Novel experiments, which were only made possible due to the unique abilities of the AFM, could be conducted, helping to understand basic cytological processes. Not only has the possibility to visualize processes on a cellular level been shown, for example, the destructive influence of enzymes on the cell membrane, but also has the difficult analysis of the electric behavior of DNA, the analysis of intermolecular forces in the pN–regime, and the transduction processes of sound into electrical currents in the inner ear. However, many processes in the micro– and nanoworld are **not fully resolved**, for example, the mechanics of **cell motility** or **cell–to–cell contact** in the conjugation of bacteria. More sophisticated setups can help to investigate these phenomena, expanding, *e.g.,* pharmaceutical possibilites.

Despite all of its versatility, the AFM has some major problems, mostly due to construction issues, which hamper the development of automated bio–nanohandling stations:

- No real–time imaging, due to low scan–rates;
- No three–dimensional force measurements, due to the nature of the laser–detection system with a two– or four–segment diode;
- Manipulation and imaging require different types of tips for optimal results.

A solution to these problems has to be found, either by changing the underlying constructional issues, or by enhancing AFMs, *e.g.,* by the use of additional sensors.

8.6.2 Outlook

For some of the problems listed above, solutions are currently being developed. For other issues, methods of resolution are proposed in the following.

For the real–time imaging problem, a solution might have arrived with the new video–rate AFMs, of which some are already commercially available systems. However, the scan range of these systems is very limited, confining the measurements to a 3×3 µm area, which is not satisfactory for most characterization and imaging tasks. Another option, which can compensate for the lack of real–time imaging is the possibility of including a haptic interface, which gives force feedback to the user while the AFM is still updating the image [45, 46]. Generally, the enhancement of the cantilever by piezoresistive or electrical sensing methods can solve the problems in measuring three–dimensional forces, as distinct forces beyond lateral ones can be registered separately. A combination of laser and piezo–based cantilever position measurements could be a meaningful enhancement to force measurements. However, the noise problems and the accuracy of piezo–based self–sensing options are unresolved issues.

The problem of needing two different types of cantilevers or AFM modes for optimal imaging and measuring or manipulating has become obvious from the previous applications. For example, in the case of measuring the virus capsid stability, a contact cantilever was necessary for the force measurements, but a slow imaging technique (jumping mode) had to be chosen, as normal contact mode applied too strong lateral forces. For measurements of the cell elasticity, a cantilever with a spherical tip is sometimes chosen for several reasons, but for imaging, this cantilever is not suitable and has to be exchanged. Other problems occur with the contamination of the AFM probe after, for instance, conducting force–distance–curves or cutting, which afterwards degrades the imaging quality. Overall as an ideal choice, imaging and manipulating usually call for different probes. A solution for this problem can be derived from the idea which has already been implemented for all of the previously described AMNS, where the visual sensor (either SEM or light microscope) and manipulator are decoupled.

Analogously, a commercially available AFM is proposed to be only used for the task of **imaging samples**, while an **additional end–effector** is used for manipulation and measurements. This setup could have another advantage: the second end–effector is not restricted to being a cantilever, but could be a manipulator of any kind, such as an injection pipette or the like. This way, injection, patch–clamp, as well as cutting or force measurements could be conducted. Several different tools could be utilized. The endeffector should ideally be **self–sensing**, which is a large development area, *e.g.,* including three–dimensional force–sensitive pipette holders. The requirements for the necessary control system are very high, as different spatial ranges from micro– to nanometers have to be covered, and visual control of the end–effector by an optical microscope and AFM control of the manipulation effects in the nm–range have to be achieved.

8.7 References

[1]   Block, S. M. 1990, *Optical Tweezers a New Tool for Biophysics in Noninvasive Techniques for Cell Biology*, Vol. 9 of Modern Cell Biology Series, S. Grinstein and K. Foskett, chapter 15, pp. 375–402.

[2]   Gascoyne, P., Huang, Y., Pethig, R., Vykoukal, J. & Becker, F. 1992, 'Dielectrophoretic separation of mammalian cells studied by computerized image analysis', *Meas. Sci. Technol.*, vol. 3, pp. 439–445.

[3]   Binnig, G., Quate, C. F. & Gerber, C. 1986, 'Atomic force microscope', *Physical Review Letters*, vol. 56, no. 9, pp. 930–933.

[4]   Tao, N. J., Lindsay, S. M. & Lees, S. 1992, 'Measuring the microelastic properties of biological material', *Biophys. Journal*, vol. 63, pp. 1165–1169.

[5]   Fritz, M., Radmacher, M. & Gaub, H. 1993, 'In vitro activation of human platelets triggered and probed by atomic force microscopy', *Experimental Cell Research*, vol. 205, pp. 187–190.

[6]   Kaupp, G. 2006, *Atomic Force Microscopy, Scanning Nearfield Optical Microscopy and Nanoscratching. Application to Rough and Natural Surfaces (Nanoscience and Technology)*, Springer.

[7]   Morris, V. J., Kirby, A. R. & Gunning, A. P. 2001, *Atomic Force Microscopy for Biologists*, Imperial Collge Press.

[8]   MPF–3D, Asylum Research, available at http://www.asylumresearch.com.

[9]   NanoMan VS, Veeco, available at http://www.veeco.com.

[10]   NanoWizard®II, JPK Instruments, available at http://www.jpk.com.

[11]   Sneddon, I. N. 1965, 'The relation between load and penetration in the axisymmetric Boussinesq problem for a punch of arbitrary profile', *International Journal of Engineering Science*, vol. 3, pp. 47–57.

[12]   Domke, J., Dannohl, S., Parak, W., Muller, O., Aicher, W. & Radmacher, M. 2000, 'Substrate dependent differences in morphology and elasticity of living osteoblasts investigated by atomic force microscopy', *Colloids and Surfaces. B, Biointerfaces*, vol. 19, pp. 367–379.

[13]   Hinterdorfer, P., Baumgartner, W., Gruber, H. J., Schilcher, K. & Schindler, H. 1996, 'Detection and localization of individual antibody–antigen recognition events by atomic force microscopy', *Proceedings of the National Academy of Science*, vol. 93, no. 8, pp. 3477–3481.

[14]   Sullivan, C., Morrella, J., Allisona, D. & Doktycza, M. 2005, 'Mounting of Escherichia coli spheroplasts for AFM imaging', *Ultramicroscopy*, vol. 105, pp. 96–102.

[15]   Humphris, A., McConnel, M. & Catto, D. 2006, 'A high–speed atomic force microscope capable of video–rate imaging', *European Microscopy and Analysis Supplement*, vol. 20, pp. 30–31.

[16]   Alberts, B., Bray, D. & Lewis, J. 2002, *Molecular Biology of the Cell*, Taylor & Francis.

[17]   Rotsch, C. & Radmacher, M. 2000, 'Drug–induced changes of cytosceletal structures and mechanics in fibroblasts – an atmoic force microscopy study', *Biophys. J.*, vol. 78, pp. 520–535.

[18]   Vereb, G., Szöllösi, J., Matkó, J., Nagy, P., Farkas, T., Vigh, L., Mátyus, L. & Waldmann, T. 2003, 'Dynamic, yet structured: The cell membrane three decades after the Singer–Nicolson model', *PNAS*, vol. 100, no. 14, pp. 8053–8058.

[19]   Gheith, M. K., Pappas, T. C., Liopo, A. V., Sinani, V. A., Shim, B. S., Motamedi, M., Wicksted, J. P. & Kotov, N. A. 2006, 'Stimulation of neural cells by lateral currents in conductive layer–by–layer films of Single–Walled Carbon Nanotubes', *Adavnced Materials*, vol. 18, no. 22, pp. 2975 – 2979.

[20]   Baikie, I. D., Smith, P. J. S., Porterfield, D. M. & Estrup, P. J. 1999, 'Multitip scanning bio–Kelvin probe', *Review of Scientific Instruments*, vol. 70, no. 3, pp. 1842–1850.

[21]   Sackmann, E. 2006, 'Haftung für Zellen', *Physik Journal*, vol. 5, no. 8–9, pp. 27–34.

[22]   Claußen–Schaumann, H. 2000, 'Adressierung und Strukturierung von Biomolekülen auf der Nanometer–Skala', Ph.D. thesis, University of Munich, Faculty of Physics.

[23]   Müller, D. J., Hand, G. M., Engel, A. & Sosinsky, G. 2002, 'Conformational changes in surface structures of isolated connexin 26 gap junctions', *The EMBO Journal*, vol. 21, no. 14, pp. 3598–3607.

[24]   Peng, L., Yi, L., Zhexue, L., Juncheng, Z., Jiaxin, D., Daiwen, P., Ping, S. & Songsheng, Q. 2004, 'Study on biological effect of La3+ on Escherichia coli by atomic force microscopy', *Jour. of Inorganic Biochemistry*, vol. 98, no. 1, pp. 68–72.

[25]   Steltenkamp, S., Rommel, C., Wegener, J. & Janshoff, A. 2006, 'Membrane stiffness of animal cells challenged by osmotic stress', *small*, vol. 2, no. 8–9, pp. 1016–1020.

[26]   Langer, M., Fink, S., Koitschev, A., Rexhausen, U., Horber, J. & Ruppersberg, J. 2001, 'Lateral mechanical coupling of stereocilia in cochlear hair bundles', *Biophys. Journal*, vol. 80, no. 6, pp. 2608–2621.

[27]   Ebner, A., Kienberger, F., Kada, G., Stroh, C. M., Geretschläger, M., Kamruzzahan, A., Wildling, L., Johnson, W. T., Ashcroft, B., Nelson, J., Lindsay, S. M., Gruber, H. & Hinterdorfer, P. 2005, 'Localization of single avidin–biotin interactions using

simultaneous topography and molecular recognition imaging', *ChemPhysChem*, vol. 6, pp. 897–900.

[28]   Sokolov, I., Subba–Rao, V. & Luck, L. A. 2006, 'Change in rigidity in the activated form of the glucose/galactose receptor from Escherichia coli: a phenomenon that will be key to the development of biosensors', *Biophysical Journal*, vol. 90, pp. 1055–1063.

[29]   Benoit, M. & Gaub, H. E. 2002, 'Measuring cell adhesion Forces with the atomic force microscope ate the molecular level', *Cells Tissues Organs*, vol. 172, pp. 174–189.

[30]   Miller, D. C., Haberstroh, K. M. & Webster, T. J. 2006, 'PLGA nanometer surface features manipulate fibronectin interactions for improved vascular cell adhesion', *Journal of Biomedical materials Research Part A*, vol. epub ahead of print.

[31]   Oberleithner, H., Ludwig, T., Riethmüller, C., Hillebrand, U., Albermann, L., Schäfer, C., Shahin, V. & Schillers, H. 2004, 'Human endothelium: target for aldosterone', *Hypertension*, vol. 43, pp. 952–956.

[32]   Ivanovska, I. L., de Pablo, P. J., Ibarra, B., Sgalari, G., MacKintosh, F. C., Carrascosa, J. L., Schmidt, C. F. & Wuite, G. J. L. 2004, 'Bacteriophage capsids: Tough nanoshells with complex elastic properties', *PNAS*, vol. 101, no. 20, pp. 7600–7605.

[33]   Cohen, H., Nogues, C., Ullien, D., Daube, S., Naaman, R. & Porath, D. 2006, 'Electrical characterization of self–assembled single– and double–stranded DNA monolayers using conductive AFm', *Faraday Discussions*, vol. 131, pp. 367–376.

[34]   Lehmann–Horn, F. & Jurkat–Rott, K. 2003, 'Nanotechnology for neuronal ion channels', *Journal of Neurology, Neurosurgery, and Psychiatry*, vol. 74, no. 11, pp. 1466–1475.

[35]   Langer, M. G., Fink, S., Löffler, K., Koitschev, A. & Zenner, H.–P. 2003, 'Investigation of the mechanoelectrical transduction at single stereocilia by AFM', *Biophysics of the Cochlea: From Molecules to Models*, Titisee, Germany 27 July – 1 August 2002, pp. 47–55.

[36]   Rubio–Sierra, J., Heckl, W. M. & Stark, R. W. 2005, 'Nanomanipulation by atomic force microscopy', *Advanced Engineering Materials*, vol. 7, no. 4, pp. 193–196.

[37]   Brufau, J., Puig–Vidal, M., López–Sánchez, J., Samitier, J., Driesen, W., Breguet, J.–M., Gao, J., Velten, T., Seyfried, J., Estaña, R. & Woern, H. 2005, 'MICRON: small autonomous robot for cell manipulation applications', *International Conference on Robotics and Automation (ICRA'05)*, Barcelona, Spain, pp. 856–861.

[38]   PIHera, Physik Instrumente, available at http://www.physikinstrumente.de.

[39]   Axiovert 200m, Zeiss, available at http://www.zeiss.com.

[40]   MM3A, Kleindiek, available at http://www.nanotechnik.com.

[41]   NaScaTec, available at http://www.nascatec.de.

[42]   Sievers, T., Garnica, S., Tautz, S., Trüper, T. & Fatikow, S. 2005, 'Microrobot station for automatic cell handling', *ICGST International Conference on Automation, Robotics and Autonomous Systems (ARAS–05)*.

[43]   Fatikow, S., Kray, S., Eichhorn, V. & Tautz, S. 2006, 'Development of a nanohandling robot station for nanocharacterization by an AFM probe', *14th Mediterranean Conference on Control and Automation*, Rhodes, Greece, pp. 1–6.

[44]   Fatikow, S., Eichhorn, V., Hagemann, S. & Hülsen, H. 2006, 'AFM probe–based nanohandling robot station for the characterization of CNTs and biological cells', *Int. Workshop on Microfactories (IWMF'06)*, Besancon, France.

[45]   Sitti, M., Aruk, B., Shintani, H. & Hashimoto, H. 2003, 'Scaled teleoperation system for nano–scale interaction and manipulation', *Advanced Robotics*, vol. 17, pp. 275–291.

[46]   Li, G., Xi, N., Yu, M., Salem, F., Wang, D. & Li, J. 2003, 'Manipulating nano scale biological specimen in liquid', *Third IEEE Conference on Nanotechnology*.

9

Material Nanotesting

Iulian Mircea[*] and Albert Sill[**]

[*] Division of Microsystems Technology and Nanohandling,
Oldenburg Institute of Information Technology (OFFIS), Germany
[**] Division of Microrobotics and Control Engineering,
Department of Computing Science,
University of Oldenburg, Germany

9.1 Instrumented Indentation

Instrumented indentation is one of the most commonly used methods to determine the mechanical properties of materials. This method is based on the penetration of a body with a known geometry into the material's surface. Both the force (or load) necessary for this penetration and the depth of indentation have to be measured, either separately or simultaneously.

Measuring these two quantities simultaneously leads to load–depth–curves. Applying further analysis to these load–depth–curves finally leads to the basic material's hardness and Young's modulus of elasticity. Not only bulk materials are under test with this method, but also coating systems consisting of two or more layers of different materials, *e.g.*, a thin film on a substrate.

While the usage of indentation tests on the macroscale has been known for decades, the method of application in the micro- or even nanoscale has been developed only recently. In the first part of this chapter, the basics of instrumented indentation and its analysis are given. The second part deals with microrobot-based nanoindentation. Some first experiments to characterize an electrically conductive adhesive are presented here.

9.1.1 Sharp Indentation

9.1.1.1 Introduction
The instrumented indentation method has been derived from the classical hardness testing method. **Hardness** has been defined as the resistance of a material against the penetration of a hard body with a known geometry, called an **indenter**. After the penetration, the residual imprint is measured and used in different relations to

calculate the hardness. Only the plastic deformation of the tested material is taken into consideration [1]. For example, the Martens hardness *HM* can be calculated by [1]:

$$HM = \frac{F}{A_i} = \frac{F}{26.43 \cdot h^2} \text{ for a Vickers indenter,} \qquad (9.1)$$

$$HM = \frac{F}{A_i} = \frac{F}{26.44 \cdot h^2} \text{ for a Berkovich indenter,} \qquad (9.2)$$

where F is the applied force, A_i is the area of the imprint remaining after the indentation test, and h is the remaining indentation depth.

The principal difference between the classical **hardness** method and the instrumented indentation method is that during an instrumented **indentation** test, the penetration depth and the indentation force are measured continuously. It is, in practice, an evolution of the contact response during the loading and unloading stages of the experiment [2]. From this test, the elastic and plastic deformation of the tested material can also be determined. The test provides load *vs.* indentation depth curves, which give helpful information about the mechanical properties of materials. Figure 9.1 shows such a typical curve. From this curve, additionally to the maximum indentation depth and force, the unloading contact stiffness S and the contact depth h_c can be determined. With these parameters, using the theory of Oliver and Pharr [3], the Young's modulus and the hardness of the tested material can be calculated. This procedure will be explained later in Section 9.1.1.5. The method can also be applied for coating systems, but there are some difficulties especially with regard to very thin coatings. In this case, in order to avoid the influence of the substrate material, special testing conditions should be considered. This will also be discussed in Section 9.1.1.5.

The method is called sharp instrumented indentation because of the shape of the **indenter**, which can be conical for Rockwell indenters or pyramidal for Vickers and Berkovich indenters [4] (Figure 9.2).

This test method has some advantages: in principle, it can be applied on specimens and components of any geometry, requiring only a small quantity of material. There are also several disadvantages in evaluating the test results from sharp indentation experiments. The extent to which these techniques can be used to quantify material properties is limited by the current understanding of the complicated material response during indentation experiments. For example, obtaining constitutive equations from indentation techniques has traditionally been limited to stress-free, perfectly elastic plastic material, not, therefore, giving critical information regarding residual stress and strain hardening.

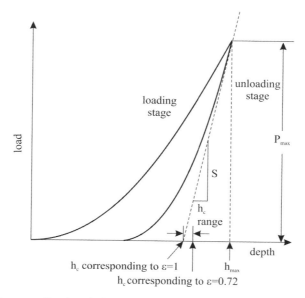

Figure 9.1. Diagram of load *vs.* indentation depth with the principal parameters needed for calculating the Young's modulus and the hardness of the tested material

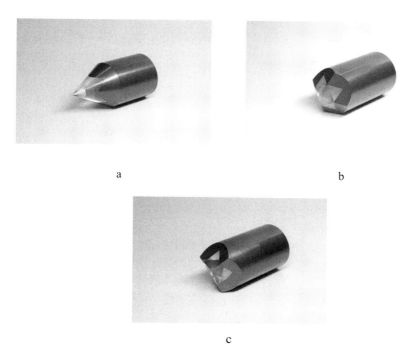

Figure 9.2. Sharp indenters: a. Rockwell indenter; b. Vickers indenter; c. Berkovich indenter (from [4], courtesy of Synton-MDP AG, Switzerland)

9.1.1.2 Basic Concepts of Materials Mechanics
The basic concepts of the **mechanics** of materials can be explained by looking at a cylindrical bar under axial forces [5]. An axial force is oriented parallel to the axis of the structural element.

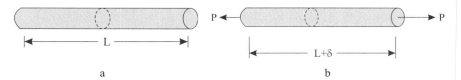

Figure 9.3. Schematic representation of a bar under axial load: a. initial length of the bar; b. bar elongated under an axial load

If the bar with an initial length L is loaded under the axial load P, it will be stretched, so the length will be $L + \delta$, where δ is the elongation of the bar. Furthermore, if a cross-sectional area A is considered, the stress σ in the bar can be defined by [5]:

$$\sigma = \frac{P}{A},$$
(9.3)

The strain is given by:

$$\varepsilon = \frac{\delta}{L},$$
(9.4)

Hook's Law expresses the linear relationship between stress and strain [5]:

$$\sigma = E \cdot \varepsilon ,$$
(9.5)

where E is the modulus of elasticity of the material. It is also called **Young's modulus**.

9.1.1.3 Similarity Between Sharp Indenters of Different Shape
Finite-element calculations have shown many similarities between sharp indenters [2]:

- there is a similarity in the stress field within the fully plastic regime;
- there is a similarity in the obtained hardness value, for a ratio between Young's modulus E and the uniaxial stress σ_r at a totally characteristic strain ε_r of 0.1, $E/\sigma_r > 150$;
- there is a similarity in the surface deformation after an indentation test with a sharp indenter; the average value around the imprint after pyramidal indentation is the same as after indentation with a conical indenter.

9.1.1.4 Indentation Ranges: Nano-, Micro-, and Macroindentation
The indentation tests can be classified as a function of the magnitude of applied
load in **nano-**, **micro-**, and **macroindentation** (Figure 9.4). It is called:

- nanoindentation, for the maximum load applied being up to 0.5 N;
- microindentation, for the maximum load applied ranging from 50–100 N;
- macroindentation, for the maximum load applied being more than 100 N.

Corresponding to these load ranges, there are differences in respect of the response
behavior of the indented material [6]:

- in the nanoindentation range, it is a response due to discrete phenomena,
  such as dislocation glide;
- in the microindentation range, it is a mesoscopic behavior, a response of
  single phases and microstructural units;
- in the macroindentation range, it is a macrosopic behavior, a response of
  the bulk material.

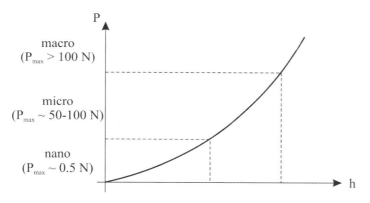

Figure 9.4. Indentation in nano, micro, and macro ranges

9.1.1.5 Analysis of Load–Depth Curves
The first researcher to use the instrumented indentation (with a spherical indenter)
was Tabor as early as 1948 [3]. Later on, in 1961, Stillwell and Tabor [3] used a
conical indenter. They tested metals and learned from their results that the
unloading stage of load–depth indentation curves can be used to calculate an elastic
modulus. In 1970, Bulychev *et al.* [3] also used instrumented indentation and
further analyzed the unloading stage of load–depth indentation curves. A typical
curve is shown schematically in Figure 9.5. They defined for the first time the
stiffness S of the upper portion of the indentation test's unloading stage [3]:

$$S = \frac{dP}{dh} = \frac{2}{\sqrt{\pi}} \cdot E_r \cdot \sqrt{A}$$

(9.6)

in which S is the stiffness of the upper portion of the indentation test's unloading stage, P is the indentation load, H is the indentation depth, E_r is the reduced modulus, defined below in the Equation 9.7, and A is the contact area between the tip of the indenter and the surface of the tested material under indentation load.

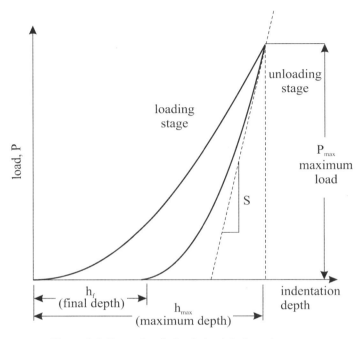

Figure 9.5. Example of a load–depth indentation curve

$$\frac{1}{E_r} = \frac{1-v^2}{E} + \frac{1-v_i^2}{E_i},$$
(9.7)

where E is the Young's modulus of the tested material, v is the Poisson's ratio of the tested material, E_i is the Young's modulus of the indenter material, and v_i is the Poisson's ratio of the indenter material.

Because determining the contact area A is difficult, in 1983, Pethica, Oliver, and Hutchins [3] defined the **indenter area function**, also called the **shape function**. The latter is the cross-sectional area of the indenter as a function of the distance from its tip [3]. First of all, it should be noted that at the maximum indentation load P_{max}, the surface of the indented material will have practically the same shape as the external shape of the indenter at the correspondent depth, the maximum indentation depth h_{max}. Consequently, if this maximum indentation depth is known from the experimentally recorded load–depth curve, it is easier to determine the contact area from such an indenter area function.

In 1986, Doerner and Nix [3] further developed this method based on their experimental observation that, in the upper portion of the unloading stage of the indentation test, the elastic behavior of the indentation contact is like that of a flat cylindrical punch. They assumed further that the unloading stage of the indentation curve can be approximated by a linear portion, and they determined the **indenter area function** by using the extrapolated depth of the "linear" **unloading stage** corresponding to zero loads. If the contact area is known, the Young's modulus can be calculated from Equation 9.6. They proposed to calculate the hardness from Equation 9.8 [3]:

$$H = \frac{P_{max}}{A},$$

(9.8)

where P_{max} is the maximum indentation load and A is the contact area at the maximum indentation load.

In 1992, indentation tests by Oliver and Pharr [3] suggested that the assumption of Doerner and Nix was too inaccurate and they further contributed to the development of this test method. In contradiction to Doerner and Nix, they took into consideration the curvature of the unloading stage of load–depth indentation curves. Additionally, they provided a new method for determining the depth, with which the area of contact should be calculated.

The necessary parameters for calculating the Young's modulus and the hardness can be identified by this method. In Figure 9.6, a cross-section of an indentation is shown.

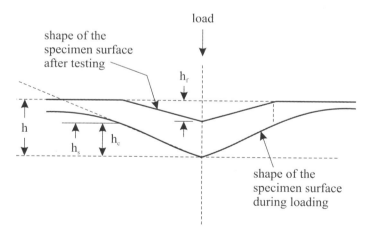

Figure 9.6. Diagram of a cross-section of an indentation

From this figure, it can be seen that the total indention depth h is [3]:

$$h = h_c + h_s,$$

(9.9)

where h_c is the contact depth, *i.e.*, the depth at which there is contact between indenter and the material surface, and h_s is the displacement of the indented surface at the outside edge of the contact between indenter and material.

After Oliver and Pharr [3], the necessary parameters for calculating the Young's modulus and the hardness of the indented material are the maximum indentation load P_{max}, the maximum indentation depth h_{max}, and the initial unloading contact stiffness S. S is only measured at the maximum indentation load. They are depicted above in Figure 9.1. The most important relationship is:

$$E_r = \frac{\sqrt{\pi}}{2} \cdot \frac{S}{\sqrt{A}},$$ (9.10)

which is Equation 9.6, but rewritten in such a way that the **reduced modulus** E_r can be calculated.

One of the most difficult problems is to determine the contact area at the maximum load A. Oliver, Hutchings, and Pethica assumed in 1986 [3] that A can be written as an **area function** $F(h)$, which takes into consideration the cross-sectional area of the indenter and the distance from the tip:

$$A = F(h_c).$$ (9.11)

This function F must be determined experimentally. Oliver and Pharr [3] proposed the following relationship for calculating the contact depth h_c:

$$h_c = \varepsilon \cdot \frac{P_{max}}{S},$$ (9.12)

where ε is a constant which depends on the geometry of the indenter: $\varepsilon = 0.72$ for a conical indenter, $\varepsilon = 0.75$ for a paraboloid of revolution, and $\varepsilon = 1$ for a flat punch [7].

In 2004, Oliver and Pharr [7] proposed a slightly modified relationship for calculating the **unloading stiffness** S:

$$S = \beta \cdot \frac{2}{\sqrt{\pi}} \cdot E_r \cdot \sqrt{A}.$$ (9.13)

If we compare this relationship with Equation 9.6, it can be seen that β, which is a correction factor, was considered as unity. This is true if it is assumed that the material is perfectly elastic and that there are only small deformations. Real indentation experiments have shown that the indenters used are non-axisymmetric and that large strains occur. Over the years, many researchers have performed numerical calculations and finite-element analysis for exactly calculating the correction factor. In 1995, Hendricks found a value of $\beta = 1.0055$ for a Vickers indenter and a value of $\beta = 1.0226$ for a Berkovich indenter.

Oliver and Pharr also presented in [7] the changes made as a result of research in order to improve the methodology and to better understand the mechanisms involved in an indentation test. For example, they investigated an error source in calculating hardness and Young's modulus, *e.g.*, the **pile-up** in the case of conical and Berkovich indenters. If during the indentation such a pile-up occurs, the contact area between the indenter and the surface of the tested material is significantly larger. This leads to relevant errors, *e.g.*, up to 50% overestimation of the Young's modulus. The material properties that affect the pile-up are the ratio of the reduced modulus to the yield stress E_r/σ_s and the work-hardening. They have found a parameter which can help to estimate the indentation behavior of materials. It is the ratio of the final indentation depth to the depth of the indentation at maximum load, h_f/h_{max}. It is possible to calculate this parameter from the experimentally recorded indentation curves. Additionally, it has been found that, if this ratio is $h_f/h_{max} > 0.7$, the indented area should be analyzed and measured in order to avoid underestimations of the Young's modulus.

Cheng *et al.* [8] developed a method that can correct the influence of the pile-up without measuring the true contact area. Analyzing indentation curves, they have found that the ratio of the **irreversible work** to the total work, $(W_{tot} - W_u)/W_{tot}$, is independent of the work-hardening of the indented material. Here, W_{tot} is the **total work of indentation**, which is equal to the area under the loading curve, and W_u is the work recovered during unloading (Figure 9.7).

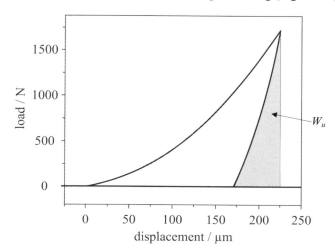

Figure 9.7. Irreversible work W_u of indentation from an indentation curve

One of the most important improvements of this method was the use of the "continuous stiffness measurement" technique [7], which means that during the loading stage of the indentation test, the stiffness is measured continuously; additionally, a frequency-specific amplifier is used for imposing a small dynamic oscillation on the recorded signals of force and indentation depth and for measuring the corresponding amplitude and phase of these signals. This was only possible through the advance achieved in measuring techniques over the years.

Based on the **"continuous stiffness measurement" technique**, Oliver and Pharr [7] improved the calibration procedures for the stiffness of the testing machine and for the indenter area function. Using this method, it is no longer necessary to perform multiple indentation tests on different reference materials for calibrations, and consequently, the necessary time for the calibrations is significantly reduced.

Myiake *et al.* [9] analyzed the influence of the pile-up on the results obtained by nanoindentation experiments. They conducted such experiments by means of a commercial AFM (atomic force microscopy) system, equipped with a rectangular stainless steel AFM cantilever with a diamond tip, on reference specimens (fused silica and single-crystal silicon). For the determination of the contact area, they analyzed the data by means of the Oliver and Pharr method and by direct measurement of this area with an AFM. Figure 9.8 shows a schematic representation of a cross-section of an indentation with pile-up.

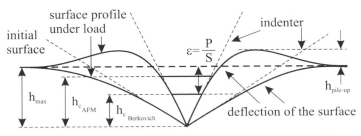

Figure 9.8. Schematic representation of a cross-section of an indentation pile-up (from [9], with kind permission of the *Japanese Journal of Applied Physics*, Institute of Pure and Applied Physics)

They found that the results obtained by the Oliver and Pharr procedure were overestimated. Therefore, it can be concluded that for very precise results, it is very useful to measure the contact area of the residual imprint after the indentation tests. As we have seen, the only difference in calculating the Young's modulus by instrumented indentation for indenters with a different shape is the factor β introduced by Oliver and Pharr [7] in Equation 9.13 for calculating the unloading stiffness S: $\beta = 1.0055$ for a Vickers indenter and $\beta = 1.0226$ for a Berkovich indenter. Furthermore, the relationship for calculating the reduced modulus E_r can be written as:

$$E_r = \beta \cdot \frac{\sqrt{\pi}}{2} \cdot \frac{S}{\sqrt{A}}. \qquad (9.14)$$

Franco *et al.* [10, 11] developed this method in order to calculate the **Vickers hardness** and to obtain the Vickers number *HV*:

$$HV = \frac{P}{A_{Vickers}}, \qquad (9.15)$$

where $A_{Vickers}$ is the contact at the maximum load for the Vickers indentation. They used the following relationship for calculating this area:

$$A_{Vickers} = \frac{d^2}{1.854368},$$
(9.16)

where d is the average diagonal of the Vickers indenter, which can be calculated by:

$$d = \beta \cdot \frac{S}{E_r} \cdot \sqrt{\frac{\pi}{2}},$$
(9.17)

where β, S, and E_r are parameters from instrumented indentation tests.

All of these considerations about the method of Oliver and Pharr for determining the mechanical properties of materials are correct for **bulk materials**. For coating systems, the method can also be used, but the influence of the substrate material should be considered [12]. Therefore, for determining the properties of a **coating**, the indentation depth should not exceed 1/10 of the total thickness of the coating.

9.1.1.6 Applications of the Sharp Instrumented Indentation
Sharp instrumented indentation is a technique widely used for determining the mechanical properties of materials ranging from metals and ceramics [2, 12–14], through polymers [15], to biological materials [16].

Zhao et al. [12, 17] determined the uniaxial **residual stress** and the mechanical properties of thermal barrier coatings (TBC) for aerospace applications by instrumented indentation. The determination of the residual stresses in a coating system is very important, because these stresses affect the integrity and reliability of such material systems. Moreover, the presence of the residual stresses (Figure 9.9) in a material system affects the contact area of indentation.

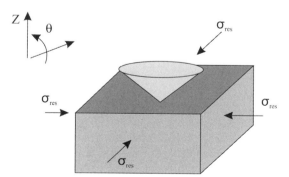

Figure 9.9. Schematic representation of a material with compressive residual stresses (reprinted from [17], with permission from Elsevier)

For example, if the residual stresses are compressive, the contact area will be smaller than in the case of a material system without residual stresses. The results have been verified by means of finite-element analysis.

Mircea *et al.* [13] investigated TBC systems that comprised a single crystal super alloy substrate, a plasma-sprayed metallic (NiCoCrAlY) layer with a thickness of 180 µm, and a ceramic coating with a thickness of 280 µm, applied by electron beam physical vapor deposition. The test method of investigation was the **Rockwell indentation**. Rockwell indentation tests generated delamination cracks of a butterfly shape, while the delaminated coating did not buckle. Such fracture behavior was in contrast to results reported earlier, which show circular delamination cracks and buckling [18]. In order to understand the specific fracture behavior, displacement-controlled indentation tests were conducted on the TBC system and on the metallic substrate, with and without metallic coating. The shape of the imprints after indentation and the fractured coatings were investigated by means of optical and scanning electron microscopy. It was found that the observed butterfly shape of the delamination was caused by anisotropy in the deformation behavior of the substrate material. Because the coating was thick compared to the indentation depth, the interaction indenter tip/coating affected the available energy for propagating delamination cracks. For example, after heat treatment, the energy for achieving a certain indentation depth increased, and shorter delamination cracks were observed. In some cases, it was possible to attribute the formation of shear cracks in the ceramic to specific portions of dissipated energy achieved from the load–displacement curves. However, during the penetration of the ceramic coating, many energy-dissipating processes are acting simultaneously. Thus, it is generally not possible to attribute certain amounts of energy to only one mechanism. For the case of thick coatings, it is proposed to modify the classical Rockwell indentation test by removing the ceramic directly underneath the indenter, in order to achieve these conditions. In this way, the computation of the energy available for propagating the delamination crack can be made.

Pharr [19] has shown that by nanoindentation it is possible to determine the **fracture toughness** of brittle materials (*e.g.*, ceramics) by measuring the radial cracks that occur after indentation and by using a simple relationship. The relationship for the calculation of the fracture toughness K_c is [19]:

$$K_c = \alpha \cdot \left(\frac{E}{H}\right)^{1/2} \cdot \left(\frac{P}{c^{3/2}}\right), \tag{9.18}$$

where P is the maximum indentation load, α is an empirical constant depending on the geometry of the indenter, and c is the length of the crack.

Other researchers [15] have mentioned the use of instrumented indentation for characterizing polymeric materials. These materials have mechanical properties between elastic solids and viscous fluids. They have shown a change in displacement at a constant force (creep). Polymeric materials are more compliant in comparison to metals and ceramics, which are normally characterized by instrumented indentation.

Schöberl *et al.* [20] investigated the mechanical properties of human teeth by instrumented nanoindentation. The teeth investigated were tested just after extraction, in order to determine their hardness and elastic modulus under conditions very close to *in vitro*. They also determined the mechanical properties of teeth after about 15 minutes; in this time, the containing water was expected to evaporate. It was found that the mechanical properties of teeth depend on the water content.

9.1.2 Spherical Indentation

The name **spherical indentation** refers to the shape of the indenter, which is a sphere (Figure 9.10). Spherical indentation began with Hertz's theory in 1880.

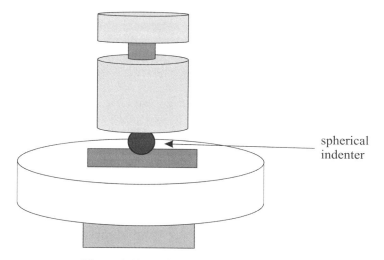

spherical
indenter

Figure 9.10. Device with a spherical indenter

9.1.2.1 Comparing Spherical and Sharp Instrumented Indentation

Spherical indentation has some advantages over sharp indentation. First of all, it has a variable contact pressure, which is suitable for studying pure elastic contact. Moreover, it has a variable **contact curvature radius** (Figure 9.11). Therefore, this method shows the ability to simulate real contact conditions. However, it is difficult to measure the contact radius experimentally, especially in the elastic regime that is characteristic for nanoindentation.

While sharp indentation depends on the indentation depth, spherical indentation depends on the sphere radius [21], the hardness increasing with a decreasing indenter radius.

In the case of spherical indentation, the roughness of the surface of the tested material is important; an asperity can lead to a false evaluation of the contact area at the maximum load [22]. It is also very difficult to manufacture indenters with a perfect spherical shape and with a required radius. Swadener and Pharr [22] developed a calibration method for spherical instrumented indentation, by perform-

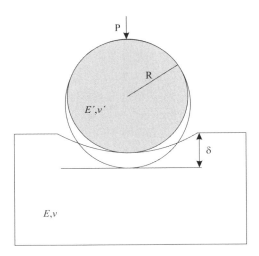

Figure 9.11. Schematic representation of the variable contact curvature radius

ing indentation tests on two different ceramic materials at the same indentation depth. Using this method, they were able to determine the **real indenter radius** and the **machine compliance**.

9.1.2.2 Analysis of Load–Depth Curves Using Spherical Indenters

Because of the different shape of the indenters, the relationship between load and displacement is also different [7, 23]. The contact depth for the spherical indenter is determined by applying Hertz's theory for a spherical indenter with a radius R_1 that penetrates a spherical hole with a radius R_2, which finally corresponds to the remaining impression in the surface of the tested material after the indentation test. The equivalent radius R may be defined by:

$$\frac{1}{R} = \frac{1}{R_1} + \frac{1}{R_2}.$$
(9.19)

The load–displacement relationship for the case of the spherical indentation is described by:

$$P = \frac{4}{3} \cdot \sqrt{R} \cdot E_r \cdot \left(h - h_f\right)^{3/2}.$$
(9.20)

All of the parameters in Equation 9.20 have already been introduced in Section 9.1.1.5. The **unloading stiffness** S for the spherical indentation can be written as:

$$S = 2 \cdot \sqrt{R} \cdot E_r \cdot \left(h - h_f\right)^{1/2}.$$
(9.21)

Taking $h = h_{max}$ for evaluating the expression above for the maximum indentation depth, and taking into consideration that the parameter ε (Section 9.1.1.5) for the case of spherical indentation is 0.75, a **contact depth** h_c follows:

$$h_c = \frac{h_{max} + h_f}{2},$$
(9.22)

which is characteristic for spherical indentation.

The considerations made in Section 9.1.2.4 for the determination of the mechanical properties of coating systems are also valid for the case of spherical indentation.

9.1.2.3 Applications of Spherical Instrumented Indentation
The applications of spherical instrumented indentation are similar to those of sharp instrumented indentation. There are only few differences in the application ranges. One of them is the application of spherical instrumented indentation to determine stress–strain curves of bulk materials and coating systems. These curves are useful for determining the transition from elastic to a plastic behavior of a material [24].

An important application field of spherical instrumented indentation is the study of the damage behavior of layered structures under concentrated load (*e.g.*, contact, impact), especially of those with brittle outer layers [25]. Such structures are: TBC for aerospace applications, cutting tools, electronic packaging devices, and biomechanical systems – natural ones such as shells and teeth and artificial ones such as dental crowns. Brittle out-layers can show cracking before the expected lifetime has been achieved; therefore, it is very important to study their damage behavior, and spherical indentation is a good tool for it. Moreover, it helps the designers of layered structures by providing guidelines for choosing the optimal material combinations.

9.2 Microrobot-based Nanoindentation of Electrically Conductive Adhesives

The use of the **microrobot-based nanoindentation** testing method is a good example of microrobotics technology helping materials research. Nowadays, new materials have been developed, and they need to be tested and investigated. The new materials have smaller dimensions, reaching the nanometer range (*e.g.*, protecting nanocoatings). These materials have more and more applications in the microsystems technology. Without microrobot-based nanoindentation, the determination of their mechanical properties (hardness and elasticity modulus) is almost impossible. The use of microrobots offers more flexibility during the tests, because they can solve tasks, which normally imply the use of many devices; moreover, their relatively small dimensions allow for the miniaturization of the testing devices. Consequently, such microrobot-based nanoindentation testing devices can be used inside an SEM for performing *in situ* nanoindentation tests. In this way, the tip of the indenter can be easily and precisely positioned on the surface of such small specimens; moreover, the occurred phenomenon (*e.g.*, cracks, plastic deformations, pile-up, sink-in, or delamination of the coating) can be observed *in situ*, which consequently can help to better understand their behavior and properties.

9.2.1 Experiments

9.2.1.1 Material System

DIN 16920 defines **adhesives** as non-metallic materials that join assembly parts by the adhesion of surfaces and internal strength [26]. They have no electrical conductivity. An **electrically conductive adhesive** (ECA) offers the possibility to join electronic components with a substrate and, at the same time, to allow their electrical connection. An ECA consists of a binder (matrix), which gives mechanical strength to the ECA, and a conductive filler providing the electrical conductivity [28]. The matrix can be a polymer, a polyamide, or a silicone; the conductive filler can be silver, gold, nickel, or copper [27]. They are relatively new materials that are receiving more and more interest in the microelectronic industry. ECAs are an alternative to the lead-based solder adhesives, which have been intensively used for the interconnections of electronic components on printed circuit boards. They are environmentally friendly, and they have a better creep resistance and flexibility in comparison to the lead-based solder adhesives. However, they also have disadvantages. For example, warm and moist environments can lead to a degradation of their performance.

In this work, an epoxy-based ECA silver-filled type PC 3002 [29] has been used. It has high electrical and thermal conductivity and high reliability. It is a thermoset polymer, which means that it is a cross-linked polymer with a three-dimensional molecular network structure [28]. The consequence is that such ECAs have a higher resistance to deformation at high temperature. Silver is the most used filler for manufacturing ECAs, because in comparison to gold, it is cheaper, and in comparison to others, it has a higher conductivity and a better chemical stability. Moreover, despite the fact that silver oxidizes (other fillers oxidize also), it shows a high conductivity [28].

Silver fillers are usually fabricated in a flake shape. The properties of the uncured adhesive are [29]:

- Ag Content   83±1.5%
- Density   4.4 g/cm^3
- Viscosity[1)]   22–28 Pa·s
- Processing life[2)] ~ 32 h
- Placing time[3)]   ~ 16 h
- Storage[4)]   6 months
- Coverage[5)]   ~ 100 cm^2/g

where
1) At shear rate $D = 50$ s^{-1}, plate-cone system with a cone 2°, temperature 23°C.
2) Time at room temperature during which the glue can be processed.
3) Maximum time between paste application and component placement, at up to 60/ R.H., at room temperature.

4) Storage in the freezer at –40°C.
5) 25 μm printed film thickness.

The properties of the cured adhesive are [29]:

- Curing (Peak temperature):
  Substrates: Al$_2$O$_3$. lead frames, PI – foil   10'/150°C
  Substrates: PES-Foil   20'/120°C
- Volume resistivity   < 0.3 mΩ·cm
- Adhesion (after DIN EN 1465)   > 8.5 N/mm
- Elasticity modulus (after ISO 527-2)   ~ 3600 N/mm^2
- Temperature stability[1)]   180°C
- Glass transition temperature   ~ 41°C
- Weight loss during the curing process at 140°C   < 0.8%
- Weight loss while being at 250°C/1h   < 0.5%
- Water absorption   < 0.19%
- Impurities:
  Cl$^-$ < 20 ppm
  Na$^+$ < 10 ppm
  K$^+$ < 10 ppm
- Thermal conductivity   > 5 W/m·K
- Shrinkage   4.4%

where
1) After 1000 hours at 180°C, the adhesion remains nearly unchanged.

9.2.1.2 Description of the Experimental Setup

The **microrobot-based nanoindentation** testing method has a setup which can be used both with an SEM and with an optical microscope. In Figure 9.12a, the principal components of the setup are shown: a **piezoresistive cantilever** is used for measuring the indentation force; the piezoresistive cantilever also plays the role of an indenter, as the indentation has been performed with the help of the cantilever tip; the cantilever is mounted in a special holder which assures the necessary electrical connections; the holder with the cantilever is carried by means of a **linear table** driven by a Nanomotor® which is also known as an NMT-motorized table; it also plays the role of a **sensor** measuring the indentation depth. The piezoresistive cantilever and the NMT-motorized table are described below (Sections 9.2.1.3 and 9.2.1.4). It should be noted that the test is performed in the horizontal direction (x-direction). Figure 9.12b indicates how a specimen is held and how the cantilever is pressed onto the specimen surface.

NMT-
motorized
table

holder of the
cantilever

cantilever

hole for the
specimen holder

a

specimen
holder

b

Figure 9.12. Experimental setup: a. principal components of the setup; b. cantilever pressing onto a specimen

A **control system** is required for performing the tests (Figure 9.13). The entire set-up is controlled by a computer, which evaluates and processes the measurement data and the input signals. This PC is connected to the controller of the NMT-motorized table. The PC sends data to the controller and reads the encoder signal from the NMT table. The output voltage of the Wheatstone bridge of the piezoresistive cantilever is AD-converted and measured by a bridge amplifier. The SEM also is equipped with a PC.

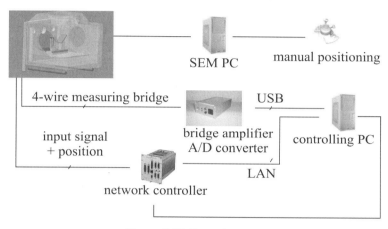

Figure 9.13. Control system

9.2.1.3 The AFM Cantilever

Nanoindentation tests require very sensitive **force sensors**. One of the recent technical solutions is to use piezoresistive cantilevers, which are most commonly used for performing atomic force microscopy, *e.g.*, for determining surface roughness or tribological properties of surfaces (like the friction coefficient) [15].

In Figure 9.14 [30], a piezoresistive cantilever and its tip can be seen. This cantilever is manufactured on a single crystal silicon substrate. The beam length is ≈ 500 μm and it is equipped with piezoresistors mounted in the form of a Wheatstone bridge. If the beam is deflected by applying an external force, a change in resistance of the piezoresistors can be measured. The change in resistance, and consequently the differential voltage of the Wheatstone bridge, can be converted into a force signal by performing a calibration described below (Section 9.2.4.2). The beam is also furnished with a tip. The pyramidal tip has a height of ≈ 17 μm; the tip radius is ≈ 10 nm; the maximum applicable load is ≈ 20 mN; the measurement resolution is ≈ 1 μN [30].

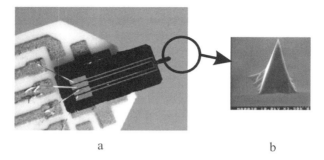

a   b

Figure 9.14. Piezoresistive cantilever: a. overview; b. AFM tip

9.2.1.4 Description of the NMT Module

Nanoindentation tests also require very sensitive **displacement sensors** and positioners. A device that can play both roles is the NMT module [31]. In Figure 9.15, such a module is shown. In principle, it is a motorized table driven by a very precise linear motor. The motor can carry loads up to 2 kg with a maximum stroke of up to 70 mm with a resolution of ≈ 2 nm. The NMT module is equipped with a very precise displacement sensor with a resolution of ≈ 10 nm, has small dimensions (*e.g.*, module type NMT-20: $50 \times 26 \times 10mm^3$), and is vacuum compatible [31].

Figure 9.15. NMT module

9.2.1.5. Experimental Procedure

Before performing nanoindentation tests, **calibrations** of the piezoresistive AFM are necessary: the stiffness calibration and the electrical calibration. They are described in Section 9.2.2. Nanoindentation tests were performed on flat specimens consisting of the ECA material system described above (Section 9.2.1.1). The ECA was applied on SEM stubs. The specimens were variously investigated after a first curing at 70°C in the oven for 120 minutes, 150 minutes, 180 minutes, 240 minutes, 300 minutes, and finally after 325 minutes at the same temperature. The nanoindentation tests were performed using the setup described above (Section 9.2.1.2). The maximum indentation depth was up to 1 µm.

9.2.2 Calibrations

For using an AFM cantilever as a **force sensor**, several **calibrations** are required. First of all, the stiffness of the beam of the cantilever must be determined. For a beam with a rectangular cross-section, the relationship between an applied load F and its deflection d is:

$$F = K \cdot d , \tag{9.23}$$

where K is the stiffness of the beam. When the beam is pressed against a specimen, it is deflected, and consequently, the differential voltage U of the Wheatstone bridge can be measured. The relationship between force and this voltage can be written as:

$$F = c_{el} \cdot U , \tag{9.24}$$

where c_{el} is the electrical constant of the AFM cantilever. Thus:

$$U = \frac{K}{c_{el}} \cdot U = c \cdot d , \tag{9.25}$$

$$c_{el} = \frac{K}{c} . \tag{9.26}$$

So, if the stiffness of the beam K is known, its electrical constant can be extracted, and the differential voltage U of the Wheatstone bridge is measured as a function of the beam deflection. Consequently, their relationship and the constant c are known. Such calibrations are described in the sections below.

9.2.2.1 Calibration of the Stiffness

There are many methods for determining the **stiffness** of an AFM cantilever. A first category is the so-called geometric method, where the dimensions of the beam are used in equations for calculating the stiffness [32]. Other methods are the so-called thermal methods, which use the acquisition of the cantilever thermal distribution spectrum, written as a function of the cantilever's resonance frequency [32].

In this study, one of the geometric test methods has been used. The dimensions of the cantilever beam were measured by means of an SEM. The resonance frequency was determined with a special setup described in [33]. With these quantities, the stiffness of the AFM cantilever can be calculated [32]:

$$K = 2 \cdot w \cdot \left(\pi \cdot l \cdot f_r \right)^3 \cdot \sqrt{\frac{\rho^3}{E}} \tag{9.27}$$

where w is the width of the cantilever, l is the length of the cantilever, f_r is the resonance frequency of the cantilever, ρ is the density of the cantilever material, and E is the elastic modulus of the material of the cantilever beam.

An example diagram of a test determining the **resonance frequency** of the cantilever is shown in Figure 9.16.

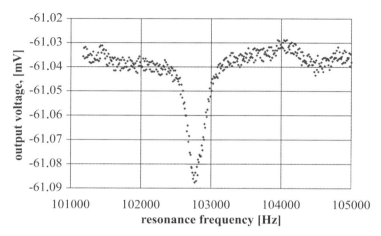

Figure 9.16. Example of a measured resonance frequency as a function of the Wheatstone bridge output voltage

9.2.2.2 Electrical Calibration

As mentioned above, besides the knowledge of the stiffness of the cantilever, an **electrical calibration** is needed. It allows the translation of the measured differential voltage of the Wheatstone bridge to a force (in newtons), by using the Equations 9.23 to 9.26. The electrical calibration was carried out by conducting measurements of the differential voltage of the Wheatstone bridge as a function of the deflection of the cantilever beam. In such tests, the end of the cantilever beam is pressed against a hard and smooth surface (*e.g.*, silicon wafer) by using the setup shown in Figure 9.12. The output voltage is measured as a function of the cantilever beam deflection. This calibration is described in detail in [33].

9.2.3 Preliminary Results

9.2.3.1 Dependency on the Hardness of the ECA on the Curing Time

The results obtained after nanoindentation tests by using the tip of the AFM on the above-mentioned ECA specimens are shown in Figure 9.18. A dependency between the hardness of the ECA and the **curing time** can be seen. The results show an increase of the hardness with the increase of the curing time at constant temperature.

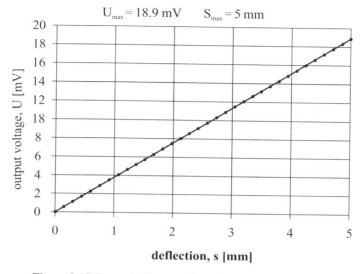

Figure 9.17. Example diagram of an electrical calibration test

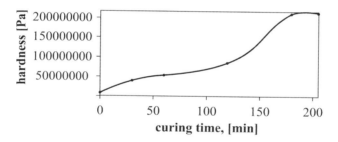

Figure 9.18. Evolution of the ECA hardness with the curing time

9.2.4 Discussion

9.2.4.1 Different Tip Shapes

A severe problem encountered when using the tip of the piezoresistive AFM cantilever as indenter is its insufficient hardness. After several nanoindentation tests, it
was worn-out (Figure 9.19). In order to avoid this, a change of the tip of the AFM
cantilever was proposed. In place of its tip, a **ruby sphere** with a diameter of 120
μm was glued to the cantilever (Figure 9.20). Ruby has a higher hardness than
silicon. Moreover, the geometry of the tip is different as the radius of the sphere is
significantly bigger than the initial radius of the **silicon AFM tip**.

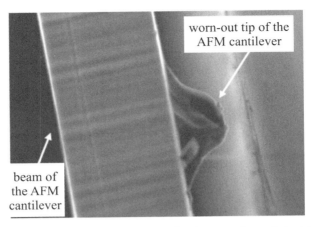

Figure 9.19. Worn-out piezoresistive AFM tip after several nanoindentation tests

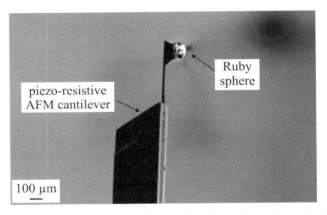

Figure 9.20. Piezoresistive AFM cantilever with a hard ruby sphere as indenter tip

First tests with this type of AFM cantilever are very promising. After an optical inspection, no wearout has been observed. However, additional **calibration tests** are necessary, and they are still in process at this time.

Finally, the initial setup for nanoindentation has been modified. The setup now uses a **Berkovich diamond tip** for performing nanoindentation tests and a highly sensitive load cell (produced by Honeywell). Moreover, the test is performed in the vertical direction (z-direction). The load cell was calibrated by using small weights. The setup also requires calibration with reference specimens (fused silica and sapphire) with known mechanical properties, in order to calculate both the hardness and Young's modulus of the tested material. At least 5 measurements at 10 different maximum loads must be performed in order to assure a good statistical base for the calibrations. They are also in process at this time. A photo of this setup is shown in Figure 9.21.

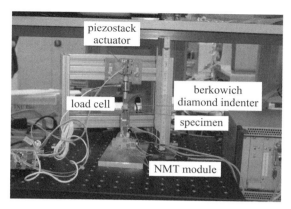

Figure 9.21. Overview of the nanoindentation setup with a highly sensitive load cell and a Berkovich indenter

In this case, the positioning of the specimen under the indenter tip is carried out by the NMT module in *x*-, *y*-, and z-directions. The indenter tip is then pressed onto the surface of the specimen by means of a **piezostack actuator** with integrated strain gages. This means that this actuator can measure the displacement of the indenter tip. First tests by using this setup are also very promising. The results obtained after nanoindentation tests by using the Berkovich tip, on ECA specimens and on ceramic substrates, are shown in Figure 9.22. The ceramic shows a steeper **loading slope** compared to the ECA, meaning that a higher load was necessary for achieving the same indentation depth for the test on the ceramic, confirming the assumption of higher hardness of ceramics compared to ECA.

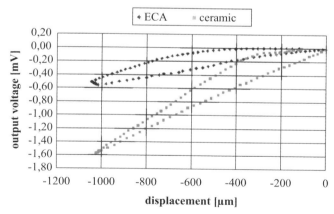

Figure 9.22. Indentation diagrams of ceramic and ECA specimens with the Berkovich diamond indenter

9.3. Conclusions

In Section 9.1, concepts of the instrumented indentation test method have been presented. Furthermore, its theoretical analysis with respect to sharp and spherical indentation has been explained. The theoretical model of Oliver and Pharr [3] for calculating the mechanical properties of bulk materials and coatings, based on the results of instrumented tests, was also described. Instrumented indentation has been recently developed for testing in the range of very low loads, up to 0.5 N, and of small depths, from a few nanometers up to 1 μm. This so-called **nanoindentation** has many applications for determining **mechanical properties** of materials, such as ceramics, metals, polymers, and human teeth.

Section 9.2 presents results of microrobot-based nanoindentation tests on specimens consisting of an ECA material after a first curing at 70°C in the oven for 120 minutes and after longer curing treatments up to 325 minutes at the same temperature. The components of the ECA material system and its properties in the uncured state are described. ECA are relatively new materials that are more and more used in the microelectronics industry. Therefore, it is very important to find reliable test methods for characterizing such materials. Material characterization is a very important issue when new materials are manufactured. Moreover, the new branch of microrobotics can help materials science in its investigations. By means of the microrobot-based nanoindentation tests, reasonable values for the hardness of an ECA can be obtained, and its evolution depending on curing time at 70°C can be monitored. An increase of the hardness with increasing curing time has been noticed.

For performing such tests, a setup that uses the tip of an AFM piezoresistive cantilever as indenter has been developed, and the required calibrations have been conducted. It was found that, despite the fact that the cantilever as a force sensor is very sensitive, the wearout of its tip after only few tests is a big disadvantage. Therefore, new solutions for overcoming this problem have been proposed. The use of a hard ruby sphere as indenter instead of a silicon tip has been proposed. First results are promising, and further calibration tests will be conducted. Finally, the setup has been modified. The advantage is now that the tip of the Berkovich indenter used is made from diamond. Moreover, for its geometry, an established theory for calculating material hardness and Young's modulus is available in the literature [3]. For this setup, it is also necessary to perform further calibration tests on reference specimens. The results obtained by using this setup are also promising. By comparing the slope of the loading stage of the nanoindentation tests on specimens with different hardness, e.g., ceramic and ECA, a difference of this slope can be noticed. This permits differentiation of the hardness of different materials.

For improving the results of the nanoindentation tests, the use of very sensitive load and displacement sensors plays an important role (Sections 9.2.1.3 and 9.2.1.4). For further experiments, it is proposed to use a novel force sensor that can measure force in all three directions. This sensor has been developed at the Technical University of Braunschweig for dimensional metrology [34]. The sensor consists of a silicon boss membrane, fabricated by chemical etching and furnished with integrated piezoresistors [34]. Because the sensor can simultaneously measure forces in all three directions, scratch tests can also be performed. It is expected that, by correlating the results of nanoindentation and scratch tests, the mechanical properties of bulk materials and the adhesion of nanocoatings can be determined in a better way.

9.4 References

[1]   NPL 2006, 'www.npl.co.uk', National Physical Laboratory UK.
[2]   Mata, M. & Alcala, J. 2003, 'Mechanical property evaluation through sharp indentations in elastoplastic and fully plastic contact regimes', *Journal of Materials Research*, vol. 18, no. 7, pp. 1705–1709.
[3]   Oliver, W. C. & Pharr, G. M. 1992, 'An improved technique for determining hardness and elastic modulus using load and displacement sensing indentation experiments', *Journal of Materials Research*, vol. 7, no. 6, pp. 1564–1583.
[4]   Synton-MDP 2006, 'www.synton-mdp.ch', Micro Diamond Points Schweiz.
[5]   Gere, J. M. 2001, *Mechanics of Materials*, Nelson Thormes, pp. 3–22. 5th SI Edition.
[6]   Alcala, J., Anglada, M. & Gonzales, P. M. 2003, 'Advanced indentation testing, Training course in the frame of the EU Project RTN2-2001-00488 "Structural integrity of ceramic multilayers and coatings (SICMAC)"'.
[7]   Oliver, W. C. & Pharr, G. M. 2004, 'Measurement of hardness and elastic modulus by instrumented indentation: advances in understanding and refinements to methodology', *Journal of Materials Research*, vol. 19, no. 1, pp. 3–20.
[8]   Cheng, Y. T. & Cheng, C. M. 1998, 'Relationships between hardness, elastic modulus, and the work of indentation', *Applied Physics Letters*, vol. 73, no. 5, pp. 614–616.
[9]   Miyake, K., Fujisawa, S., Korenaga, A., Ishida, T. & Sasaki, S. 2004, 'The effect of pile-up and contact area on hardness test by nanoindentation', *Japanese Journal of Applied Physics*, vol. 43, no. 7B, pp. 4602–4605.
[10] Franco, A. B., Pintaude, G., Sinatora, A., Pinedo, C. E. & Tschiptschin, A. P. 2004, 'The use of a Vickers indenter in depth sensing indentation for measuring elastic modulus and Vickers hardness', *Materials Research*, vol. 7, no. 3.
[11] Pintaude, G., Cuppari, M. G. V., Schön, C. G., Sinatora, A. & Souza, R. M. 2005, 'A review on the reverse analysis for the extraction of mechanical properties using instrumented Vickers indentation', *Zeitschrift für Metallkunde*, vol. 96, pp. 1252–1255.
[12] Zhao, M., Chen, X., Yan, J. & Karlsson, A. M. 2006, 'Determination of uniaxial residual stress and mechanical properties by instrumented indentation', *Acta Materialia*, vol. 54, pp. 2823–2832.
[13] Mircea, I. & Bartsch, M. 2005, 'Modified indentation test to estimate fracture toughness of thick thermal barrier coating systems', *11th International Conference on Fracture*, on CD of the Proceedings.
[14] Bartsch, M., Baufeld, B., Dalkilic, S., Mircea, I., Lambrinou, K., Leist, T., Yan, J. & Karlsson, A. M. 2007, 'Time-economic lifetime assessment for high performance thermal barrier coating systems', *Key Engineering Materials*, vol. 333, pp. 147–154.
[15] VanLandingham, M. R. 2003, 'Review of instrumented indentation', *Journal of Research of the National Institute of Standards and Technology*, vol. 108, no. 4, pp. 249–265.
[16] Barthelat, F., Li, C. M., Comi, C. & Espinosa, H. D. 2006, 'Mechanical properties of nacre constituents and their impact on mechanical performance', *Journal of Materials Research*, vol. 21, no. 8, pp. 1977–1986.

[17] Chen, X., Yan, J. & Karlsson, A. M. 2006, 'On the determination of residual stress and mechanical properties by indentation', *Materials Science and Engineering A*, vol. 416, pp. 139–149.

[18] Vasinonta, A. & Beuth, J. L. 2001, 'Measurement of interfacial toughness in thermal barrier coating systems by indentation', *Engineering Fracture Mechanics*, vol. 68.

[19] Pharr, G. M. 1998, 'Measurement of mechanical properties by ultra-low load I ndentation', *Material Science and Engineering A*, vol. 253, pp. 151–159.

[20] Schöberl, T. & Jäger, I. L. 2006, 'Wet or dry – hardness, stiffness and wear resistance of biological materials on the micron scale', *Advanced Engineering materials*, vol. 8, no. 11, pp. 1175–1179.

[21] Swadener, J. G., George, E. P. & Pharr, G. M. 2002, 'The correlation of the indentation size effect measured with indenters of various shapes', *Journal of Mechanics and Physics of Solids*, vol. 50, pp. 681–694.

[22] Swadener, J. G. & Pharr, G. M. 2000, in R. Vinci, O. Kraft, N. Moody, P. Besser & E. Shaffer (eds), *Materials Research Society Symposium Proceedings 594*, pp. 525–530.

[23] Swadener, J. G., Taljat, B. & Pharr, G. M. 2001, 'Measurement of residual stress by load and depth sensing indentation with spherical indenters', *Journal of Materials Research*, vol. 16, no. 7, pp. 2091–2102.

[24] CSM & Instruments 1999, 'Nanoindentation with spherical indenters for characterization of stress-strain properties', *Applications Bulletin*, no. 11.

[25] Pajares, A., Wei, L., Lawn, B. R., Padture, N. T. & Berndt, C. C. 1996, 'Mechanical characterization of plasma sprayed ceramic coatings on metal substrates by contact testing', *Materials Science and Engineering A*, vol. 208, pp. 158–165.

[26] Fatikow, S. 2006, 'Mikroboter weisen den Weg', *Mikroproduktion*, vol. 4, pp. 60–62, in German.

[27] Orthmann, K., Dorbath, B., Klatt, H., Richly, W. & Schmidt, J. 1995, *Kleben in der Elektronik*, number 472 in Kontakt&Studium, Expert Verlag, Kapitel 4: Grundlagen der Leitklebstoffe, pp. 71–94, in German.

[28] Xu, X. S. 2002, 'Evaluating thermal and mechanical properties of electrically conductive adhesives for electronic applications', Ph.D. thesis, Virginia Polytechnic Institute and State University Blacksburg.

[29] W.C. Heraues GmbH, Circuit Material Division, 'www.4cmd.com', Data sheet of the conductive adhesive PC3002.

[30] Nascatec GmbH, 'www.nascatec.de'.

[31] Klocke Nanotechnik GmbH, 'www.nanomotor.de'.

[32] Burnham, N. A., Chen, X., Hodges, C. S., Matei, G. A., Thoreson, E. J., Roberts, C. J., Davies, M. C. & Tendler, S. J. B. 2003, 'Comparison of calibration methods for atomic-force microscopy cantilevers', *Nanotechnology*, vol. 14, pp. 1–6.

[33] Kray, S. 2006, 'Internal Report, University of Oldenburg, Division Microbotics and Control Engineering', in German.

[34] Phataraloha, A., Büttgenbach, S., 2004, 'A novel design and characterization of a micro probe based on a silicone membrane for dimensional metrology', *Proceedings of the 4th euspen International Conference*, pp. 310–311.

Nanostructuring and Nanobonding by EBiD

Thomas Wich

Division of Microrobotics and Control Engineering,
Department of Computing Science,
University of Oldenburg, Germany

10.1 Introduction to EBiD

Since 1974, when Taniguchi coined the expression **nanotechnology** [1] as a description of manufacturing processes, a lot of different techniques for manufacturing on this small scale have been developed. Until that time, manufacturing processes on the micrometer scale used to be the limit. Conventional semiconductor-processing technologies are mostly limited by the achievable resolution in lithography. However, this resolution depends on the wavelength of light – or, in general, on electromagnetic waves. In order to process materials on the nanometer scale, it is either necessary to develop a new approach in materials structuring (often referred to as the bottom-up approach) or to extend the possibilities of common techniques, for example by using electromagnetic waves with considerably shorter wavelengths.

Although it is possible – and not extraordinary anymore using today's research facilities – to structure materials by handling single atoms or molecules with scanning tunneling microscopes (STM), this processing technology is not economically viable due to its long processing time (approx. 10 ms per atom) [2]. Still, the most widely used structuring processes for micro- and nanotechnology are based on lithography. Especially with regard to 3D structuring, **focused beams** with high-energy density on the impact spot represent a promising technology. Common irradiation technologies fulfilling the demands for high energy on the impact spot and of short wavelength are UV-laser, ion beams (*e.g.,* focused ion beams – FIB), electron beams, or even X-rays. The basic principles of these processing technologies for either the deposition or ablation of materials bear on the chemical activation of a precursor on the impact spot, that is where the focused beam interacts with the substrate.

Electron beam-induced deposition (EBiD) can be regarded as a **local chemical vapor deposition (CVD)** technology inside an electron microscope [3]. CVD is a (semiconductor) processing technology, where a vaporized precursor is chemically activated by means of thermal energy. This energy is commonly induced by light, direct heating, or a plasma source (plasma-enhanced CVD – PECVD). For CVD processes, the substrate is heated up inside a chamber filled with the precursor vapor at a certain pressure. Due to the thermal energy, a chemical reaction sets in on the substrate surface and leads to deposition of the non-volatile reaction products. Thus, the substrate surface is coated with these reaction products. In semiconductor technology, CVD processes are widely used for deposition of uniform thin films over a silicon wafer.

In contrast, the EBiD process is a **very localized** CVD process, occurring on an area of the substrate defined by the electron beam. Most EBiD setups use an electron microscope, because the electron beam can be used for localizing the deposition site, triggering the chemical reactions and observation of the deposit. In most cases, either scanning electron microscopes (SEM) or scanning transmission electron microscopes (STEM) are used for the EBiD process, as their beam can be focused on a very small area. Positioning and scanning of the beam allows the deposition of dots, lines, areas, and even of three-dimensional figures. In [4], an overview of three-dimensional nanostructures is given.

The EBiD process must be seen as a process defined by the **interactions** between the electron beam, the substrate, and the chemical precursor. In Section 10.2, a detailed description of electron beam generation and its relevant parameters for EBiD processes will be given.

In and around the spot where the SEM's electron beam interacts with the surface, the chemical reaction is triggered and a deposition is formed. In the early years of EBiD, the primary electron beam of the SEM was seen as the trigger for the chemical reaction. More recent results show that there are other types of triggers, such as the secondary electrons generated through the primary beam [5], which might have more influence than the primary electrons. The **basic equation** for the layer growth rate of a deposition described in [6] is still seen as a sufficient model to describe the reaction. In Section 10.2.3 this model will be explained and the role of the secondary electrons will be considered. The possible heating effects of the primary electron beam on the deposition will also be discussed.

A **simplified setup** for EBiD in an SEM consists of the electron column, providing the electron beam, and the vacuum chamber, where the substrate is positioned. The precursor for the chemical reaction is provided by a gas injection system (GIS). There, the precursor is evaporated inside a reservoir with a small vapor outlet. Different types of gas injection systems and evaporation stages are explained in Section 10.3. The gas injection system influences the chemical reaction significantly, as parameters like precursor flux density determine the growth rate to a considerable extent. Different methods of controlling the precursor flux are described in Section 10.4.

In Section 10.5, **control concepts** for the EBiD process will be presented. EBiD can serve as a technology for bonding objects on the nanometer scale and thus be a very important task in automated nanomanufacturing in the SEM. Basic control approaches will be discussed, enabling the user to control the geometry of the deposits and the process itself.

From the beginning of EBiD, a lot of research has been done regarding the application of EBiD deposits for lithographic masks or for the manipulation of electronic circuits. Even so, very little knowledge has been gathered for the application of EBiD as a manufacturing technique for mechanical elements on the nanometer scale, that is, with respect to the deposit's **mechanical properties**. Known values from the literature and own measurements will be presented in Section 10.6.

10.1.1 History of EBiD

The history of EBiD is strongly linked to the history of electron microscopy. As soon as specimens have been imaged using electron microscopes, thin **contamination layers** on the specimens' surface could be observed. This contamination effect is still a problem in today's electron microscopy, even though its extent could be reduced. The formation of contamination dots, lines, or layers on a specimen surface is due to the presence of small amounts of vapor in the vacuum chamber. The first researcher known to describe the effect of specimen contamination under electron impact in vacuum was R.L. Stewart in [7]. In his experiments, he used an electron gun fixed in a vacuum tube with a pressure of about 5 mmHg ($1.33 \cdot 10^{-5}$ mbar). He pointed the electron beam, of approximately 200 V, on different target materials, to find them contaminated with a thin film where the beam hit the target. He proved that these films were carbon compounds and assumed that the precursor for these deposits was organic vapor in the vacuum chamber. Furthermore, he observed self-deflection effects of the electron beam after deposition which he blamed on the charging of the deposit. He went on to formulate the first description of electron-beam-induced contamination deposition: The electron beam triggers the **dissociation of organic vapor** where the beam hits the target.

In 1953, a systematic investigation of organic vapor sources [8] with regard to contamination experienced in scanning electron microscopy was performed. The researchers measured the thickness of the deposited carbon layer on different metal surfaces for different sources of organic vapor, such as diffusion **pump oil**, vacuum grease, and different types of gaskets. Additionally, they discovered two interesting effects regarding thermal treatment. Firstly, heating up the target leads to a significant reduction in contamination deposition. Secondly, using a cold trap close to the specimen also leads to a significant drop in deposition rate.

In [9], possible applications of EBiD are introduced. As in the previous publications, the precursor used was of organic composition (silicone pump oil). Its major contribution was the proposing of a phenomenological theory about the growth rate of the deposited thin films. Based on the equations in [9], it has been shown that the growth rate of the deposit depends on the electron current density, which is one of the major beam parameters. The second parameter determining the growth rate is the presence of the precursor on the substrate's surface, which is mainly determined by the evaporation stage, a function of the gas injection system. In accordance with [8], it has been proven that the substrate's temperature influences the growth rate. However, the theory was based on the electron density; and did not distinguish between primary and secondary electrons, which are struck out of the substrate by the primaries. In [10], this theory has been refined, taking into account the secondary electrons.

In 1976, Broers [11] described in the formation of 8 nm metal structures, using EBiD in a high-resolution SEM. He suggested the use of his specialized deposition technique on 10 nm gold-palladium films supported by 10 nm carbon foils for mask application in microfabrication processes. He achieved these very thin contamination structures due to the reduction of **electron scattering** effects in these very thin substrates. The protecting film on the metal was formed by conventional contamination lithography.

Since the 1980s, H. W. P. Koops has contributed many pieces of research on EBiD. In [6], pattern generation in the submicron range has been investigated, using metallo-organic substances – ruthenium carbonyl ($Ru_3(CO)_{12}$) and osmium carbonyl ($Os_3(CO)_{12}$) – as precursors. A very interesting contribution was the determination of the gas flux from the **gas injection system** to the substrate. At that time, the main application for EBiD was seen as the generation of lithography masks [6, 12]. In the following publications, further possible applications have been found; AFM supertips [13], three-dimensional lithography [14], field-emitter devices [15], and even magnetic-flux sensors [16].

10.1.2 Applications of EBiD

As has been shown in the previous section, EBiD started as a troublesome effect in SEM imaging. However, over the years, it developed into a highly sophisticated processing technology on the nanometer scale. Typical applications are described in the following paragraphs.

Lithography: As already mentioned, [9] proposed EBiD as a coating technology for producing insulating and dielectric thin films on substrates. With the first experiments using metallo-organic precursors it was possible to deposit conducting structures. For example, [17] proposed to use EBiD for the generation of etch masks, for mask repair, and also for direct rewiring on IC-prototypes for test applications. Due to the high theoretical resolution of the electron beam, especially when compared to common lithography tools, it offers a wide range of possibilities, at least for non-high-throughput systems. [18] proposed an online nanolithography system using EBiD technology in an e-beam lithography system. However, the minimum feature size was as high, some 25 nm.

3D structuring and nanomanufacturing: Based upon the early experiments on the three-dimensional deposition of nanostructures [14], this field developed into one possible application for EBiD. A typical example of structuring is the generation of supertips for atomic force microscopy [16, 19], where comparatively simple geometries (pin-like deposits) are deposited. However, more and more complex deposits have been generated, showing the possibilities of this technique [20], but also generating new devices [21]. The combination of deposits with different mechanical properties has the potential to generate new mechanical devices. An example for flexible hinges will be given in Section 10.4. Another field of application in nanomanufacturing is the use of EBiD as a **bonding technology** for attaching objects like carbon nanotubes to each other. Typical examples can be seen in [22] and in [23].

10.2 Theory of Deposition Processes in the SEM

In this section, the aim is to describe the state-of-the-art regarding the theory behind electron-beam-induced deposition and the **interactions** between electron beam, substrate, and precursor, which lead to material deposition. For the understanding of the deposition process, it is necessary to recognize the effects of secondary electron generation. Major electron beam parameters influence the deposition significantly, which will be described in the following section, with the single effects such as the generation of secondary electrons and their range. Based on this, a description of a deposition growing process can be formulated. Afterwards, the rate equation of EBiD and its relevant parameters will be presented. After discussing the single parameters, the influence of heat during deposition will be taken into account.

10.2.1 Scanning Electron Microscopy for EBiD

For most applications of EBiD, SEMs are used, although TEMs (transmission electron microscopes) and STEMs are also possible instruments. In this chapter, the application of EBiD will be restricted to SEMs, but the fundamentals can be applied for other types of electron microscopes as well.

10.2.1.1 *Generation of the Electron Beam*
Inside the electron column of an SEM, the electron beam is generated, shaped, and deflected. The extraction of electrons out of the cathode can be achieved using three physical effects. These are thermionic emission, field emission, and Schottky emission.

In Figure 10.1, a schematic sketch of a **thermionic electron gun** is shown [24]. The cathode, a hairpin-like wire usually made of tungsten, is heated with an electric current to a temperature T_c of 2500-3000 K [24]. At that temperature, the electrons' kinetic energy is high enough to overcome the work function of 4.5 eV for tungsten. The electrons are emitted from the tungsten wire into the vacuum and form a spatial charge. Due to the acceleration voltage U_0 between cathode and anode, the electrons are accelerated towards the anode. For common SEMs, the acceleration voltage is between several hundred volts and up to 50 kV.

The Wehnelt cup between cathode and anode bundles the exiting electron cloud to an electron beam. The current of the electron beam is designated emission current I_E with a current density J_E of approximately 1.8 A/cm^2 for tungsten hair-pin cathodes [25]. Close to the Wehnelt cup, the cathode, the Wehnelt cup, and the anode generate the crossover, which is the beam's smallest cross-section between cathode and anode. After [24], the current density at the crossover has a Gaussian profile, which can be calculated using the following equation:

$$J(r) = J_E e^{-(\frac{r}{r_0})^2},$$
(10.1)

where r_0 ranges from approximately 10 μm to 50 μm, depending on the electron column parameters.

Within the last couple of years, **field emission guns (FEG)** for the generation of the electron beam have become more and more popular, due to their improved resolution for scanning electron microscopy. Compared to thermionic guns, the electrons are extracted out of the cathode by a very high electrical field between the cathode and the first anode. This electrical field enables the electrons to tunnel through the potential wall. The second anode is used for accelerating the electrons. The major advantage of FEGs compared to thermionic guns is the higher current density J_E at a much smaller crossover size. This leads to better resolution and a better **signal-to-noise ratio**.

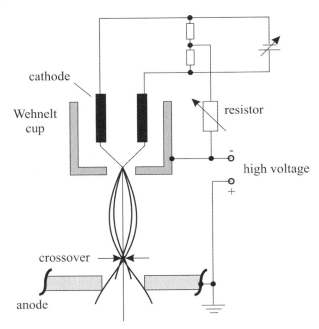

Figure 10.1. Schematic of a thermionic electron gun as widely used in SEMs. The high tension between Wehnelt cup and anode leads to the formation of an electron cloud. The crossover is referred to as the virtual source.

The third effect used for electron extraction is the so-called **Schottky effect**, which can literally be seen as a combination between thermionic emission and field emission. At a temperature of about 1800 K [24], the electron's kinetic energy is high enough to overcome the potential wall of 2.7 eV, which is lowered by the electric field.

The **current density** J_E mainly depends on the emission type of the cathode. As will be shown in the next section, the minimum spot size of the beam on the substrate has an important influence on the spatial resolution of the EBiD deposit and depends on the emission current density.

10.2.1.2 General SEM Setup

In Figure 10.2, a typical setup of an SEM is shown [26]. The two main parts are the vacuum chamber, where the specimen and the electron detectors are positioned, and the electron column. The chamber is evacuated to a pressure around 10^{-5} mbar by a turbo-molecular pump and a rotary pump. The **electron column** holds the necessary parts for the beam generation, consisting of the cathode and anode and for the shaping and deflection of the beam (apertures, condensor lenses, and scan coils). From the cathode, the primary electrons are released and accelerated by the potential difference U_0 between cathode and anode. The potential difference determines the electron's kinetic energy, which is mostly quoted in keV.

The generated beam is bundled by the condensor lens, which leads to the so-called crossover. This is also called the **virtual source**, which is then demagnified by the condensor lens(es) and the objective lens on the specimen. The objective lens focuses the beam on the specimen. The size of the demagnified crossover image is the so-called spot size d_p, which determines the SEM's resolution. The spray diaphragms prevent a contamination of the lenses and the scan coils.

The final aperture determines the current I_p passed on the specimen and thus has to be set in accordance with the resolution. The aperture diameter and the **working distance** WD, which is the distance between the aperture and the specimen, determine the aperture angle α:

$$\alpha = \tan^{-1}\left(\frac{D}{2WD}\right).$$

(10.2)

The aperture angle α determines the depth of focus and the resolution. A lower aperture diameter D leads to a reduced probe current and to a smaller aperture angle. The consequence is an increase in resolution and in focus depth. However, the aperture angle can only be reduced to the point where the signal-to-noise ratio gets unacceptable for imaging.

The generated beam is scanned by magnetic fields, which are generated by the scan coils. A scan generator controls the deflection and scans the beam line by line over the specimen's surface for generating an image of the specimen. A reduced scan speed leads to a better **signal-to-noise ratio**. The smaller the scanned area, the higher the magnification, which can be as high as 300,000 times for an SEM.

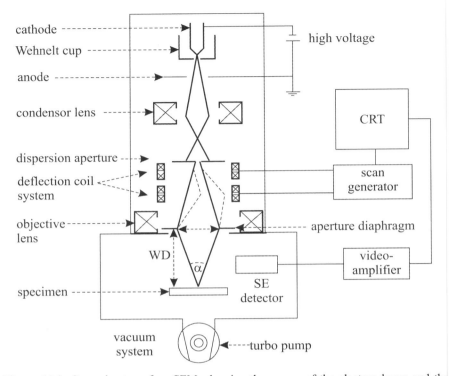

Figure 10.2. General setup of an SEM, showing the course of the electron beam and the relation between the aperture angle, the aperture, and the working distance. Further information regarding the image processing can be found in Chapters 4 and 5.

By positioning the electron beam on a spot, it is possible to generate pin-like deposits with EBiD, whereas scanning of the beam can generate line or layer deposits.

10.2.1.3 *Secondary Electron Detector*

When the high-energy **primary electrons (PE)** hit the substrate, low-energy **secondary electrons (SE)** are generated; the mechanisms for this will be explained in more detail in the next section. However, the number of generated secondary electrons strongly depends on the substrate material and geometry. The generated SEs are counted by an Everhart-Thornley detector, whose collector accelerates the emitted SE towards the detector with a voltage up to 400 V. The high voltage between collector and scintillator of about 10 kV accelerates the electrons on the scintillator, where a luminescent layer emits photons when being hit by electrons. The emitted photons are guided by a light conductor to a photomultiplier, where the light signal is amplified and converted into a current conducted through a resistor. The voltage drop on the resistor corresponds to the brightness value of the spot where the primary electron beam hits the substrate. By correlating the SE detector signal with the beam position on the substrate by reading the scan generator, an image of the specimen's surface can be constructed.

10.2.2 Interactions Between Electron Beam and Substrate

For understanding the deposition mechanisms in EBiD, deeper investigations of the reactions between the primary electron beam and the substrate are necessary. As will be shown, the primary electron beam parameters have a major influence on the deposition of materials. Furthermore, different substrate materials lead to different results in growing experiments.

10.2.2.1 *Energy Spectrum of Emerging Electrons*
Due to the high vacuum in the vacuum chamber, the primary electrons hit the specimen with approximately the same energy as that with which they have been accelerated. Depending on the substrate's material parameters such as composition, density, and geometry, the penetration depth of the PEs vary. When penetrating the substrate, the electrons are scattered elastically and inelastically, which leads to a characteristic energy distribution within the substrate and on its surface. During this scattering process of the PEs in the substrate, the PEs lose their energy. The lost energy is used for the generation of electrons, electromagnetic radiation, and heat. In Figure 10.3 [24], an overview of the emitted radiation is given and the electrons with their characteristic energies are described.

Backscattered and low-loss electrons: Due to elastic and inelastic scattering of the primary electrons on the substrate atoms, the electrons lose part of their energy and change their direction until their energy is lost completely or until they escape at the substrate's surface. A distinction is made regarding the electrons' energy when they leave the substrate. Electrons with energies higher than 50 eV are referred to as backscattered electrons, whereas low-loss electrons are defined as electrons with an energy close to that of the primary electrons. In Figure 10.4, the energy distribution of electrons emerging from the substrate surface is given.

Auger electrons and X-rays: When inelastic scattering of a primary electron leads to ionization on an inner electron shell of the substrate atom, this shell is filled up by an electron from one of the outer shells. During this process, energy is released in the form of X-rays and so-called Auger electrons. The energy of the X-rays and of the Auger electrons is very characteristic of the substrate material and can be measured using specialized detectors (energy dispersive X-ray analysis = EDX-analysis or Auger electron detectors). The energy spectrum of the Auger electrons is between 50 eV and 2 keV.

Secondary electrons: Inelastic scattering of the primary electrons can lead to the emission of electrons from the substrate material itself. These emitted electrons are mainly SE that are generated when PEs strike electrons out of the outer shell of substrate atoms by lifting up the shell electrons with an energy level high enough to overcome the Fermi level. Depending on the energy of the primary electron, it can trigger many ionization effects.

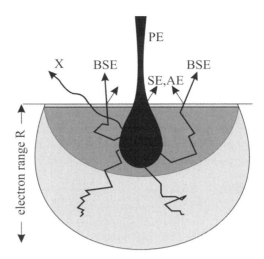

Figure 10.3. Interaction between primary electrons (PE) and substrate, which leads to electron radiation (secondary electrons (SE), backscattered electrons (BSE), Auger electrons (AE), and X-rays (X))

Based on their origin, secondary electrons only have very limited energy (less than 50 eV) and can only leave the substrate when they are generated close to the surface. Secondary electrons and backscattered electrons are differentiated with respect to their energy. About 70% of secondary electrons only have an energy of less than 15 eV [24, 25] (Figure 10.4).

Specimen current and charging effects: Due to the electron beam impacting on the substrate, different electrons are generated in the substrate, as explained above. However, not all primary electrons leave the specimen. These are adsorbed and dissipated to the ground through the specimen holder. The current flowing between specimen and ground, known as the specimen current, can be measured

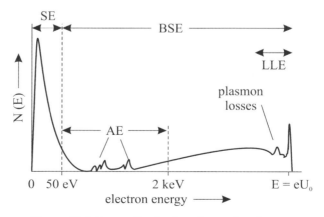

Figure 10.4. Energy distribution of emerging electrons

using a pico-amperemeter. However, charging effects of the substrate may occur if the resistance between substrate and ground is too high. This can either be due to a low-conductivity specimen or to a bad electrical connection between specimen and ground. If fewer electrons leave the specimen – as SEs, BSEs, or specimen current – than are being injected by the primary beam, charging also occurs. In this case, optimizing the beam parameters can reduce charging effects.

10.2.2.2 *Range of Secondary Electrons*

In the early years of EBiD, it seemed to be common sense to consider the electrons of the primary electron beam as responsible for triggering the chemical reaction of the precursor. However, within the last decade, several observations indicated that the reaction is triggered by the secondary electrons. In 1993, [27] noticed that the diameter of a pin-like deposit, which is fabricated in spot mode, is about one to two magnitudes bigger than the diameter of the PE beam. [28] identify the high influence of the secondary electrons on the deposition process with a higher cross-section of the secondaries (energies below 50eV) compared to the primary electrons (25 keV). [5] confirm this by simulating the growth process of EBiD deposits with the Monte-Carlo method. The simulation results are in good agreement with their experiments. Based on these considerations, further investigation regarding the secondary electrons and their different generation effects, intensity, and energy spectrum, is necessary.

SE1 and SE2: Secondary electrons are defined as low-energy electrons with an energy below 50 eV. Especially with regard to EBiD, not only is their energy of considerable interest, but also their spatial distribution when emitted from the substrate. The generation of SEs is attributed to the inelastic scattering of the high-energy electrons. Secondary electrons generated due to inelastic scattering of the primary electrons on the substrate atoms are called secondary electrons 1 (SE1).

However, due to the elastic scattering of primary electrons (backscattered electrons and low-loss electrons), these electrons can diffuse several μm away from the impact spot of the primary beam (Figure 10.3). During their travel, these high-energy BSE can generate secondary electrons as well, which are then called SE2. In Figure 10.5, an overview over the different generation spots for secondary electrons is given.

Furthermore, the backscattered electrons can again leave the substrate and generate secondary electrons (SE3) far away from the first impact (*e.g.,* on the vacuum chamber). However, this effect will be neglected here, as it does not contribute to the deposition on the substrate.

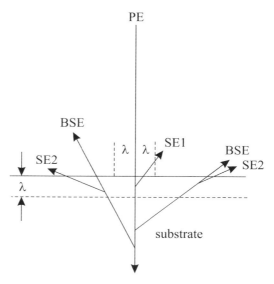

Figure 10.5. Generation of SE1 and SE2 [29]: The primary electrons (PE) hit the substrate generating backscattered (BSE) and secondary electrons (SE1). The backscattered electrons can again generate secondaries (SE2).

Spatial distribution of the SE1: Secondary electrons are limited by their low energy, and they can, therefore, only leave the substrate when they are generated within a depth λ of the surface [24, 29]. A formal description of the most probable escape depth can be found in [30]:

$$\lambda = 2.67 \frac{A_0 I}{\rho Z^{\frac{2}{3}}},$$ (10.3)

Where λ is denoted in Å. A_0 is the atomic weight, I the first ionization energy in eV, ρ the density denoted in g·cm^{-3} and Z the atomic number. Based on this formula, escape depths ranging from approximately 0.5 nm for cerium to 10 nm for carbon can be calculated. Reimer reports [24], escape depths between 1 and 30 nm for metal oxides and alkali halides. Based on this, Seiler delivered in 1983 an approximation for the spot diameter of the primary excited secondary electrons [29]:

$$d_{SE1} = \sqrt{d_0^2 + \lambda^2},$$ (10.4)

where d_0 is the diameter of the primary beam and λ is the mean escape depth of the SE1. The diameter of the focused primary beam d_0 on the substrate can be calculated using an equation from [31]:

$$d_0 = \sqrt{\frac{4 \cdot k_B \cdot T_C \cdot I_P}{\pi \cdot \alpha^2 \cdot J_E \cdot e \cdot U_0}} \,, \tag{10.5}$$

where k_B is the Boltzmann constant, T_C is the cathode temperature, I_P is the beam current, α the aperture angle, J_E the current density, e the elementary charge and U_0 the acceleration voltage.

Spatial distribution of the SE2: The spatial distribution of the emitted SE2 is determined by the range of the backscattered electrons. The range of the BSE is furthermore limited by the amount of energy lost per impact, the primary energy and the density of the material. The dependence of the relative range R_x on the primary energy can according to [24] be formulated as:

$$R_x = aE^n \,, \tag{10.6}$$

where a and n are material- and energy-dependent parameters which can again be taken from [24], e.g., $a = 11.5$ and $n = 1.35$. However, [24] notes that the dependence on materials is low and is thus neglected for calculations here. E denotes the energy of the primary electrons in keV. The result R_x is denoted in $\mu\text{g}\cdot\text{cm}^{-2}$. The absolute range R can be formulated as the fraction of the relative range R_x and the material density ρ in $\text{g}\cdot\text{cm}^{-3}$ [24]:

$$R = \frac{R_x}{\rho} \,. \tag{10.7}$$

Based on this equation, the maximum range of a backscattered electron can be estimated and consequently the maximum distance from the impact spot of the primary beam where SE2 can be generated is defined as $\frac{R}{2}$. When the distance between backscattered electrons and the primary beam increases, its energy decreases and thus its ability to ionize substrate atoms. Seiler established in [29] that the spatial distribution of the SE2 released by BSE can be approximated by Gaussian distributions. Based upon measurements from [32], the full width at half maximum (FWHM) intensity of the SE2 yield can be estimated to approximately 3 μm on silicon at 20 keV primary beam energy. This value rises to 5.8 μm at 30 keV.

Secondary electron yield: According to [24], the relationship between the number of primary electrons N_{PE} and the total number of secondary electrons can be defined through the yield factor δ :

$$N_{SE} = \delta N_{PE} \,. \tag{10.8}$$

The yield factor δ is dependent on the energy of the primary electrons E_{PE} and a factor δ_{max}, which can be taken from tables (e.g., [30, 33]). The reason for the decreasing SE-yield with increasing energy, that is, acceleration voltage, is the increasing penetration depth of the primary electrons, which handicaps the deeper-generated secondaries from reaching the surface.

Seiler worked from published experimental data in [29] and derived an equation, with which the **secondary electron yield** δ can be approximated:

$$\delta = 0.86 \cdot \delta_m^{1.35} \cdot (E_{PE})^{-0.35}. \tag{10.9}$$

The energy of the primary electrons is denoted in keV. The equation above can be used for primary electron energies E_{PE} greater than E_{PEmax}: 1000 eV. Especially with regard to the formation of the deposit, the intensity distribution of the secondary electrons is of major interest, because it determines the spatial resolution of the deposition. In Figure 10.6 [29], a typical intensity distribution of the secondary electrons is shown. It is noticeable that the emission distance of the SE2 can be much wider compared to the emission distance of SE1. In order to find out about the correlation between the spatial resolution of the deposition and the spatial distribution of the secondary electrons, a comparison of the secondary electron yield density between the SE1 and SE2 is necessary.

The secondary electron yield's mean density for an electron beam of 20 kV acceleration voltage, and a beam diameter of 130 nm must therefore be compared for silicon and gold as substrate material with respect to the contributing SE1 and SE2. The overall secondary electron yield δ can be calculated by summing up the contribution of the SE1, reflected by δ_{PE}, and the contribution of the SE2 δ_{BSE} multiplied by **the backscattering coefficient** η [29]:

$$\delta = \delta_{PE} + \eta \delta_{BSE}, \tag{10.10}$$

where δ_{BSE} can be calculated from the following equation [29]:

$$\delta_{BSE} = \beta \delta_{PE}. \tag{10.11}$$

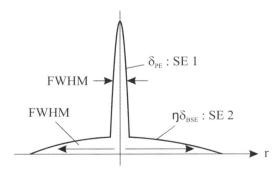

Figure 10.6. Schematic intensity distribution of the secondary electrons (SE1 and SE2) against the distance from the point where the primary beam hits the substrate

And, according to [24], δ_{PE} is proportional to:

$$\delta_{PE} \propto E_{PE}^{-0.8} \cdot \sec(\phi) \cdot \lambda , \tag{10.12}$$

Where ϕ is the incident angle of the electron beam and the area surface normal of the substrate, and λ is the exit depth of the SE. The backscattering coefficient η is a material-dependent parameter. Values can, for example, be found in [29] and [24]. From [29], the value taken for silicon is 0.22, and for gold 0.45. According to the calculations of [29], the ratio β, which describes the number of secondary electrons released per backscattered electron, can be assumed as 2 for electron energies higher than 5 kV. Using these values, the average SE1 yield density can be calculated for silicon as $1.47 \times 10^{13}\,\mathrm{m}^{-2}$, and for gold as $1.77 \times 10^{13}\,\mathrm{m}^{-2}$. Accordingly, the average SE2 yield density for silicon is $1.48 \times 10^{10}\,\mathrm{m}^{-2}$ and for gold $1.01 \times 10^{12}\,\mathrm{m}^{-2}$. It is obvious that the SE1 yield densities for silicon and gold are in the same region. One reason for this is the low influence of the **escape depth** λ (assumed 30 nm for both materials) compared to the high beam diameter of 130 nm, if typical values for thermionic guns are assumed. However, the average SE2 yield density of silicon is about two orders of magnitude smaller than for gold. The main reason for this is the reduced **range** R of the backscattered electrons in gold (approx. 0.34 µm) compared to that of silicon (approx. 2.8 µm). From these simple calculations, the following conclusions can be drawn:

1. The SE1 yield density is very similar for both substrates and therefore the influence of the substrate material should be low for beams generated with thermionic guns.
2. The influence of the substrate material on the SE2 yield density is quite strong. The result is the effect of a socket formation for depositions on high-density materials like gold, as it can be seen in Figure 10.19, where pin-like depositions have been made on silicon substrate coated with about 40 nm of palladium-gold, which is bigger than the exit depth of the SE. In contrast to these sockets being formed while depositing on high-density materials, thin circular contamination layers several µm in diameter form when depositing pins on light materials such as silicon, as shown in Figure 10.7.

10.2.2.3 *Results*

Based on the assumption that secondary electrons play the major role in the deposition process, the relevant generation processes for SE1 and SE2 have been explained. The lower limit for the lateral dimension of an EBiD deposit can thus be approximated by the diameter of the primary electron beam, corrected by the range of the SE1 (λ). Depending on the major beam parameters, which are user-adjustable on a common SEM, *i.e.,* the acceleration voltage U_0, the probe current I_P, and the working distance WD, this limit can be calculated using Equations 10.3-10.5. However, the range enhancement of the SE1 due to scattering of the primary beam is very small compared to the beam diameter of thermionic SEMs.

This additional material parameter λ only comes into account in **high-resolution imaging** using field emission guns or a substrate with low density, where the parameter λ cannot be neglected against the beam diameter d_0. In our experiments, the probe current at constant acceleration voltage and the acceleration voltage at constant probe current, respectively, have been varied. For acceleration voltages above 10 kV and probe currents below 10 nA, the difference between beam and pin diameter is approximately between 200 and 400 nm.

The generation of SE2 due to the inelastic scattering of backscattered electrons certainly is a major drawback for high-resolution depositions, as these SE2 can be generated up to several μm away from the spot where the primary beam hits the substrate. This effect can only be minimized when the substrate is very thin, *e.g.,* a **foil** [5]. This results in a very low backscattering coefficient. However, if **bulk** substrates are used, the spatial intensity of the SE2 is very low compared to the SE1. This can lead to the deposition of a socket, which can be seen in Figure 10.19.

10.2.3 Modeling the EBiD Process

In Figure 10.7, typical EBiD deposits are shown. These pin-like deposits are generated when the SEM's electron beam is pointed on a spot and allow comparative experiments for the process parameters.

10.2.3.1 *Rate Equation Model*

In 1986, [6] developed a **rate equation model**, which describes the growth of planar layers by EBiD, depending on precursor parameters and electron beam parameters. This phenomenological model has, until now, been the most widely used theoretical approach for describing this growth. In Figure 10.8 [34], the EBiD process is sketched. The precursor with flux density F_{Pre} in molecules $/(\mathrm{s \cdot m^2})$ strikes the surface and is adsorbed with a probability given by the sticking coefficient s. The adsorbed precursor molecules stay on the substrate surface with a mean stay time of τ in seconds. The molecule density in one monolayer is designated by the parameter N_0 in molecules $\cdot \mathrm{m^{-2}}$. The cross-section for dissociation under electron bombardment is given by σ in $\mathrm{m^2}$. N_{Pre} denotes the density of adsorbed precursor molecules on the substrate surface in molecules $\cdot \mathrm{m^{-2}}$. Finally, the influence of the electrons is expressed by the electron flux density J_e in electrons $/(\mathrm{s \cdot m^2})$.

Figure 10.7. Example of EBiD deposits on a silicon wafer from a tungsten-hexacarbonyl precursor. These "pins" have been deposited by pointing the SEM's electron beam on one spot for several minutes.

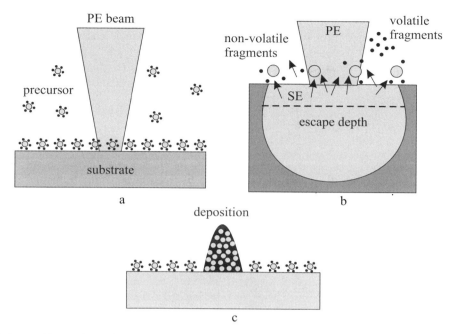

Figure 10.8. Schematic illustrating the EBiD process, showing the different interactions between precursor molecules and substrate, secondary electrons and precursor and the deposit. a. The gaseous precursor is adsorbed on the substrate surface. b. The primary electron beam interacts with the substrate and leads to the emission of secondary electrons, which dissociate the adsorbed precursor molecules into volatile and non-volatile fragments. c. The non-volatile fragments form the deposit.

Based upon these parameters, [6] proposed the following equation for the change in adsorbed precursor molecules:

$$\frac{dN_{Pre}}{dt} = s\,F_{Pre}\left(1 - \frac{N_{Pre}}{N_0}\right) - \frac{N_{Pre}}{\tau} - \sigma\,N_{Pre}\,J_e\,. \tag{10.13}$$

On the right side of the equation, the first term describes the density of adsorbed precursor molecules, the second term the density of desorbed precursor molecules, and the third term the density of dissociated molecules, *i.e.*, the density of the deposited molecules.

Assuming a **steady state** in Equation 10.13, *i.e.*, $dN_{Pre}/dt = 0$, the density of the adsorbed precursor molecules in the equilibrium state N_{Pre} can be calculated [12]:

$$N_{Pre} = N_0 \left(\frac{s\,F_{Pre}}{N_0}\right)\bigg/\left(\frac{s\,F_{Pre}}{N_0 + \frac{1}{\tau} + \sigma\,J_e}\right). \tag{10.14}$$

This equation reveals that the maximum surface coverage is one monolayer [12], unless the precursor pressure in the gas phase is very high (condensation) [35].

Furthermore, Scheuer gives a description of the growth rate R_{Dep} of the deposited layer, which is proportional to the volume V_m of the deposited molecule in m^3, the dissociation cross-section σ, the adsorbed precursor molecule density N_{Pre} and the electron flux density J_e [6]:

$$R_{Dep} = V_m \, \sigma \, N_{Pre} \, J_e .$$ (10.15)

Substituting N_{Pre} in the equation above with the steady-state precursor density N_{Pre} from Equation 10.14, the growth rate can be calculated:

$$R_{Dep} = V_m \, s \, F_{Pre} \Big/ \left(1 + \tfrac{N_0 + s \tau F_{Pre}}{N_0 \tau \sigma J_e} \right).$$ (10.16)

Based upon this rate equation model, two cases **similar to chemical vapor deposition processes** can be distinguished:

- $\tau s F_{Pre} < N_0$: in this case, the surface is never covered with a monolayer. The chemical reaction happening in the EBiD process is **diffusion-limited**, analogous to low-pressure CVD processes (LPCVD). Thus, increasing the precursor flux should result in a significant increase in the growth rate.
- $\tau s F_{Pre} \geq N_0$: the surface is always covered with a precursor monolayer. The limiting factor is the electron flux. A further increase in precursor flux leads to only a small increase in the deposition rate. This regime is comparable with atmospheric pressure CVD (APCVD), where the process is **energy-limited**.

The influence of the molecular flux on the deposition rate is discussed in Section 10.4 and can be seen from the experiments shown in Figure 10.14 and Figure 10.16. These diagrams indicate that the EBiD process using a heated gas supply in the high-vacuum chamber shows characteristics of a diffusion-limited process. Thus, an increase in the deposition rate can be achieved by increasing the molecular flux. However, this is mostly limited by the maximum pressure the SEM can tolerate in its chamber. Another approach is the use of environmental electron beam-induced Deposition (EEBD), as proposed by [36].

10.2.3.2 Parameter Determination for the Rate Equation Model

The main benefit of the rate equation model is the fact that the deposition rate can be calculated from a set of parameters. The main parameters, which can be influenced in a simple way, are the precursor flux density F_{Pre}, which will be discussed in Section 10.4, and the electron flux density J_e, discussed in the following section. However, only a little information is given in the literature about many of the parameters in the rate equation. One of the reasons for this is the challenge of measuring them, *e.g.,* for the dissociation cross-section σ for metallo-organic precursors under electron impact.

Dissociation cross section σ: For this parameter, values from $1.2 \pm 0.2 \times 10^{-20}\,m^2$ at 30 keV [17] to $0.2 \pm 0.15 \times 10^{-20}\,m^2$ at 40 keV [6] for tungsten-hexacarbonyl $(W(CO)_6)$ are found in the literature. The dissociation cross-sections given were measured originally with respect to the energy of the primary electron beam. However, the dissociation cross sections peak at low energy (below 100 eV), which has been determined by fitting EBiD growth measurements. Other values in the literature are mostly found for hydro-carbon complexes, e.g., $4.8 \times 10^{-20}\,m^2$ at 25 eV for C_2H_5 [34].

Sticking coefficient s: The sticking coefficient s is often found in the literature set to one, due to a lack of data. This coefficient is dependent on the substrate surface coverage Θ, i.e. [35]:

$$s = s_0\, f(\Theta).$$

(10.17)

Langmuir proposed a simple model, where the sticking coefficient drops linearly with the surface coverage. A further approach assumes that the sticking coefficient is constant until a surface coverage of 0.3 to 0.4 and then drops linearly [35]. Under this assumption, dissociation is not yet implemented, which might play a significant role in the EBiD process, especially as the sticking coefficient cannot be modeled without temperature dependence.

Mean stay time τ: The mean stay time describes the desorption rate of precursor molecules from the substrate surface. This parameter strictly depends on temperature, i.e. [35]:

$$\tau = \tau_0\, e^{\frac{E_{des}}{R\,T_{sub}}},$$

(10.18)

where τ_0 is the pre-exponential factor, E_{des} is the desorption energy, R is the molar gas constant, and T_{sub} the substrate temperature. The mean stay time is usually considered to be 10^{-13} s [37, 35], whereas measured values for the pre-exponential factor τ_0 range from 10^{-16} s to 10^{-9} s [35]. [6] determined the mean stay time during the EBiD process for $Ru_3(CO)_{12}$ and $Os_3(CO)_{12}$ by fitting measured values to 1.4 ± 1 s.

Deposited molecule volume V_M: The volume of a deposited molecule depends on the morphology of the deposit, i.e., the crystal structure of the deposit. Once the density and the atomic composition of the deposit is known, the volume can be calculated easily. Density measurements have been carried out in [38].

Molecule density in a monolayer N_0: The molecule density in a monolayer can be estimated based on the assumption that molecules on a surface are packed as densely as possible. If the molecule is considered to be spherical with a radius r, the molecule density can be calculated as follows [35]:

$$N_0 = \frac{1}{2\sqrt{3}r^2}.$$

(10.19)

Applying this equation to the tungsten-hexacarbonyl precursor with a molecule diameter of 7.2×10^{-10} m, the monolayer density is 2.23×10^{18} m^{-2}. A value determined from EBiD measurements is 7.4×10^{17} m^{-2} [6].

10.2.3.3 Influence of the SE

Based on the assumption that the **secondary electrons** are responsible for triggering the reaction, the electron flux density J_e should be substituted by the secondary electron flux density J_{SE}. For pin-like deposits, the deposition rate R_{Dep} should then be proportional to secondary electron flux density.

By contrast, the width of the pins should be equivalent to the minimum resolution. The secondary electron flux density J_{SE} can then expressed as:

$$J_{SE} = \frac{\delta I_P}{(\frac{d_0}{2})^2 \pi},$$
(10.20)

where δ is the secondary electron yield, I_P is the probe current and d_0 is the diameter of the e-beam on the substrate (spot size).

Taking into account the fact that the total SE-yield δ is within the spot size dominated by the contribution of the SE1 (Figure 10.6), the SE-yield δ can be substituted by the SE1-yield δ_{PE}. However, there is no simple method for

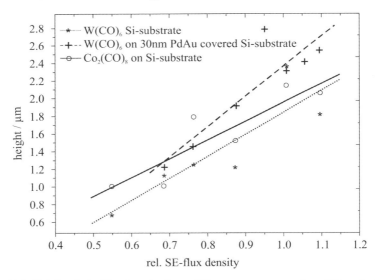

Figure 10.9. Height of pin-like deposits from different precursors and on different substrates *vs.* the relative flux densities of the SE1. The relative SE-flux density has been increased by increasing the acceleration voltage of the primary beam. Although this leads to a decrease in SE-yield (Equation 10.12), the reduction in spot size (Equation 10.5) is more influential on the SE-flux density. All other parameters, *e.g.,* beam current, apertures, working distance, and those of the GIS, have been kept constant.

determining the SE1-yield, and so it is assumed that the secondary electron flux density is related to the secondary electron flux density of the single experiment at acceleration voltages of 20 kV. The deposited height and thus the vertical growth rate shows linear behavior for an increasing secondary electron flux density in Figure 10.9. This indicates that the growth rate is also **energy-limited**.

10.2.3.4 *Heat Transfer Calculations*

In the previous section, the fundamental equation for describing the deposition process in EBiD has been shown. Many of the parameters are approximated or calculated from deposition experiments. One of the main parameters determining the deposition rate, which is strongly dependent on temperature, is the mean stay time τ. This parameter describes the average time a molecule rests on the substrate surface until it is either desorbed into the gas phase or disassociated. The mean stay time strongly depends on temperature, as shown in Equation 10.18. A raised temperature of the substrate will lead to a reduction in the deposition rate.

The influence of heat generation on a sample under observation in the SEM, and the possible damage through heating, have been estimated in [24]. Based upon his calculations, the effects of heat generation on the deposition are restricted to the case when the generated heat cannot dissipate due to a reduction in the heat dissipating cross-section or an increase in the insulating length. This is, for example, the case for pin-like deposits.

Assuming a pin-like deposit of diameter D, length L and heat conductivity λ, the temperature difference ΔT between the cold end of the substrate and the hot end irradiated by the electron beam can be calculated using the following equation [24]:

$$\Delta T = \frac{4 P_H L}{\pi \lambda D^2},$$ (10.21)

where P_H is the power dissipated into heat in the deposit. The heating power P_H can be estimated from the electron beam power P_{PE} by a factor f between 40 and 80% for plain substrates according to [24]. Simulations based on the **Monte-Carlo method** showed heat dissipation factors f approximately between 10 and 20% for pin-like structures, due to the scattering of electrons through the pin. Within these simulations, the number and energies of electrons entering a pin-like structure have been balanced against the number and energies of electrons leaving the structure again.

Simulations from [39] showed that temperature increases of 30 K for pin-like deposits from a TEOS (tetra-ethoxy-silane) precursor are possible for certain structures. Experimental data for deposits of other precursors is not available, mostly due to the missing data on the **thermal conductivity** λ of the deposited material.

10.3 Gas Injection Systems

In this section, an overview of gas injection systems will be given. These systems allow the user to inject a gaseous precursor stream into the SEM's vacuum chamber, out of which the deposition on the substrate's surface will be made. Depending on the application, different gas injection systems will be discussed. In order to determine the mass flow and its flow characteristics, a model will be described, explaining the different flow characteristics and their influence on deposition parameters. The need for a mobile GIS will, therefore, be discussed. Additionally, techniques and methods for pressure control will be described. As an application showing the possibilities of these systems, combined depositions are presented.

10.3.1 Introduction

Gas injection systems deliver the evaporated precursor to the deposition spot. Gas injection systems can be distinguished as being of mainly two types – internal systems situated inside the vacuum chamber and external systems which are flanged on the outside of the vacuum chamber.

Requirements for gas injection systems: For the deposition, one major parameter is the density of the precursor flux j_{Pre}. This parameter determines the amount of precursor molecules per solid angle and time unit (molecules $\cdot (\mathrm{sr} \cdot \mathrm{s})^{-1}$) provided for the chemical reaction. As will be shown in the following sections, this is one of the most crucial process parameters. An increased molecular flux of the precursor also leads to a rise in the system vacuum, and so the mass flow cannot be allowed to rise above a certain level determined by the vacuum system. In order to allow high precursor density on the spot at low increase in vacuum chamber pressure, the molecular beam should be focused or "peaked". As the flux decreases with increasing distance between the capillary outlet and the deposition spot, another attribute is the ability to minimize the distance. However, this distance is also dependent on the geometry of the capillary and cannot be reduced to zero for most applications. Especially for **handling and assembly applications**, where bonding with EBiD is one process in a chain, it is necessary to switch off or to reduce the precursor flux in order to prevent contamination while imaging with the electron beam.

Precursors: In the experiments presented here, only metallo-organic precursors like tungsten-hexacarbonyl ($W(CO)_6$) or di-cobalt-octacarbonyl ($Co_2(CO)_8$) have been used. These metallo-organic precursors are sublimates, *i.e.,* at room temperature and ambient pressure they are available as a solid powder, although they have to be kept under a protective gas atmosphere in order to prevent dissociation. These sublimates evaporate directly from the solid state, and the evaporation rate can be controlled by temperature for a given pressure. The metallo-organic precursors can easily be dissociated by the EBiD process, which leads to depositions composed of metal, carbon, and oxygen.

10.3.2 The Molecular Beam

10.3.2.1 *Modeling of the Mass Flow Between Reservoir and Substrate*
In Figure 10.10, a schematic drawing of an evaporation system for metallo-organic precursors is shown. This evaporation system is situated in the SEM's vacuum chamber. The precursor fills the reservoir and evaporates through a capillary of length l and diameter d into the vacuum chamber. The precursor flow leaving the capillary outlet forms a molecular beam pointed towards the deposition spot on the substrate.

The characteristics of the precursor flow through the capillary can be described using the **Knudsen number** Kn, which is the ratio between the molecular mean free path λ of the precursor and the characteristic dimension of the capillary – the diameter d :

$$Kn = \frac{\lambda}{d}. \qquad (10.22)$$

Depending on the Knudsen number, three flow regimes can be distinguished [37]:

- For $Kn > 0.5$, the **flow** is of **molecular** type, and the mean free path of the molecules is very long compared to the capillary's geometry. Thus, the interactions between the single molecules are very low and collisions of the molecules on the capillary's surface affect the flow. This flow regime is present, therefore, in the SEM's (high) vacuum chamber and in the outlet of the capillary.
- The other extreme is a very low Knudsen number, *i.e.,* $Kn < 0.01$. In this regime, the interactions between molecules dominate the flow. This can be the case for high pressures in the capillary inlet.
- Between those two extremes, the so-called **transitional regime** is situated (*i.e.,* $0.01 < Kn < 0.5$). This is the case in the capillary before the molecular flux at its outlet is reached.

The **mean free path** λ can be calculated using an equation formulated by Maxwell [35]:

$$\lambda = \frac{1}{\sqrt{2}\pi\sigma^2 n}, \qquad (10.23)$$

where σ is the diameter of the molecule and n is the molecule density. The mainspring for the precursor flux between the reservoir and the vacuum chamber is the difference in pressure, *i.e.,* the reservoir pressure p_{res} and the pressure inside the vacuum chamber p_{vac}. The pressure along the capillary, starting from the inlet and ending at the outlet, decreases from p_{res} to p_{vac} continuously. Thus, the mean free path of the molecules changes as well as the flow regimes. However, for different flow regimes, the conductance of the capillary also changes. In the

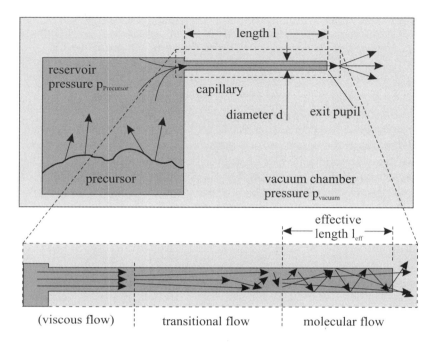

Figure 10.10. Schematic of an evaporation system for metallo-organic precursors

following paragraphs, these conductances are determined in order to calculate the molecular flux from the reservoir to the vacuum.

Figure 10.10 shows how the flow regime can change from the capillary inlet to the outlet. The sketched regimes are strongly dependent on the pressures along the capillary, and also at the borders, when flow regime changes cannot be strictly defined. For the calculation of the mass flow, it is reasonable to start the calculation with the outlet part of the capillary, because its conductivity is the dominant factor on the mass flow.

Molecular flow on the capillary outlet: In 1961, Becker [40] published his results on the formation of molecular beams with channel sources, which emit into vacuum. Based upon the theory in [41], the effective length l_{eff} of a long circular tube, which defines the distance between the capillary outlet and the section where the collision probability of a molecule is 50%, has been introduced. The effective length can be calculated using the following equation, which considers the blocking of the molecular flow at the capillary outlet due to the formation of a precursor cloud [40]:

$$l_{eff} = \frac{4}{3} \cdot d \cdot \left(\left(1 + \frac{3}{64} \frac{d \cdot v}{\dot{N}_{Pre} \cdot \sigma^2} \right)^{\frac{1}{2}} - 1 \right), \tag{10.24}$$

where d is the capillary diameter, v is the mean gas velocity, $\dot{N}_{Pre}$ is the molecular flux and σ is the molecule diameter. Based on this, the molecule density n_0 at the capillary section at position $l - l_{eff}$ can be calculated [40]:

$$n_0 = \frac{24\dot{N}_{Pre}}{2\pi d^3 v} l_{eff} + \frac{16\dot{N}_{Pre}}{\pi d^2 v}. \qquad (10.25)$$

From the equation above, the pressure in the capillary at the position where the flow changes from the transitional to the molecular regime $p_{l=l_{eff}}$ can be calculated by using the ideal gas law for ideal gases.

Transitional flow: Between the capillary inlet and the capillary section in a distance l_{eff} from the outlet, the flow can be considered transitional. For this flow regime, the pV-throughput can be calculated from [35]:

$$q_{pV} = \frac{\pi}{128} \cdot \frac{1}{\eta} \cdot \frac{d^4}{l - l_{eff}} \cdot \frac{p_1^2 - p_{l-l_{eff}}^2}{2} + Z \cdot \frac{\pi}{12} \cdot \bar{c} \cdot \frac{d^3}{l - l_{eff}} \cdot (p_{Res} - p_{l=l_{eff}}), \qquad (10.26)$$

where Z is a dimensionless correction factor, first established by Knudsen [35], and $\bar{c}$ is the mean molecule velocity. The q_{pV}-throughput can be transformed into the molecular flux $\dot{N}_{Pre}$ for ideal gases using the ideal gas law.

Flux calculation algorithm: Based upon this calculation scheme, the interesting parameters molecular flux $\dot{N}_{Pre}$ and effective length l_{eff} can be calculated using an iterative algorithm. In the first iteration ($i = 1$), the molecular flux $\dot{N}_{Pre}^{i=1}$ between capillary inlet and outlet is calculated using the inlet pressure p_{res} and the outlet pressure p_{vac}, which can be assumed to be 0 (capillary exhausts into vacuum). Then the effective length $l_{eff}^{i=1}$ and the pressure $p_{l-l_{eff}}^{i=1}$ at capillary position $l - l_{eff}$ are calculated. Using these values, in the second and n-th iteration step, the molecular flux $\dot{N}_{Pre}^{i=n}$ in the transitional regime is calculated for a capillary of length $l - l_{eff}^{i=n-1}$, inlet pressure p_{res}, and outlet pressure $p_{l-l_{eff}}^{i=n-1}$. The iteration steps are repeated, until the change in the values becomes negligible.

The above scheme is for capillaries where the Knudsen number at the inlet is in the transitional regime, which has been the case for the GIS and precursors described above. Thus, possible viscous flow, which might be the case for high inlet pressures or very small capillaries, is neglected. However, the algorithm can easily be extended for this case.

Directivity of the molecular beam: In the previous section, a method for calculating the molecular flux $\dot{N}_{Pre}$ between the precursor reservoir and the deposition spot has been described. In general, the deposition rate is proportional to the quantity of precursor molecules on the deposition spot and thus to the molecular flux. However, the maximum molecular flux is limited by multiple parameters. Generally speaking, the molecular flux should be kept as low as possible, in order to reduce contamination of the SEM, namely the vacuum system and the electron gun. These devices can be damaged by precursor reaction and contamination. In order to achieve high precursor density on the deposition spot at a minimum molecular flux, it is necessary to peak the molecular beam, generating a beam with very high directivity and thus a high molecular flux density j_{Pre}.

Lambert emitter: The molecular flux density or intensity j_{Pre} relates to the solid angle emitted from an aperture of diameter d and thickness l, where the thickness is very small compared to the diameter. This is referred to as the Lambert emitter. The flux density is strongly dependent on the angle θ measured perpendicularly from the aperture and can be described using the following equation [41]:

$$j_{Pre}(\theta) = cos(\theta) j_{Pre}(\theta = 0°) .$$ (10.27)

The Lambert emitter is, due to its cosine dependence, also known as the cosine emitter. The unit of $j_{Pre}(\theta)$ is $molecules \cdot (sr \cdot s)^{-1}$. The molecular flux density at angle $\theta = 0°$ can be calculated from the molecular flux $\dot{N}_{Pre}$ [40]:

$$j_{Pre}(\theta = 0°) = \frac{\dot{N}_{Pre}}{\pi} .$$ (10.28)

Peaking factor: In order to compare a molecular beam source regarding its directivity, a comparative factor χ is defined, which is the ratio of the peaked center line intensity $j_{Pre_{Peaked}}(\theta = 0°)$ to the intensity of a cosine emitter $j_{Pre_{cosine}}(\theta = 0°)$ with the same leak rate [42]:

$$\chi = \frac{j_{Pre_{peaked}}(\theta = 0°)}{j_{Pre_{cosine}}(\theta = 0°)} .$$ (10.29)

After [40], the maximum possible directivity χ_0 of a capillary with a diameter d and length l can be calculated using the following equation:

$$\chi_0 = 1 + \frac{6l}{8d} .$$ (10.30)

If the capillary length l is substituted with the effective length l_{eff} from Equation 10.24, the peaking factor χ for a given gaseous precursor flowing through a capillary with dimension and d with a molecular flux $\dot{N}_{Pre}$ can be calculated.

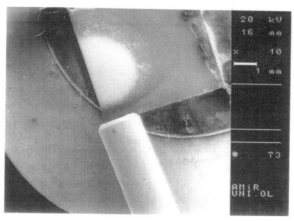

Figure 10.11. SEM image showing the substrate being partly coated with a circular deposition. The precursor was evaporated through the capillary looming from the lower image edge to the substrate edge. The circular deposition spots show the cosine-dependence of the molecular flux evaporating through the capillary outlet.

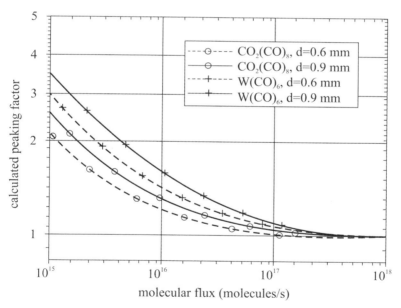

Figure 10.12. Calculated peaking factor for tungsten-hexacarbonyl and di-cobalt-octacarbonyl precursors *vs.* molecular fluxes for a capillary of 40 mm length and different inner diameters. Reasonable peaking factors can only be achieved by a reduction of molecular flux.

In Figure 10.12, the calculated peaking factor χ *vs.* the molecular flux for different precursors and capillary diameters is shown. The calculations demonstrate that high peaking factors can only be achieved if the molecular flux is reduced, or that for a given molecular flux, the diameter has to be widened. The main reason for this is the **blocking of the capillary outlet**, as concluded in [40]. Consequently, a **compromise between molecular flux and peaking factor** has to be found.

10.4 Mobile GIS

The precursor flux density on the deposition spot J_{Pre}, i.e., at **distance** r from the capillary outlet, is strongly dependent on the distance r:

$$J_{Pre}(r) = \frac{j_{Pre}}{r^2},\qquad(10.31)$$

where the distance r is denoted in m and the precursor flux density $J_{Pre}(r)$ at distance r is denoted in $\text{molecules}\cdot(s\cdot m^2)^{-1}$. Hence, the distance between the capillary outlet and the deposition spot has a major effect on the precursor flux density and thus on the deposition rate. A system for adjusting this distance and controlling the precursor flux will be presented in the following sections.

10.4.1 General Setup

In Figure 10.13 the setup for the EBiD is shown. **Inside the vacuum chamber**, one or multiple gas injection systems are positioned.

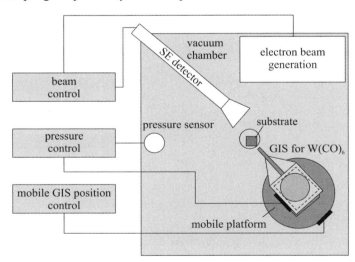

Figure 10.13. Schematic of the EBiD setup. In the SEM's vacuum chamber, the substrate and a mobile GIS are situated. The GIS can be positioned with the mobile platform and its controller. Closed-loop pressure control is achieved by using the pressure sensor for feedback, heating and cooling, respectively, of the GIS. The deposition on the substrate can be controlled by the beam controller, i.e., positioning the beam and reading the SE-detector signal.

These systems are positioned on mobile platforms and so the capillary outlets can be positioned close to, and oriented to, the deposition spot. The pressure in the vacuum chamber can be measured by a separated full-range **pressure gage**, which is flanged to the vacuum chamber door. The substrate is positioned, rotated, and tilted using the SEM's stage in all three dimensions for observation with the electron beam. In principle, the gas injection systems are positioned so that they do not conceal the SE-detector.

10.4.2 Position Control of the GIS

A major parameter for the deposition is the distance between the capillary outlet and the deposition spot. In order to allow high flexibility, the GIS has to be transportable, which ensures that this distance can be adjusted with respect to the deposition task. As has been shown previously, the molecular flux density decreases with increasing distance between capillary outlet and deposition spot. In this way, the influence of the distance on the deposition rate has been measured. In Figure 10.14, the dependence of deposition height on distance between capillary outlet and deposition spot is shown. During this experiment, all other parameters have been kept constant.

As can be seen from Figure 10.14, the dependence between deposition height and distance from the GIS outlet is very strong. Thus, it is necessary to **minimize**

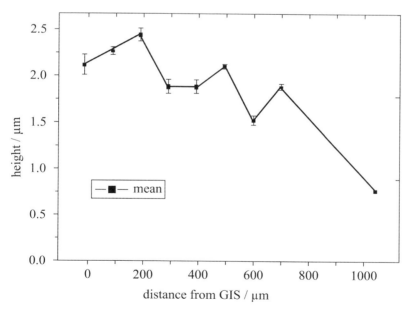

Figure 10.14. Dependence of the height of the deposited pin-like structures from the distance between capillary outlet and deposition spot for tungsten-hexacarbonyl. All pin-like structures have been deposited with the same process parameters. Only the distance has been varied.

the distance between the deposition spot and the capillary outlet. In most cases, this can only be achieved by using a mobile gas injection system, which is adjustable in three axes. Furthermore, by withdrawing the capillary from the deposition spot, the precursor flux on the deposition spot can quickly be reduced to zero, which is an effective method for minimizing contamination.

10.4.3 Pressure Control

According to Equation 10.31, the precursor flux density at distance r from the capillary outlet depends on both, the distance r and the precursor flux density j_{Pre}. The latter can be calculated using the flow calculation algorithm described in Section 10.3.2. From Equation 10.26, it is obvious that the pressure difference between the precursor reservoir and the vacuum chamber is the driving force for the molecular flux. In this section, different types of evaporation systems will be presented and compared. Based on this, methods for controlling the precursor flux will be shown.

10.4.3.1 *Constant Evaporation Systems*
One of the simplest setups for evaporating a metallo-organic precursor in the vacuum chamber of the SEM is a small precursor container, which emits a precursor flux through a small capillary on the substrate. If the container is thermically well connected to a heat reservoir (*e.g.,* the vacuum chamber walls), it can be assumed that the container's temperature does not change significantly during the evaporation of the precursor. Thus, the pressure in the container only depends on the **vapor pressure** of the precursor at the given temperature of the container and on the conductance of the capillary.

This setup is easy to realize and may serve for deposition experiments where process parameters are evaluated. In order to interrupt the precursor flux to the substrate, the reservoir and capillary can be moved away from the substrate using a mobile platform or similar actuating device. However, the precursor will still be evaporated and **contaminate the vacuum chamber**. Apart from this disadvantage, the molecular flux from constant evaporation stages for a given precursor is determined by the geometry of the capillary. The molecular flux density on the deposition spot can only be controlled by the distance between capillary outlet and deposition spot.

The diameter of the capillary has to be chosen small enough to allow the vacuum system of the SEM to reach the system's operating pressure, but wide enough to allow sufficient precursor flux. If the diameter chosen is too wide, the vacuum pumps will evacuate the reservoir to a considerable extent, until operating pressure is reached.

10.4.3.2 *Heating/Cooling Stages*
In order to overcome the above disadvantages of constant evaporation stages, gas injection systems, using Peltier elements for **heating and cooling** the precursor reservoir, have been developed. In Figure 10.15, a gas injection system for metallo-

organic precursors is shown. The precursor fills the reservoir, which is sealed by a rubber o-ring between reservoir and cap. A steel capillary with length 40 mm and an inner diameter of 0.6 mm is connected to the reservoir. Heating and cooling, respectively, of the reservoir can be achieved by shifting heat between the thermal mass and the reservoir. In order to control the temperature of the reservoir, a temperature sensor has been built between the reservoir and the Peltier element. This sensor also prevents the overheating of the precursor during the heating-up process. A second temperature sensor has been built into the column carrying the capillary, for measuring the temperature difference between the reservoir and the capillary. In order to prevent possible condensation effects in the capillary, it is supported by a column, which ensures a low temperature gradient along the capillary and reservoir even during the heating-up process.

10.4.3.3 *Control of the Molecular Flux*

The precursor is evaporated in the reservoir of the gas injection system by applying a positive voltage on the Peltier element, which heats up the reservoir and cools down the thermal mass. For fast disruption of the molecular flux from the reservoir to the deposition spot, the polarity of the Peltier element's driving voltage is switched, and the reservoir is cooled. This allows **fast temperature cycles** and prevents precursor leakage through the capillary if no deposition is wanted. For controlling the molecular flux from the reservoir to the deposition spot, the gas injection system can be operated in two modes, either by **closed-loop control** of the reservoir temperature or by closed-loop control of the vacuum chamber pressure. These two modes will be explained in the following paragraphs.

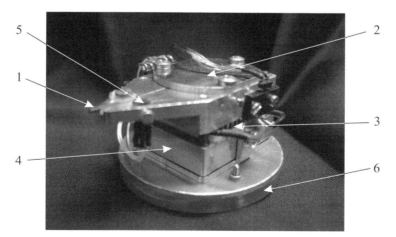

Figure 10.15. Gas injection system for metallo-organic precursors. The precursor is evaporated through the capillary (1) pointing to the deposition spot. The reservoir (2) holds the precursor and is heated and cooled, respectively, by the Peltier element (3). It shifts the heat between reservoir and thermal mass (4); the temperature can be controlled by a sensor (5). The GIS hole can be positioned and oriented towards the substrate using the mobile platform (6).

Temperature control: In temperature control mode, the temperature sensor between the Peltier element and the precursor reservoir is used as closed-loop sensor input. If the heating-up process to a distinct temperature is finished and the evaporation system can be considered isothermal, the molecular flux and thus the pressure in the vacuum chamber and on the deposition, can be considered constant. However, experiments showed that a constant pressure is very hard to achieve using temperature control. There are a couple of reasons for this effect. As the vapor pressure of metallo-organic compounds is strongly dependent on temperature (vapor pressure can be approximated by an exponential law), a small change in temperature can lead to a dramatic change in molecular flux and thus vacuum pressure. These small changes in temperature can also be based on the tolerances of the temperature sensors. Furthermore, change in pressure can be caused by the inertia of the control system itself, as the heat is slowly distributed in the evaporation stage compared to the velocity of the evaporation process.

Pressure control: During the experiments with temperature control, it became obvious that a pressure control system is better suited for EBiD. For sensor feedback in the control loop, a full-range pressure gage has been flanged to the SEM's door, which enables the measuring of the pressure in the vacuum chamber. Furthermore, the SEM's vacuum gage values have been gathered, in order to reveal pressure differences. However, the pressure gage is **gas-dependent** and thus the measured values can only be correlated to the true pressure by means of a correction factor. Although the pressure sensor is far away from the deposition spot, compared to the distance between capillary outlet and deposition spot, it can be assumed that a high pressure correlates to a high precursor flux.

For controlling the pressure in the vacuum chamber, the pressure gage value is used as sensor in the feedback loop. Based on the difference between actual pressure and pressure set point, the voltage was adjusted on the Peltier element using a PID controller.

10.4.3.4 *Pressure Dependence of the Deposition Rate*

Changing the vacuum chamber pressure by introducing the precursor is a parameter that is very simple to use for closed-loop control of the evaporation stages. However, the measured pressure can only be evaluated as a relative value, because the pressure sensor is calibrated against nitrogen. Unfortunately, correction values are not available for the precursors used.

A pressure increase is caused by an increase in precursor flux, which leads to a higher adsorption rate of precursor molecules on the substrate surface. This again leads – according to the rate equation model – to a higher deposition rate, as long as the EBiD process is in the **diffusion-limited region**. However, the possible pressure range is limited by two boundaries: the lower boundary is the minimum pressure the vacuum system of the SEM is able to generate; the upper limit is the maximum pressure before the vacuum switches off the SEM for safety reasons, when, for example, there is overheating of the turbo-molecular pump or danger of short circuits in the cathode of the e-beam system.

Figure 10.16 shows measured values for heights of pin-like deposits. The experiments were carried out on a plain silicon wafer as substrate, under the same conditions and parameters, except for pressure. All deposits clearly indicate that with rising precursor flux and thus pressure in the vacuum chamber, the deposition rate rises as well.

10.4.4 Multimaterial Depositions

Nanostructuring with EBiD is very flexible with respect to the geometric dimension. Pin-like deposits can be generated by pointing the electron beam on one spot on the substrate, line deposits by scanning the beam in a line, and layer deposits by scanning an area. Out of these basic geometric elements, more complex systems can be built up. By combining deposits with different physical or chemical properties, sensory and actuator elements can be built. In the following list, examples for possible applications are given:

- **Flexible hinges**: combination of three pin-like deposits on top of each other, where the one in the middle is made of a more flexible material.
- **Thermal actuators**: two pin-like deposits made of materials with different thermal elongation coefficients, situated beside each other and linked by a bridge on top of both, where one of the pins is deposited on a circuit path for heating.

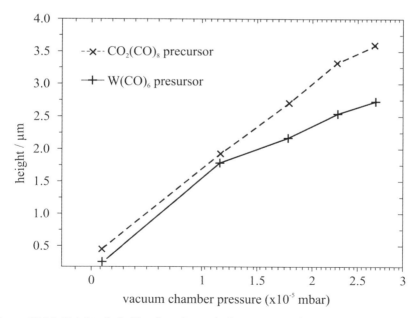

Figure 10.16. Height of pin-like deposits made from tungsten-hexacarbonyl and di-cobalt-octacarbonyl precursor at varying vacuum chamber pressures

- **Thermocouples**: two line-deposits with two materials crossing each other in one point.
- **Strain gages**: laminar deposit of a non-conducting material, on top of which lines of a metallic material are deposited.
- **Sensory elements**: two laminar deposits on top of each other.

Three-dimensional deposits can also be realized by scanning the electron beam slowly while depositing. For generating sensor or actuator elements, it is advantageous to combine deposits of different precursor materials. Therefore, it is necessary to introduce multiple precursors alternating on the deposition spot. This can be achieved by exchanging the precursor, but this is not very flexible. For this reason, a setup has been developed that allows almost seamless change in the precursor flux. Two independent gas injection systems are positioned around the substrate, one containing tungsten-hexacarbonyl, the other di-cobalt-octacarbonyl. An **EDX-microanalysis** of the deposited structure can be seen in Figure 10.17. This combined deposition forms a flexible hinge. In bending experiments, it has been found that the middle part is much more flexible than the bottom and top depositions.

The gas injection system provides the precursor flux, which is the most important and influential part in the EBiD setup of the primary beam. Based upon theoretical considerations regarding precursor flux, its density and directivity, it has been shown that mobile evaporation stages that are heated and cooled,

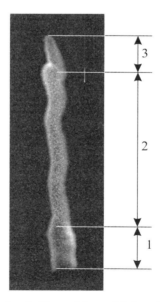

Figure 10.17. EDX-analysis of a pin-like multimaterial deposition of tungsten-hexacarbonyl (1 and 3) and di-cobalt-octacarbonyl (2). The deposited pin has a length of approx. 4 μm. This combination forms a flexible hinge, as cobalt deposits are much more flexible than tungsten deposits.

respectively, are helpful for optimizing the deposition process with respect to the growth rate and low contamination of the vacuum chamber.

One of the most promising technologies for increasing the precursor flux density seems to be the application of miniaturized **multichannel arrays**, which promise a better directivity.

10.5 Process Monitoring and Control

In recent years, EBiD has developed into a promising nanostructuring technique. Various applications have been reported, ranging from rapid prototyping on the nanometer scale to nanolithography. Many examples have been published, showing the application of EBiD for three-dimensional structuring. In [27], the manufacturing of supertips on top of an AFM tip has been shown. This application has been extended by many others, *e.g.,* the manufacturing of high-resolution magnetic supertips [19] and submicron hall devices [43]. Further examples for three-dimensional structuring are given in [20]. In the field of nanolithography, EBiD has several advantages compared to FIB techniques, because no damage is caused in the observed substrate [17]. This fact means that EBiD is especially suitable for the repair of lithography masks, which was proposed by Koops *et al.* as early as 1987 [12]. Mask repair technology based upon EBiD developed into full-scale wafer technology [44], which can be integrated into today's semiconductor-processing technologies.

These more and more complex applications of the EBiD process demand sophisticated methods of process control. In this section, methods for controlling the EBiD process with respect to automation will be given.

10.5.1 Time-based Control (Open-loop Control)

The most simple method for controlling the deposition process is a time-based technique. In Figure 10.18, the deposited tip height *vs.* the deposition time is shown. From this diagram, the growth process can be divided into two parts. At first, the deposition starts with a very high rate which drops continuously to a constant growth rate in the second part. The slope of the deposition rate *vs.* time in the first and in the second part is strongly dependent on process parameters such as the precursor, its flux density, and electron flux density. However, guaranteeing constant process parameters enables the user to control the deposition rate.

For two- and three-dimensional deposits, respectively (*e.g.,* deposits generated through line scans and area scans), the **residence time** t_R on a spot can be calculated from the scan speed between electron beam and substrate v_{Scan} and the diameter d_{Pin} of pins deposited with equal process parameters:

$$t_R = \frac{d_{Pin}}{v_{Scan}}.$$

(10.32)

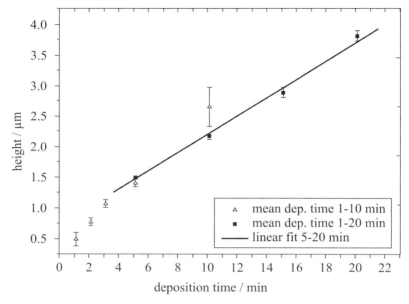

Figure 10.18. Tip height *vs.* deposition time. The single pins have been deposited using the same process parameters, and only the deposition time has been changed.

Based upon this equation and a constant growth rate, the deposited height can be approximated. However, this method only allows open-loop control of the deposition and strongly depends on the knowledge and accuracy of the process parameters for accurate depositions.

10.5.2 Closed-loop Control of EBiD Deposits

In order to overcome the disadvantages of time-based control, especially for deposition heights which are in the non-linear growth period (Figure 10.18), a closed-loop control method has been developed. This method comprises the evaluation of the SE-detector signal, which is a standard detector for any SEM. In [45] the correlation between probe current, SE-signal, and deposition height for pin-like structures is described.

Based on the time variable-change of the detector signal during 1-D deposits and its correlation with the deposition height, a method has been developed to control the deposition height during 2D depositions. Furthermore, the evaluation of the **detector signal** enables the user to track down certain process failures during the process.

10.5.2.1 *Growth of Pin-like Deposits and SE-signal*

Pin-like deposits are grown with EBiD, when the SEM's electron beam is positioned on one spot of the substrate's surface and a gaseous precursor is provided. In Figure 10.19, the results from typical growing experiments are shown.

The single deposits with a deposition time of up to 5 minutes have a **Gaussian-like shape**. After approx. 5 minutes, the tip starts growing out of the surface and the deposit has the shape of a needle. However, the tip geometry is still the same. The growing process can be separated into two distinctive parts. In the first interval (in Figure 10.18 from 0 to 5 min), horizontal and vertical growth takes place, until the tips' Gaussian-like shape is completely developed [45]. In this interval, the height growth rate is non-linear. It starts with a high growth rate, which decreases, until a constant rate is reached. The second interval is characterized by linear growth. The time when the non-linear growth changes to linear growth depends on several process parameters such as substrate material, beam parameters, and precursor flux density, when the tip profile is formed.

In Section 10.2, a description of the interactions between the specimen and the electron beam has been given. The quantity of secondary electrons generated on a sample depends, among other parameters, on its geometry. During the growing process of an EBiD deposit, the geometry changes, as can be seen in Figure 10.19. This leads to a change in the SE-detector signal and in the probe current signal. However, the signal change is only measurable until the deposit's tip shape is formed and only until vertical growth of the pin happens [45]. This correlation between the SE-detector signals can be used as an ***in situ* process control method** during the deposition. The SE-signal $S_{SE}(x,t)$ for a pin-like deposit at time t at a position x on the substrate can be determined using the following function:

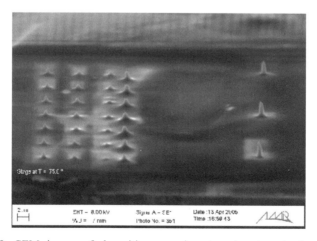

Figure 10.19. SEM image of deposition experiments using tungsten-hexacarbonyl as precursor on silicon substrate coated with palladium-gold. The rows show different deposition times (3, 2, 1, 5, and 10 minutes from left to right). Within a row, the deposition time was equal.

$$S_{SE}(x,t) = S_{UG}(x) + S_{Dep}(t),$$
(10.33)

where $S_{UG}(x)$ is the signal from the substrate during observation at position x. This can, for example, be achieved through imaging the deposition spot. However, the imaging time should be small compared to the deposition time. The signals are strongly dependent on the detector parameters, *e.g.*, brightness and contrast, but also on beam parameters. $S_{Dep}(t)$ is the time-dependent signal, while the pin is growing. This signal can be determined using an exponential law [45]:

$$S_{Dep}(t) = S_0 \cdot (1 - e^{-\frac{t}{\tau}}),$$
(10.34)

where S_0 is the maximum signal achievable while growing the pin for certain detector parameters and τ stands for process-dependent parameters. They reflect, for example, the substrate material, the precursor parameters like chemical composition and flux density, as well as the beam parameters. The growth function for a pin-like deposit in the non-linear growth regime can then be denoted by:

$$h(t) = L \cdot S_{Dep}(t),$$
(10.35)

where L is a process-dependent constant, which has to be evaluated by parameter variation.

10.5.2.2 *Application for 2D Deposits*
Based upon the previously described relationship between the SE-detector signal and the growth of one-dimensional (*i.e.*, pin-like) deposits, a method for **closed-loop control** of two-dimensional (*i.e.*, line) deposits has been developed. If the electron beam is scanned along a certain trajectory X on the substrate, and the SE-signal along this trajectory is recorded, the function $S_{UG}(X)$ is established. In the simplest case, this can be achieved by using the "line scan" functionality of the SEM control. For a defined process parameter set, the correlation between the SE-signal $S_{Dep}(t)$ and the deposited height $h(t)$ is known from deposition experiments. The residence time t_R can be calculated from Equation 10.32 and thus be controlled by the scan speed v_{Scan} using a control algorithm.

In Figure 10.20, an SEM image of line deposits using the algorithm described above is shown. Three lines have been deposited as a bundle with the same height. On the right edge of the image deposited pins can be seen, which have been used for recording the growing function $h(t)$ and the corresponding SE-signal function $S_{Dep}(t)$. From these functions, the relationship between sensor signal and deposited height $S(h)$ can be calculated.

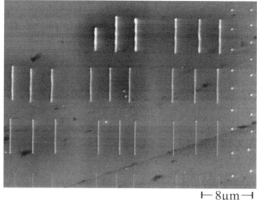

⊢—8µm—⊣

Figure 10.20. SEM image of deposited lines of different heights using the height control algorithm

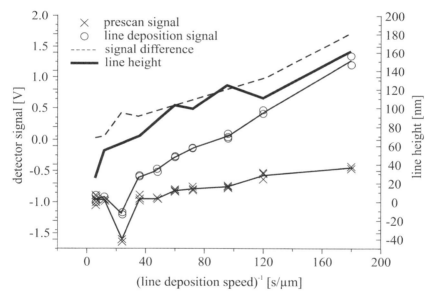

Figure 10.21. Signals during line deposits. Points on the line marked with a cross represent the mean value of the detector signal during the line prescan. Points marked with a circle on the line represent the mean value of the detector signal during deposition with a scan speed given by the reciprocal of the x-axes. The thick line marks the mean height of the deposited line with respect to the right y-axes. The dashed line represents the difference in the detector signal between prescan and deposition, which shows the proportionality between detector signal during deposition and line height.

The relationship between scan speed, SE-signal, and deposited line height is shown in Figure 10.21. The signals along each deposited line and the height of the line have been averaged. The prescan signals have been gathered by imaging the line. The distinct relationship between deposition speed and the difference in the detector signal between prescan and deposition is very clear. The **prescan** can be seen as a fast scan ($t_R << \tau$) along the given trajectory where a line is to be deposited. In the simplest case, this can be the recording of an image which contains the trajectory.

Based upon the algorithm described above, many relevant applications become realizable in closed-loop mode. One example is the **height control** during line deposition, even if the substrate is not plain or has a strong SE-signal contrast. Another typical nanomanufacturing application is the *in situ* control of EBiD bonds during deposition. In this case, two objects are connected to each other by depositing a line perpendicular over the flange where both objects are in touch. In the SE image, this usually gives a strong contrast to the image. Using the algorithm described, constant height of the bonding line can be achieved.

10.5.3 Failure Detection

By evaluating the detector signal during deposition with EBiD, distinct process failures can be evaluated and the process can be stopped or restarted. Typical process failures are:

- **Drift effects** during deposition: these occur if the substrate or the deposition are charged electrically through the primary beam and thus cause an electrostatic deflection of the primary beam. Another possibility is drift due to thermal elongation effects (heat transfer due to heated precursor gases, heating of the deposition) or mechanical instabilities of the setup.
- **Beam generation problems**: due to flashover, short circuits or other failures in the beam generation, the primary beam is disturbed during deposition.

Both failure scenarios show distinct characteristics in the SE-detector signal during deposition. In the case of drift effects, the signal slowly drifts away from its reference, as shown in Figure 10.22a. The inlaid picture shows the effect of drift while depositing a pin-like structure. For evaluating the signal characteristics, upper and lower limits can be defined, between which the signal is allowed to float during deposition. If the signal strays outside these limits, the deposition process should be restarted, taking into account the fact that that the beam possibly has to be repositioned automatically or manually.

In the case of beam generation failures, the SE-detector signal shows a distinct behavior, as shown in Figure 10.22b. The signal characteristics are formed by cracks and peaks. In this case, the deposition process usually has to be stopped for maintenance on the beam generation system.

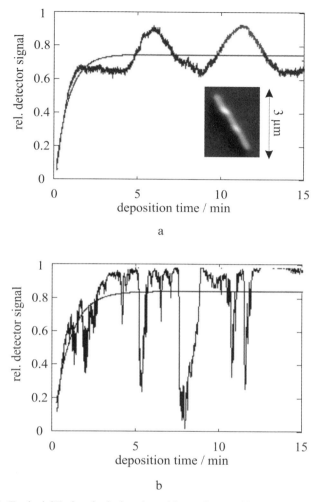

a

b

Figure 10.22. Typical SE signals during deposition, when a. drift or charging happens and when b. the beam generation system is disturbed, *e.g,.* through short circuits

Based upon the evaluation of the SE-detector signal, it is possible to control the height of one- and two-dimensional deposits, *i.e,.* pin-like and line deposits. Out of two-dimensional objects, more complex three-dimensional objects like layers, combined depositions, or highly structured objects can be formed using the same algorithm. However, the major process parameters must be accurate for the exact control of the deposited geometry. The influence of the single parameters can be evaluated using pin-like deposits. As has been shown, the control of the SE-signal can be used to detect process failures and thus regulate, restart, or stop the process.

10.6 Mechanical Properties of EBiD Deposits

EBiD is often used to build mechanical devices, to connect objects on the nanometer scale to each other, or for electrical connections. Especially with regard to bonding tasks in manufacturing processes inside the SEM, it is necessary to observe the mechanical properties of the EBiD deposits. In recent years, the mechanical properties of deposits have become more interesting. The main reason for this is the technology's potential use for setting up micro- or nanomechanical systems. The values determined in the literature and from own experiments are presented below.

Young's modulus: In [46], EBiD layers from paraffin were deposited. These were tested using nanoindentation in order to determine the Young's modulus. The values where measured as 34.3 ± 3.4, 46.3 ± 2.3 and 59.5 ± 2.5 GPa for acceleration voltages of 3, 12, and 20 keV, respectively.

Utke *et al.* [47] estimated the Young's modulus for EBiD deposits from $W(CO)_6$- and $Co_2(CO)_8$-precursors in the range between 10 and 100 GPa. Wich *et al.* [48] measured the Young's modulus of pin-like deposits, which were deflected using an AFM-cantilever until fracture of the pin for the $W(CO)_6$-precursor, of between 21.6 and 50.4 GPa.

Tensile strength: In [47], the tensile strength for deposits from a $W(CO)_6$-precursor has been determined to be 2 ± 0.5 GPa and for deposits from a $Co_2(CO)_8$-precursor to be 1 ± 0.5 GPa, using an AFM-cantilever-based method.

Hardness: The hardness of EBiD layers has been investigated in [46] for paraffin precursors and in own unpublished experiments for tungsten-hexacarbonyl and di-cobalt-octacarbonyl. In both experiments, Berkovich-shaped indentation-tips were used. The hardness values were determined for the paraffin precursor to be 3.6 ± 0.3, 4.0 ± 0.2, and 4.4 ± 2 GPa for acceleration voltages of 3, 12, and 20 kV, respectively. The values for the $W(CO)_6$-precursor were measured to be $7.6^{+2.4}_{-0.6}$ GPa at 20 kV acceleration voltage. The measured hardness for EBiD layers from a $Co_2(CO)_8$-precursor was $3.7^{+1.5}_{-1.3}$ GPa at 20 kV acceleration voltage.

10.7 Conclusions

10.7.1 Summary

Based on the working principle of an SEM, the **interactions** between electron beam and substrate have been described. The generated secondary electrons can be distinguished in SE1 and SE2; methods for calculating their range and the yield have been described for both. Typical EBiD phenomena, such as socket formation and rings around pin-like deposits, have been explained, with the different secondary electron flux densities for the single species.

The **rate equation model** described in [6] has been discussed with regard to single equation parameters. In the literature, less knowledge applicable for EBiD is known for these parameters, complicating the exact determination of the EBiD growth rate. Based on the assumption that the secondary electrons play a major role in EBiD processes [5], the rate equation model has been adapted to the secondary electron flux density. The height of pin-like deposits made from different precursors and on different substrates with different acceleration voltages, and thus secondary electron flux densities, showed a linear behavior when plotted against relative SE flux density.

The **GIS** that delivers the precursor to the deposition spot is a very important part of the EBiD process. The crucial parameter for the EBiD process is the **precursor flux density**, which can be optimized by optimizing the capillary geometry. A model was introduced for calculating the precursor flux density and the peaking factor of the precursor emission. Based on this model, it has been shown that a tradeoff has to be accepted for capillary-type GIS between peaking factor and molecular flux.

For optimizing the achievable results with capillary-type GIS, the distance between capillary outlet and deposition spot should be minimized. Further improvements with regard to process control can be achieved by closed-loop pressure control of precursor evaporation.

Further **process control** can be achieved by evaluating the SE-detector signal during deposition; this is of special interest for line deposits, as the deposited height can be controlled *in situ*. Bonding applications in the SEM can profit from this closed-loop control method. Additionally, process failures can be identified early by evaluating the SE-detector signal.

Finally, the **mechanical properties** of EBiD deposits found in the literature and from own measurements have been presented, emphasizing the importance of EBiD for micro- and nanomechanical applications.

10.7.2 Outlook

The EBiD process has developed over the last 40 years from a disturbing effect in electron microscopy to a promising deposition technology. Today, it offers many different **applications**, such as mask repair, nanomanufacturing, or nano-structuring. The main reason for this can be seen in the progressively shrinking structural size in the semiconductor industry and in nanotechnology. This development process will lead to a wider field of application for EBiD, due to its miniaturization potential. Especially with regard to the evolution of nanotechnology and the resulting need for analysis methods (*e.g.,* CNT characterization), EBiD is one of the most promising technologies for setting up a bonding system on the nanometer scale in the SEM.

The process itself is defined by the interaction of the primary electron beam with the substrate and the precursor molecules and therefore a wide variety of process parameters arises from these interactions. Although a basic phenomenological description of the deposition process is known, until recently it was not possible to describe the EBiD process precisely and in detail. A major field of research today still is the influence of **heat** on the chemical reaction and the **chemical and structural composition** of (and influence on) the deposits.

Additionally, for a wider field of applications in industry, the growth rates are too slow; further improvements on the gas injection systems seem to be a promising approach.

Utilizing the improvements of the primary electron beam systems, *e.g.*, cold field emission guns, further reductions in the lateral size of deposits below 20 nm have already been achieved in SEMs. Further improvements in this field will certainly increase the application of EBiD in the semiconductor industry.

10.8 References

[1]   Taniguchi, N. 1974, 'On the basic concept of nanotechnology', *Proc. Intl. Conf. Prod. Eng. Tokyo*, vol. 2, pp 18-32.

[2]   Dujardin, G., Mayne, A., Robert, O., Rose, F., Joachim, C. & Tang, H. 1998, 'Vertical manipulation of individual atoms by a direct STM tip-surface contact on Ge(111)', *Phys. Rev. Lett.*, vol. 80, no. 14, pp. 3085–3088.

[3]   Utke, I., Bret, T., Laub, D., Buffat, P., Scandella, L. & Hoffmann, P. 2004, 'Thermal effects during focused electron beam induced deposition of nanocomposite magnetic-cobalt-containing tips', *Microelectron. Eng.*, vol. 73-74, no. 1, pp. 553–558.

[4]   Hoffmann, P., Utke, I. & Cicoira, F. 2002, 'Limits of 3D nanostructures fabricated by focused electron beam (FEB) induced deposition', *10th International Symposium on Nanostructures: Physics and Technology*, vol. 5023, pp. 4–10.

[5]   Silvis-Cividjian, N., Hagen, C. W., Leunissen, L. H. A. & Kruit, P. 2002, 'The role of secondary electrons in electron-beam-induced-deposition spatial resolution', *Microelectronic Engineering*, vol. 61-62, pp. 693–699.

[6]   Scheuer, V., Koops, H. & Tschudi, T. 1986, 'Electron beam decomposition of carbonyls on silicon', *Microelectron. Eng.*, vol. 5, no. 1-4, pp. 423–430.

[7]   Stewart, R. L. 1934, 'Insulating films formed under electron and ion bombardment', *Phys. Rev.*, vol. 45, no. 7, pp. 488–490.

[8]   Ennos, A. E. 1954, 'The sources of electron-induced contamination in kinetic vacuum systems', *British Journal of Applied Physics*, vol. 5, no. 1, pp. 27–31.

[9]   Christy, R. W. 1960, 'Formation of thin polymer films by electron bombardment', *Journal of Applied Physics*, vol. 31, no. 9, pp. 1680–1683.

[10]   Ling, J. 1966, 'An approximate expression for the growth rate of surface contamination on electron microscope specimens', *British Journal of Applied Physics*, vol. 17, no. 4, pp. 565–568.

[11]   Broers, A. N., Molzen, W. W., Cuomo, J. J. & Wittels, N. D. 1976, 'Electron-beam fabrication of 80-A metal structures', *Applied Physics Letters*, vol. 29, no. 9, pp. 596–598.

[12]   Koops, H. W. P., Weiel, R., Kern, D. P. & Baum, D. P. 1988, 'High-resolution electron-beam induced deposition', *Journal of Vacuum Science and Technology B: Microelectronics and Nanometer Structures*, vol. 6, no. 1, pp. 477–481.

[13]   Griesinger, U. A., Kaden, C., Lichtenstein, N., Hommel, J., Lehr, G., Bergmann, R., Menschig, A., Schweizer, H., Hillmer, H., Koops, H. W. P., Kretz, J. & Rudolph, M. 1993, 'Investigations of artificial nanostructures and lithography techniques with a scanning probe microscope', *Proceedings of the 16th international symposium on electron, ion, and photon beams*, vol. 11, pp. 2441–2445.

[14]   Koops, H. W. P., Kretz, J., Rudolph, M. & Weber, M. 1993, 'Constructive three-dimensional lithography with electron-beam induced deposition for quantum effect

devices', *Proceedings of the 16th international symposium on electron, ion, and photon beams*, vol. 11, pp. 2386–2389.

[15] Weber, M., Rudolph, M., Kretz, J. & Koops, H. W. P. 1995, 'Electron-beam induced deposition for fabrication of vacuum field emitter devices', *7th International Vacuum Microelectronics Conference*, vol. 13, pp. 461–464.

[16] Koops, H. W. P., Kretz, J. & Rudolph, M. 1994, 'Characterization and application of materials grown by electron-beam-induced deposition', *Jpn. J. Appl. Phys.*, vol. 33, no. Part 1, 12B, pp. 7099–7107.

[17] Kohlmann-von Platen, K. T., Buchmann, L.-M., Petzold, H.-C. & Brunger, W. H. 1992, 'Electron-beam induced tungsten deposition: Growth rate enhancement and applications in microelectronics', *Proceedings of the 36th International Symposium on electron, iron, and photon beams*, vol. 10, pp. 2690–2694.

[18] Hübner, U., Plontke, R. & Blume, M. 2001, 'On-line nanolithography using electron beam-induced deposition technique', *Microelectronic Engineering*, vol. 57-58, pp. 953–958.

[19] Utke, I., Hoffmann, P., Berger, R. & Scandella, L. 2002, 'High-resolution magnetic Co supertips grown by a focused electron beam', *Applied Physics Letters*, vol. 80, no. 25, pp. 4792–4794.

[20] Liu, Z., Mitsuishi, K. & Furuya, K. 2004, 'Three-dimensional nanofabrication by electron-beam-induced deposition using 200-keV electrons in scanning transmission electron microscope', *Applied Physics A: Materials Science & Processing*, vol. 80, no. 7, pp. 1437–1441.

[21] Utke, I., Luisier, A., Hoffmann, P., Laub, D. & Buffat, P. A. 2002, 'Focused-electron-beam-induced deposition of freestanding three-dimensional nanostructures of pure coalesced copper crystals', *Applied Physics Letters*, vol. 81, no. 17, pp. 3245–3247.

[22] Mølhave, K., Madsen, D. N., Dohn, S. & Bøggild, P. 2004, 'Constructing, connecting and soldering nanostructures by environmental electron beam deposition', *Nanotechnology*, vol. 15, no. 8, pp. 1047–1053.

[23] Wich, T. & Sievers, T. 2006, 'Assembly inside a scanning electron microscope using electron beam induced deposition', *Proceedings of 2006 IEEE/RSJ International Conference on Robots and Intelligent Systems*.

[24] Reimer, L. 1998, *Scanning Electron Microscopy – Physics of Image Formation and Microanalysis*, Vol. 45 of Springer Series in Optical Sciences, 2nd edn.

[25] Balk, L. J., Blaschke, R., Bröcker, W., Demm, E., Engel, L., Göcke, R., Hantsche, H., Hauert, R., Krefting, E. R., Müller, T., Raith, H., Roth, M. & Woodtli, J., *Praxis der Rasterelektronenmikroskopie und Mikrobereichsanalyse*, Bartz, W. J.

[26] Fuchs, E., Oppolenzer, H. & Rehme, H. 1990, *Particle Beam Microanalysis (Fundamentals, Methods and Applications)*, VCH Weinheim.

[27] Schiffmann, K. I. 1993, 'Investigation of fabrication parameters for the electron-beam-induced deposition of contamination tips used in atomic force microscopy', *Nanotechnology*, vol. 4, no. 3, pp. 163–169.

[28] Utke, I., Cicoira, F. & Jaenchen, G. 2002, 'Focused electron beam induced deposition of high resolution magnetic scanning probe tips', *Mat. Res. Soc. Symp. Proc.*, vol. 706.

[29] Seiler, H. 1983, 'Secondary electron emission in the scanning electron microscope', *Journal of Applied Physics*, vol. 54, no. 11, pp. R1–R18.

[30] Ono, S. & Kanaya, K. 1979, 'The energy dependence of secondary emission based on the range-energy retardation power formula', *Journal of Physics D: Applied Physics*, vol. 12, no. 4, pp. 619–632.

[31] Reimer, L. 1999, 'SEM/TEM Hypertext: per Mausklick (fast) alles über Elektronenmikroskopie'. CD-ROM.

[32]   Hasselbach, F. & Rieke, I. 1982, 'Spatial distribution of secondaries released by backscattered electrons in silicon and gold for 20-70 keV primary energy', *10th International Conference on Electron Microscopy, Hamburg*, vol. 1, pp. 253–254.

[33]   Kanaya, K. & Kawakatsu, H. 1972, 'Secondary electron emission due to primary and backscattered electrons', *Journal of Physics D: Applied Physics*, vol. 5, no. 9, pp. 1727–1742.

[34]   Silvis-Cividjian, N. 2002, 'Electron beam induced nanometer scale deposition', Ph.D. thesis, Technische Universiteit Delft.

[35]   Wutz, M. 2004, *Handbuch Vakuumtechnik*, 8th edn, Vieweg Verlag.

[36]   Mølhave, K. 2006, 'Tools for *In situ* Manipulation and characterisation of nanostructures', Ph.D. thesis, MIC- Department of Micro and Nanotechnology, Technical University of Denmark.

[37]   James M. Lafferty (editor) 1998, *Foundations of Vacuum Science and Technology*, John Wiley and sons.

[38]   Utke, I., Friedli, V., Michler, J., Bret, T., Multone, X. & Hoffmann, P. 2006, 'Density determination of focused-electron-beam-induced deposits with simple cantilever-based method', *Applied Physics Letters*, vol. 88, no. 3, p. 031906.

[39]   Randolph, S. J., Fowlkes, J. D. & Rack, P. D. 2005, 'Effects of heat generation during electron-beam-induced deposition of nanostructures', *Journal of Applied Physics*, vol. 97, no. 12, p. 124312.

[40]   Becker, G. 1961, 'Zur Theorie der Molekularstrahlerzeugung mit langen Kanälen', *Zeitschrift für Physik A*, vol. 162, no. 3, pp. 290–312.

[41]   Giordmaine, J. A. & Wang, T. C. 1960, 'Molecular beam formation by long parallel tubes', *Journal of Applied Physics*, vol. 31, no. 3, pp. 463–471.

[42]   Jones, R. H., Olander, D. R. & Kruger, V. R. 1969, 'Molecular-beam sources fabricated from multichannel arrays. I. Angular distributions and peaking factors', *Journal of Applied Physics*, vol. 40, no. 11, pp. 4641–4649.

[43]   Boero, G., Utke, I., Bret, T., Quack, N., Todorova, M., Mouaziz, S., Kejik, P., Brugger, J., Popovic, R. S. & Hoffmann, P. 2005, 'Submicrometer Hall devices fabricated by focused electron-beam-induced deposition', *Applied Physics Letters*, vol. 86, no. 4, p. 042503.

[44]   Edinger, K., Becht, H., Bihr, J., Boegli, V., Budach, M., Hofmann, T., Koops, H. W. P., Kuschnerus, P., Oster, J., Spies, P. & Weyrauch, B. 2004, 'Electron-beam-based photomask repair', *The 48th International Conference on Electron, Ion, and Photon Beam Technology and Nanofabrication*, vol. 22, pp. 2902–2906.

[45]   Bret, T., Utke, I., Bachmann, A. & Hoffmann, P. 2003, '*In situ* control of the focused-electron-beam-induced deposition process', *Applied Physics Letters*, vol. 83, no. 19, pp. 4005–4007.

[46]   Ding, W., Dikin, D. A., Chen, X., Piner, R. D., Ruoff, R. S., Zussman, E., Wang, X. & Li, X. 2005, 'Mechanics of hydrogenated amorphous carbon deposits from electron-beam-induced deposition of a paraffin precursor', *Journal of Applied Physics*, vol. 98, no. 1, p. 014905.

[47]   Utke, I., Friedli, V. & Fahlbusch, S. 2006, 'Tensile strengths of metal-containing joints fabricated by focused electron beam induced deposition', *Advanced Engineering Materials*, vol. 8, no. 3, pp. 137–140.

[48]   Wich, T., Kray, S. & Fatikow, S. 2006, 'Microrobot based testing of nanostructures inside an SEM', *Proceedings of the 10th International Conference on New Actuators (Actuator)*.

Index